MARSHALL HALL (1790–1857): SCIENCE AND MEDICINE IN EARLY VICTORIAN SOCIETY

THE WELLCOME INSTITUTE SERIES IN THE HISTORY OF MEDICINE

Forthcoming Titles

The Correspondence of James Jurin (1684–1750)
Physician and Secretary of the Royal Society
Andrea Rusnock

History of Cardiology
P. Fleming

Operative Chymist: A Biography of Thomas N. R. Morson
Anthony F. P. Morson

Fantasy Surgery
Ann Dally

Accidents in History
Edited by R. Cooter and B. Luckin

Academic inquiries regarding the series should be addressed to the editors W. F. Bynum and Roy Porter at the Wellcome Institute for the History of Medicine, 183 Euston Road, London NW1 2BE, UK

MARSHALL HALL (1790–1857): SCIENCE AND MEDICINE IN EARLY VICTORIAN SOCIETY

Diana E. Manuel

Rodopi

Amsterdam – Atlanta, GA 1996

First published in 1996
by Editions Rodopi B. V., Amsterdam – Atlanta, GA 1996.

Design and Typesetting by Christine Buckley, the Wellcome Trust.
Printed and bound in The Netherlands by Editions Rodopi B. V.,
Amsterdam – Atlanta, GA 1996.

British Library Cataloguing in Publication Data
A catalogue record for this book is available from the British Library
ISBN 90-5183-921-9 (Bound)

Manuel, Diana E.
Marshall Hall (1790–1857): Science and Medicine in Early Victorian
Society/Diana E. Manuel, – Amsterdam – Atlanta, GA: Rodopi. – ill.
(Clio Medica 37/ ISSN 0045-7183;
The Wellcome Institute Series in the History of Medicine)

CIP-GEGEVENS KONINKLIJKE BIBLIOTHEEK, DEN HAAG

Front cover: Marshall Hall portrait
by James Luntley, signed and dated 1856.
By kind permission of the Wellcome Institute Library, London

Met index, lit. opg.
ISBN 90-5183-921-9 geb.
Trefw.: Hall, Marshall / geneeskunde; Engeland; geschiedenis;
19e eeuw

Printed in The Netherlands

I dedicate this book to my late parents Godfrey Manuel and Ellen Byrne and to my daughter Emma Szelepet.

Contents

	Acknowledgements	viii
	Preface	x
1.	Biography	1
2.	Hall's Nottingham Period: 1817–26	44
3.	Medical Education in London	108
4.	Marshall Hall and the Royal Society: 1830–50	156
5.	Hall's Work on the Nervous System	232
6.	Hall's Final Years	303
	Bibliography	333
	Index	364

Plates

I	Marshall Hall as a boy facing a school bully	24
II	Young female with symptoms of one of the *Mimoses*	69
III	Young female with symptoms of one of the *Mimoses*	70
IV	Cartoon of Man-Midwife	81
V	Responses in sectioned newt	245
VI	Diagrammatic Illustration of the True Spinal or Excito-Motory System	247
VII	Table showing Hall's Three Divisions of the Nervous System	250
VIII	Book plate – Royal Society of Medicine	316
IX	Hall's Memorial and Tomb – Nottingham	326

Acknowledgements

The person to whom I am most indebted is Professor W. F. Bynum, Head of the Academic Unit of the Wellcome Institute for the History of Medicine. He has been extremely generous with his time and his knowledge. I am most grateful to the Institute itself for providing me with a Research Fellowship which enabled me to take a year's Sabbatical Leave from the University of Durham where I was formerly an academic. The University of Durham also provided financial support as did the Royal Society through its numerous travel grants which enabled me to search for manuscripts and other material in the United States and France, two of the countries with which Marshall Hall had professional contacts throughout his life.

The Libraries and their Librarians and other staff who have provided me with invaluable help over the years, include the British Library where I must pay a special tribute to the unfailing help and courtesy of the staff of the North Library Early Collections Service. Other London Libraries include those of the Wellcome Institute, University College London, the Royal Society, the Royal Institution, the Royal Society of Medicine, the Zoological Society, the Royal Society of Physicians, the Royal Society of Surgeons, St Thomas's Hospital, the British Medical Association and the Library of the Society of Friends. In Edinburgh, the libraries where I received help include those of the Department of Manuscripts, Edinburgh University Library, the Royal Infirmary, the Royal College of Physicians, the Public Record Office and the student Royal Medical Society. In Nottingham, Hall's birthplace, the University of Nottingham, the Nottingham Infirmary and the small elegant Bromley House Library, all provided me with help that enabled me to make the most of my short visits to Nottingham.

I would like to thank the staff of the Bibliothèque Nationale, the Institut de France, Académie des Sciences, the Académie Nationale de Médecine and the Archives Nationales in Paris. The Libraries in the United States include those of the Library of Congress, and the Folger, Shakespeare Library Washington, the Massachusetts Historical Society in Boston, the American Philosophical Society in Philadelphia and above all the Francis A. Countway Library of Medicine, Harvard, where I received especially generous help on a number of occasions from Mr Richard R. Wolfe, Curator of Rare Books and Manuscripts and Garland Librarian of the Boston Medical Library. I have also received invaluable help from a number of other individuals. These include Professor Roy Porter of the Wellcome Institute for the History of Medicine, Professor Arnold W. Wolfendale, FRS, Astronomer Royal (1991–5), now Emeritus Professor of Physics, University of Durham, and Dr Amalie M. and the late Professor Edward H. Kass, both of Harvard, who made available to me their collection of manuscript material on Thomas Hodgkin, a contemporary and one-time colleague of Marshall Hall.

Finally, I am grateful for the opportunity to thank the individuals who were most important in helping me actually produce the disk and hard copy. At the Wellcome Institute, Sally Bragg has throughout provided me with unfailing support and has most admirably facilitated my regular access to Bill Bynum. Christine Lavery, of the Wellcome Trust, has given excellent guidance regarding the final preparation of the manuscript. In Durham, Mrs Betty Fairley and Mrs Lynn Carrington did the initial typing and were always immensely good humoured. They carried on despite finding some passages relating to Hall's animal experiments somewhat difficult! On leaving Durham my material had to be converted from one computer package to another and the individual who coped with this task and who has consistently helped me through to completion is Mr Michael Maltby, a friend and colleague at University College London. Michael is a real computer guru who enjoyed the challenges in organizing my material. Amusingly, despite my background and his research in chemistry we both almost missed the slight misspellings of the names of two eminent nineteenth-century chemists. Nevertheless, errors of fact and other shortcomings in this work are entirely my own.

Whilst I dedicate this book to my late parents Godfrey and Ellen and to my daughter, it is to Emma that I owe a special debt for her support and above all for her tolerance of my maternal neglect. Undoubtedly more than a little of the time spent on this book should have been hers, particularly in view of the fact that I was a weekend mother commuting between Durham and our home in London. Thank you Emma.

Preface

Hall was brought up among the chimneys, smells and smoke of his father's cotton manufacture and bleaching factory in Basford, Nottingham. He was reminded daily of the positive results of his father's interest in the chemistry of bleaching and of his older brother's interest and inventiveness in the technology of gassing lace to clear it of excess fibres and in a host of other technological applications. Thus, Hall was surrounded by the serious application of science to a range of processes associated with some of the city's industrial activities. In a sense, this was an extension of the earlier application, in England, of contemporary science to technological problems concerned with the ancient activities of navigation, mining, agriculture and war. This was a far cry from the manner in which science was then being pursued as a hobby among sections of the leisured classes. There it could be regarded as an additional conversational accomplishment for a gentleman – rarely for a gentlewoman.

Hall was frustrated by this amateur status accorded to science. He believed in the Victorian ethos that progress was inevitable and for him science was the key to this advance. He wished to take it as seriously as did his father, to apply it to medicine and to raise the status of the profession. Already as a student he had begun to address the issue of diagnosis and had tackled its key concept of classification, in an attempt to re-order diseases on the basis of a more rigorously scientific process of observation.

Hall made no major breakthrough with this work on diagnosis – knowledge of the germ theory of disease was still some way in the future. So, Hall moved on to a new set of medical problems concerning the body's physiological systems, which he investigated through a series of painstaking and elegant experiments. He worked first on the circulatory system where he questioned the very existence of the fine ultimate capillaries within the tissues. He also devised a rule to promote and guide safe blood-letting, which was still being used therapeutically. Finally, he embarked on his major experimental studies of that most difficult area in physiology – the nervous system.

To have been born into a manufacturing family, albeit one with a genuine interest in pure science and its technological applications, in the Midlands at the end of the eighteenth century, was not the most propitious beginning for one determined to join the ranks of the seriously renowned in medicine. Since, unsurprisingly for the period, the family manufacturing metier was combined with an allegiance to Methodism, the least prestigious of the dissenting religious groups, such burning ambition might seem to have been doomed from the start. What Hall had needed were Anglican parents associated with one of the noble professions of law or the church. Added to these disadvantages had been the influence of a father, probably distracted with his business affairs, who neglected to ensure that his ambitious third son Marshall was studying Latin, the generally essential prerequisite for entry to the professions. He was not. This omission only added to the obstacles which had to be overcome in pursuit of a medical qualification, gained at Edinburgh rather than Oxbridge. Furthermore, as a short person, to have suffered from the stereotyping of being perceived as also bombastic increased both Hall's burden and his determination to succeed.

Personal experience as a child of illness which, according to his mother, was at the root of his being short for the rest of his life, could have motivated Hall to study medicine rather than follow his father and at least two brothers into the lace industry. The shock of losing one of his three sisters who died in childbirth could have prompted his early interest in women's diseases, including puerperal fever, which almost certainly carried her off. His *Eupædia* or *Letters to a Mother on the Watchful Care of Her Infant* could be seen as a private homage to the lost sister and little Ellen and, in its early drafts, provided a practical guide for the sister who had taken on the care of the infant.

In Edinburgh Hall had spent more time than any of his student contemporaries in the dissecting room. Little wonder that he acquired brilliant manipulative skills. He would clearly have made a splendid surgeon. But, despite his quasi-egalitarian views and Whiggish politics, he chose the more socially prestigious role of the physician. Hall seemed genuinely not to have coveted the golden cane image of the physician. But despite his apparently simple lifestyle – modest diet, little entertaining of friends and colleagues – Hall did enjoy some of the privileges associated with being a physician. These included a smart house and, as a horse lover, fast horses and a good carriage. What Hall valued immensely was the recognition that a physician's task was an intellectual one. A physician used his head rather than his hands to solve medical and clinical problems;

he could remain well-groomed – although this did not exactly apply to Hall – and never become well-grimed on the job.

Nevertheless, at the drop of a hat, had there been encouragement and funding from institutions such as the Royal Society, to which he applied for help, Hall would have put aside these benefits to become a medical research scientist. And this, despite the as yet un-established and altogether more insecure nature of such a position. He coveted with great intensity the prospect of initiating a revolution in ideas about human physiology and in particular about reflex action. The renown for such a change, including its contribution to the establishment of scientific medicine, would, he anticipated, endure beyond his lifetime.

A centrist politically at home – he was seen by some as a radical but his radicalism was generally to promote his own ends – Hall was an ardent Francophile, likely to have been perceived as 'not one of us' by Establishment figures looking across the channel at post-revolutionary France. Towards the end of his life a tour of the United States somewhat lessened Hall's Whiggishness and hence by implication his support for the colonists and their cession from the United Kingdom. His considerable linguistic ability in French, in which he eventually published work and which gave him entrée to the great French institutions of science and medicine, would certainly not have endeared Hall to xenophobic individuals and groups at home. One of Hall's most enduring friendships was with the physician Pierre C. A. Louis, a Republican sympathizer who strove to introduce a quantitative methodology to medical research. Louis, like Hall an unsuccessful applicant for a prestigious chair, was also something of a loner in the Paris medical fraternity.

Hall had read the work of the French philosophes. He imbibed something of what Roszak describes as their concept of citizenship, an ideal that synthesizes what is best in the seemingly irreconcilable traditions of service and scholarship. Citizenship for the philosophe embodied a moral vocation as well as having a legal status and was concerned with influencing and educating people through shaping knowledge into a basis for action. For the philosophe, intellectuality began at the point where one undertook to make knowledge work and, according to Roszak, the intellectual was one who intervened in society for the defence of civilized values – free speech, free thought, free enquiry for the sake of reform. Hall's own professional life has to be set against the broader canvas of a society and its institutions, including those associated with science and medicine, undergoing major professional, social and political reform. And while he was not prepared to throw himself wholeheartedly into medical politics like, for instance, Wakley of the *Lancet,* or Hastings who was deeply involved in the origin of the British Medical Association, because he was not prepared to devote the time to it, Hall was not afraid to be an individual protestor. He preferred the freedom of the individualist but without group support this was a largely ineffectual way of trying to bring about change. Nevertheless he had made positive contributions towards the greater professionalization of medicine especially relating to its education and training procedures and its practices.

In Hall's day the patient did not generally occupy centre stage. But for Hall, his patients were not entirely invisible. His Edinburgh training provided first-class, first-hand bedside experience and his early work *On Diagnosis* is based on this close proximity and the fruits of his detailed observational work at those bedsides. It is unlikely that despite his fight for a more science-based medicine, which was to come but not in his day, and his personal struggle for professional prestige based on his laboratory research, Hall would uncritically have approved of the highly technological, research-oriented present-day medicine. With its pursuit of machine-based therapies, radical surgical procedures including key-hole surgery and research strategies, modern medicine, it is claimed by some, gives primacy to scientific over clinical objectives and has lost sight of the sick individual patient as a person needing care and treatment for a disease condition. It has, say its critics, lost much of its humanity and of this trend Hall would not have approved.

1

Biography

A life-long passion for science, seen as an essential underpinning to medicine, and a burning ambition not for money but for an enduring reputation in the world of scientific medicine, ruled Marshall Hall's life. Both the passion and the ambition had their roots in his home background as the son of a Nottingham manufacturer who was a Methodist and in his scientifically oriented medical course at Edinburgh University. Born 18 February 1790 Marshall Hall was the sixth of eight children and the fourth son of Robert Hall a cotton manufacturer and bleacher at Basford, Nottingham.[1] Robert and at least one older son Samuel were role models of men of ideas and action in applied science or technology.[2] These men were part of the manufacturing cadre which had arisen in the Midlands and the North during the late eighteenth and early nineteenth centuries. A younger brother John, who became a surgeon, was in his turn encouraged by his brother Marshall to study in Edinburgh, which at the end of the eighteenth century had begun to pioneer the practice of university education for surgeons as well as for physicians.

Family Background

Marshall Hall was devoted to his family, especially to his three sisters, two of whom were older and who no doubt spoilt him. They were Sarah, often called Sally, Anne and a third who died in childbirth during his student days leaving a young niece Ellen. She was for her uncle always a reminder of her lost mother. One sister remained a maiden aunt, probably to care for Ellen and it was to her, 'Miss Hall' as Hall's wife called her, that Hall and his wife went following their return from a 15-month tour of the United States of America a few years before Marshall Hall's death. Always anxious to please his father, fears of failure and concern about the cost of his education during his student days were constantly expressed in letters to these sisters.[3] Regular assurance was sought from them about their father's attitude to Hall's progress in Edinburgh. From them too he also looked forward to little extras such as silk stockings and above all supplies of tea which he described as his 'bon bon'.[4] Later in life a particular fondness for coffee with plenty of cream replaced that for tea and during his final days, when swallowing was almost impossible, this was sometimes the only nourishment taken.

His father Robert Hall had a knowledgeable interest in chemistry and its applications to his own industrial concerns. Familiarity with current work on chlorine and its bleaching properties, associated with individuals such as Joseph Black (1728–99) and Charles Scheele (1742–86), Antoine Lavoisier (1743–94), Claude Berthollet (1748–1822) also put him in communication with Joseph Priestley (1733–1804), with whom he corresponded. As a regular subscriber to *Annales de Chimie* he could read French, a language which Marshall Hall came to read, write and speak with great flair and fluency. He also subscribed to *Nicholson's Journal* a scientific publication to which his son, during his student years, contributed papers on chemistry.[5]

The French chemist Berthollet had discovered that chlorine, called oxymuriatic acid gas, possessed the power of destroying vegetable colours. Robert Hall contacted Priestley and Thomas Henry (1734–1816) who in those days of imprecise titles for medical men, especially below the level of physician, was sometimes described as a chemist and sometimes as a surgeon–apothecary.[6] Henry became Secretary, then President of the Manchester Literary and Philosophical Society.[7] He was made FRS in 1775 on the recommendation of Sir John Pringle and Joseph Priestley. His patent for milk of magnesia was extremely lucrative and with his son William he actively pursued several aspects of applied chemistry. Like Robert Hall he was interested in the relevance of chemistry to manufacturing industries and lectured on bleaching and dyeing in the Manchester College of Arts and Science, which became incorporated in the University, an institution which he had helped to establish.[8]

Robert Hall raised with Henry the question of using chlorine on a large scale as a bleaching agent in industries such as his own.[9] Priestley and Henry did not encourage Robert Hall; Henry's reluctance was probably due to the fact that he himself was engaged in the same activity.[10] In a published exchange of letters with Mr Samuel Parkes in 1816, Henry's son claimed that his father and not

Messrs Milnes of Aberdeen who had been named by Parkes, was the most important individual among the first users of oxymuriatic acid for bleaching.[11]

Despite lack of encouragement, Robert Hall went ahead with the bleaching process, with considerable success and was at least one of the first to use chlorine for bleaching on a large scale.[12] Liebig made reference to this new process as one of the nationally important activities on which Britain's enormous manufacturing system depended and which had enabled her to compete favourably with France and Germany at the time,

> ... but for this new bleaching process, it would scarcely have been possible for the cotton manufacture of Great Britain to have attained its present enormous extent, – it could not have competed in price with France and Germany. In the old process for bleaching, every piece must be exposed to the air and light during several weeks in the summer, and kept continually moist by manual labour. For this purpose, meadow land, suitably situated, was essential. But a single establishment near Glasgow bleaches 1400 pieces of cotton daily, throughout the year. What an enormous capital would be required to purchase land for this purpose! .... Besides the diminished expense, the cotton stuffs bleached with chlorine suffer less in the hands of skilful workmen than those bleached in the sun; and already the peasantry in some parts of Germany have adopted it, and find it advantageous.[13]

Marshall Hall was extremely proud of his father's achievement; his own son was often reminded of it.[14] Hall senior and his second son Samuel were also interested in technology. A prize from the Society of Arts was awarded to the father for the invention of a new crane[15] whilst Samuel, described in the *Dictionary of National Biography* as an Engineer and as 'a prolific inventor' applied his creative skills to the cotton and allied industries. Patents were taken out in 1817 and 1823 for 'gassing' lace and net. The process involved passing the machine-produced fabric rapidly through a row of gas flames which removed all the loose fibres without damage to the lace fabric itself. According to Hall 'The sheet of lace passed to the flame opaque ... and issued from it bright and clear ....'[16] Gassing had an important influence on the lace trade and was in itself financially very rewarding to Samuel whose works were visited by the Duke of Sussex, later President of the Royal Society and someone who had a genuine interest in science.[17]

But Samuel's fortune was dissipated in bringing out about 20 other less successful inventions.[18] These included a surface condenser in which the steam was condensed by passing it through a number of small tubes cooled on the outside. This condenser was chiefly intended for use at sea and it was hoped that the disadvantages associated with the presence of salt in ships' boilers would be obviated by charging them with fresh water at the beginning of a voyage and recycling the water. The patent was taken out in 1838 and although the trials during 1839–41 were not successful the principle of Samuel Hall's tubular condensers was subsequently used for cooling purposes.

Other patents included the bleaching of starch, the consumption of the smoke of furnaces and improvements in the action of paddle-wheels as well as other inventions related to steam engines and boilers.[19] Considerable originality and inventiveness were displayed in Samuel's application of science to manufactures and the useful arts. Like his younger brother, Samuel Hall moved to London and eventually survived him by some seven years, dying in very reduced circumstances in Bow, East London. Marshall Hall's last words on his own death-bed expressed concern for the happiness of a brother,[20] probably Samuel, from whom he had obtained financial help as a student[21] and from whom he is most likely to have gained valuable motivation and skill in designing the numerous pieces of equipment which he came to construct for his own experimental work.

Like many eighteenth- and nineteenth-century entrepreneurs, including Thomas Henry who had abandoned the Church of England to become a Unitarian,[22] Robert Hall was a dissenter and a follower of John Wesley who died the year following Marshall Hall's birth and with whom Robert had personally corresponded. In a letter to Marshall Hall's wife following her husband's death, Hall's brother-in-law Mr Higginbottom explained that in 1797 when the Methodists split, Robert Hall was attached to the 'New Party', 'Reformers' or 'New Connexion Methodists' as they were called. It was claimed, even in the House of Commons that this new group was disaffected to Government. However, on hearing that Robert Hall was among the group, Mr Robert Smith, later Lord Carrington, then one of the MPs for Nottingham rose in the House to say, 'As long as that gentleman (Mr Hall) is connected with the disaffected body, all will be safe and right; there will be nothing wrong.'[23] At the time of the Luddite activities in Nottingham, Robert Hall received a letter signed 'Ned Lud', the assumed name of the leader of the group, assuring him that he was in no danger.

Marshall Hall's Education and Training

Early Education

Writing to his sisters from Edinburgh in 1812 Hall spoke of having been only a visitor to his own home over the previous twelve years.[24] He must, therefore, have become a boarder at school in 1800 at the age of ten. He had attended the Academy of the Rev. J. Blanchard in Nottingham where his older brothers had been educated. This had been the pattern of education of Thomas Henry's son William. Whereas Hall's brothers had received instruction in the classics, Hall did not, although he was taught French. Charlotte Hall attributes this deficiency in Hall's education partly to his father being preoccupied at the time and failing to check on his son's curriculum and partly to the fact that, as she claimed, the classics were not so universally taught in the earlier part of the century as they were in the 1850s.[25] According to Peterson, it was not uncommon for sons destined to enter the medical profession, rather than going into law or the church, not to study the classics.[26]

Some form of medical career was definitely Hall's wish. Maybe his father had consulted Thomas Henry who, following his apprenticeship to an apothecary in Wrexham, had later attended anatomical lectures in preparation for becoming an apothecary–surgeon. On the other hand he had sent his own son William, 16 years older than Marshall Hall, to Edinburgh where he studied first chemistry and eventually gained an MD in 1807. The son had submitted two papers, on carbonated hydrogenous gas and on muriatic acid to the Royal Society of which he was elected an FRS in 1808.[27] He was also awarded the Royal Society's Copley medal. Thomas Henry's son could have provided yet another model whom Hall wished to surpass in terms of scientific attainment.

Early Training and University

Departure from school at 14 was followed by five years of private study during which, initially, Watson's *Chemical Essays*[28] and the works of Lavoisier were Hall's main sources. There followed a placement with Mr Moor, described by Charlotte Hall as 'a chemist of some repute and wealth at Newark'[29] and here began Hall's friendship with Robert Cook who was lodging in the same accommodation and studying for the medical profession. Cook eventually became a practitioner at Gainsborough.

At the beginning of the nineteenth century three kinds of doctor were recognized – physicians, surgeons and apothecaries, each with its own central institution the College (later designated Royal Colleges) of Physicians, Surgeons, and the Society of Apothecaries.[30] These were responsible for setting examinations and awarding qualifications. Of the three groups, the physicians were the oldest and the most respected and privileged. It was said that physicians used their heads not their hands and advised rather than performed. They were also reimbursed rather than paid for their services. The number of physicians was small.[31] In 1800 there were in England fewer than 200 Fellows, Licentiates and Extra-Licentiates of the College of Physicians and by 1847 this number had risen to only 683, three-quarters of whom were in London. Even though the Royal College of Physicians represented less than five per cent of all medical practitioners their ancient Collegiate foundation, their tradition of classical learning and the absence of trade or craft function in the physician's work, all contributed to the more prestigious position of the College and its membership.

In theory there was a clear division of labour between the three grades of medical men – apothecaries were perceived as plying the trade of selling drugs, surgeons as practising the craft of surgery and physicians the profession of physic. In practice all three groups engaged in general practice and throughout Hall's life there was considerable demand both within the profession and between the profession and the government for reform. Numerous bills were introduced into Parliament concerned with defining medical qualifications, establishing an institutional structure and method of registration within the profession and removing quackery. There was hope of achieving some degree of unity at least between physicians and surgeons. A measure of success was reflected in the Medical Act passed in 1858, the year following Hall's death. But by this time a new hierarchy had arisen, that of general practitioners and consultants.

It would seem that Mr Moor was a drug seller, possibly the career initially envisaged for Marshall Hall and it is highly likely that Robert Cook, his fellow lodger either qualified as an apothecary or as both an apothecary and a surgeon and then went into general practice. The apprenticeship with Mr Moor did not please Hall who prevailed upon his father to move him. He was developing a genuine interest in medicine and as well as studying chemistry had started to make up for his deficiency in Latin. Going to University and becoming a physician were now his goals.

As the son of a dissenter it would have been difficult for Marshall Hall to enter other professions concerned with civil office where, until

1828, there was a legal requirement to take the Sacrament of the Church of England as a necessary qualification. Similarly, gaining admission to Oxford and Cambridge would have presented problems, even if Hall's father could have afforded it, since entry was usually restricted to members of the Church of England and generally excluded Catholics, Dissenters and Jews. And so the University of Edinburgh was chosen. Like other Scottish Universities, Edinburgh offered a good education 'even to men of moderate means'.[32] More importantly Edinburgh, unlike Oxford and Cambridge, was non-ecclesiastical and though professors nominally took the oath of allegiance to the Church of Scotland most classes were taught by laymen. As Morrell has noted,[33] this decidedly secular ethos drew a cosmopolitan array of students from Scotland, England and Ireland and from the colonies, as well as from Europe and America, a former colony. They attended courses in what was widely regarded as the best medical school in the English speaking world.[34]

Study in Edinburgh

The years 1809–14 were spent in Edinburgh where the annual costs, for someone as careful as Hall, could have been one-third of those at Oxford.[35] Towards the end of this period he held a paid Physician's Clerkship at the Royal Infirmary and did not attend lectures. On the contrary he himself lectured on diagnosis and laid down some of the framework for *On Diagnosis* published in 1817.[36] Chemistry which along with Anatomy was popular, vocational and compulsory for medicine, was studied during his first year. Thereafter, according to the *List of Students*[37] and Medical Matriculation Lists[38] he followed the Edinburgh medical curriculum, for someone intent on graduating, of courses in the Practice of Physic (1810–13), the Practice of Medico-Chirurgical Obstetrics (1810–11), Dr James Home's course in Materia Medica (1811–12) and a course in Midwifery. In pursuance of an MD degree, each class would have been followed for six months after which a thesis was prepared and a *viva voce* undertaken.

Andrew Fyfe, a contemporary and the son of the Professor of Anatomy at Edinburgh, was his classmate in the Anatomy course and special permission was obtained from Fyfe's father for additional access to the dissecting rooms at a very early hour in the morning. Extra time was thus devoted to practical anatomy.[39] Higginbottom, a younger contemporary who married one of Hall's sisters, recalls Hall actually helping Professor Fyfe and remembers that at an evening Anatomy class, one of Hall's preparations was shown to the class as a most beautiful specimen of dissection.[40] Hall had entered Edinburgh just after the phase when the medical faculty could boast a bevy of illustrious, charismatic figures among its professoriate including one of the Monros and Joseph Black. Hall was fortunate to coincide with some of them still in the final phase of their Edinburgh careers. The remuneration of Professors was mainly, or in some cases exclusively, derived from student fees and hence there was considerable incentive to recruit large classes.[41] There were about 311 students on the Practice of Physic course and Emblen notes the low pass rate among students attending the Medical Faculty around 1800.[42] This was related to the fact that at the time there were no admission requirements and many of the students would be taking single classes out of interest rather than whole degree programmes in medicine, which at the end of the eighteenth century was almost the only degree offered at Edinburgh. Unsurprisingly, Hall was not among those who failed. Perseverance and dedication as a student were recurring themes among other members of his Edinburgh network who provided Charlotte Hall with reminiscences of her husband during his Edinburgh period.

Andrew Fyfe recalled that little or no social time was spent with contemporaries since this would have consumed time which ought to be devoted to study. Invitations to Professor Fyfe's home were declined.[43] An undated MS reveals how highly organized Hall was as a student, how tightly structured and arduous his personal work programme compared to that of the stereotype of fun-loving and lazy medical students, and how he had already focused on diagnosis as an area of particular interest. This was important groundwork for one determined to excel as a clinician. The document also reveals Hall's early acquaintance with classical writers on medicine and his high regard for contemporary French sources. It records his intention,

1. To spend *two* hours in the morning in dissection and the study of operations.
2. Then *two* or *three* in the wards of the hospital; inquiring particularly into the history, symptoms, treatment and effect of remedies on each patient; but especially making a particular study of diagnosis.

It also sets forth,

3. The plan of studying diagnosis:
   (i) The formation of a diagnostic arrangement by bringing together those diseases which, being *most similar*, are most apt to be mutually mistaken; and
   (ii) The collection of diagnosis from every *source* of distinction, in the history, symptoms, causes, effects of remedies, etc.

(iii) This plan embraces *all diseases* – medical, surgical, puerperal, etc.

Hall further intended,

4. To go through a course of study, comprising chemistry, physiology, and materia medica, in as *practical* a manner as possible.
5. To study the Latin and French languages, reading Celsus, *Heberden*, and Gregory; Corvisart, Chardel, Pinel, Dessault, Bichat, etc. etc.[44]

Hall had continued to study Latin for his thesis, a requirement for graduation in Edinburgh. His thesis, *De Febribus Inordinatis*[45] was on fever, an issue of great importance at the time although the systematic study of temperature change in disease was developed much later in the century.[46] The thesis, dedicated to his father and to two friends A. Boutflower and Francis Whaley, was defended and published in 1812. Reference to the graduation ceremony appeared in the *Edinburgh Medical and Surgical Journal*,

> On the 24th of June, the Senatus Academicus, of the University of Edinburgh conferred the degree of Doctor in Medicine on the following gentlemen, after having gone through the appointed examinations, and publicly defended their respective inaugural dissertations: From England, Marshall Hall, *De Febribus Inordinatus* ...[47]

Royal Medical Society

Towards the end of the eighteenth century, Edinburgh could boast a whole range of thriving societies for medical and science students. These included the Royal Medical and Royal Physical Societies, the Agricultural Society of Edinburgh, the Chirurgico-Physical Society, the Hibernian Medical Society, the Natural History Society of Edinburgh, a short-lived Chemical Society and the Academy of Physics. Reflecting its popularity with North American students, Edinburgh also had an American Physical Society. For Hall it was his membership of the Royal Medical Society which he treasured most. Here he had gained invaluable experience for the presentation of his thesis. This student Society, which still functions today,[48] was founded in 1737 having originated as an informal association in 1734 of six of Professor Alexander Monro's students, following Monro's founding of the Faculty of Medicine in Edinburgh in the 1720s. The Society received its Royal Charter in 1778 and included among its membership many who became eminent figures, including Joseph Black, Oliver Goldsmith, Astley Cooper, Richard Bright, William Sharpey and Charles Darwin. It also had as a member in 1796 Peter Mark Roget who became Secretary of the Royal Society and, Hall felt, one of his main opponents within that institution.

The Medical Society owed much to the continuing encouragement of the professors and other senior members of the University.[49] It was said that only those students who had distinguished themselves by diligence, capacity and conduct were initiated into its assembly and election to the Society was by ballot.[50] Hall was elected on 8 December 1809,[51] only a few months after he began his medical studies, clear evidence that he had made his mark early.[52]

The founding group of students of the Royal Medical Society had initiated the practice of reading and discussing student papers written in English or Latin and circulated to other members. This was the practice which provided an opportunity for students to rehearse the defence of their theses. The papers presented to the Society by Hall were,

> On Chemical Attraction.[53]
>
> On the Dispersive and Refractive Powers of the Human Eye, and on Some Motions of the Iris.[54]
>
> On the Application of the Analytical and Synthetical Methods in the Prosecution of Medical Science.[55]

These titles confirm Hall's early interest in chemistry. They suggest an interest in the sensory motor system in which he came to specialize and they also indicated as did his work programme, the beginnings of his life-long concern that medicine should be firmly based on science. Dr Thomas Charles Hope, Professor of Chemistry performed large-scale experiments before his large, admiring classes and his work had stimulated the formation in 1796 of an Apparatus Committee on which Hall served. This Committee continued until 1835 when it was combined with the Museum Committee, which had been started in 1819, to build up a collection of anatomical preparations.[56]

Hall served the Medical Society at the highest level. In his day, four student Presidents were elected annually and since the membership was continually changing, offices were seldom held for more than two years. Of the four Presidents, one was nominated Senior President. For the Session LXXV in 1811/12, that is the session in which he graduated, Hall was elected from 13 candidates as Senior President by a margin of 15 votes over his nearest rival.[57] Hall deemed it a great honour when, in March 1857, five months before his death, he learned that he had been elected an honorary member of the Edinburgh Royal Medical Society. In his early days he had undoubtedly impressed his fellow students; this final honour was some indication of contemporary esteem.[58]

Clerkship to the Edinburgh Royal Infirmary

The Senior Presidency of the Medical Society was a coveted position. So were the Clerkships to the Royal Infirmary in Edinburgh. Hall's wife and some of the brief biographies suggest that Hall had the unexpected honour of being recalled to Edinburgh after his graduation,

> Hardly had he arrived (in Nottingham) when he was recalled to fill the post of Senior Clerk, otherwise Resident House Physician, to the Edinburgh Royal Infirmary.[59]

In fact the Minutes of meetings of the Infirmary Managers, who appointed the Clerks on behalf of the Infirmary, reveal that Hall had already unsuccessfully applied for such a post, to follow his graduation, one and a half years earlier.[60] However, in August 1812 another Clerkship became vacant when his friend Francis Whaley resigned prematurely.

These Clerkship posts were for unmarried individuals, lasted two years and carried a salary of twenty pounds a year.[61] The hospital regulations reveal that the hospital physicians and surgeons placed onerous duties upon the Clerks who were also subjected by the Managers to strict rules governing their entitlement to leave of absence from the hospital.[62] Since most Clerks were taking advanced medical courses at the University or were completing surgical apprenticeships, this curtailment of their independence sometimes impeded attendance at these courses. It also prevented them from indulging in moonlighting through some private practice on the side.[63]

The Medical Clerks' duties were clerical, medical and financial. Their clerical duties included listing all patients in a general admissions book. Separate ledgers were used for service men and servants. The Medical Clerks also took down dictation from attending physicians and transferred this information to the ward journals, ledgers and General Register of Patients. Their medical duties consisted of accompanying the physicians on the daily rounds and also making their own visits to patients, especially the emergency cases. They prepared progress reports for the physicians. Their financial duties included the somewhat unpleasant task of collecting money from those patients who were admitted to ward annexes because no bed was available in the ordinary wards. This task was sometimes difficult and to motivate them, Clerks were offered a bonus on completion of their contract, based on their collecting efficiency.

Surgical Clerks had similar duties but, very importantly, in Edinburgh they played a greater role in the treatment programmes of patients, being involved in procedures such as blood-letting and cupping and were allowed to prescribe in emergency cases where there was no surgeon available. The Surgical Clerks were also involved in operations in the hospital, carried out autopsy dissections and prepared the reports on them. They also trained and supervised hospital dressers, on whose performance they reported to the attending surgeons.[64]

As Risse notes, given their multiplicity of tasks, there were bound to be shortcomings. Some refused to tolerate the workload and restrictions and, like Whaley, they resigned prematurely. Following Whaley's resignation the Managers held an extraordinary meeting and Hall was the only candidate considered and appointed,

> The Managers having taken into consideration a letter laid before them at last meeting from Mr Francis Whaley resigning his situation as one of the Clerks of the House, Elected Mr Marshall Hall to be Clerk in his room.[65]

As a Physician's Clerk, Hall held a post which could be a step towards obtaining an established hospital appointment. He worked in the Ordinary Medical Wards for Drs Andrew Duncan senior and Thomas Spens, the Ordinary Physicians at the Infirmary.[66] The corresponding Clinical Clerkships were Professors' appointments but both were reserved for students of high calibre. Since Hall did not subsequently obtain a permanent hospital appointment in London, it would be interesting to know if he ever regretted his departure from Edinburgh where he had made such a promising beginning and where, difficult though the posts were to obtain, he might have gained a Chair.

Like Whaley, Marshall Hall resigned his Clerkship prematurely. The reasons are unknown, but perhaps he had decided that despite his promising beginning the competition – and it was extremely difficult to get established in Edinburgh at the time – was just too great for a southerner. His letter of resignation was read out at the Managers' Meeting of 7 February 1814 together with 'various applications from Gentlemen wishing to be appointed his successor ....'[67] These included Robert Lister, but at the Extra-Ordinary meeting of 14 February the successful candidate was Dr George Craigie. Hall's resignation provoked increased concern and a discussion on the inconvenience caused by premature departure from any of the Clerkships, especially when due notice was not given. The minutes recorded that,

> The Managers desire that in future when any of their Clerks mean to leave the house before the expiry of their time ([V.C.] two years from the date of their appointment) they will give as early information as possible to the Treasurer.[68]

The Managers then considered other breaches of certain clauses in the Clerks' contracts such as absenting themselves from the Infirmary without leave of the Managers and appointing strangers to stand in for them. In future such behaviour was to be penalized.

Departure from Edinburgh – Plans for the Future

Charlotte Hall makes no reference to the fact that Hall left Edinburgh prematurely. While in Edinburgh he had originally determined to establish himself in London which was seen as the only place where a reputation such as he coveted could be made. However, already one year before departure from Scotland his father was informed about a change of plan. He was reminded that 'my ruling passion (is) ambition', even though this affirmation was tempered with, 'it *shall* be the ambition of being *good* as well as *great*'. Having considered the pros and cons he had 'been induced to give up the idea of London'.[69]

Continuing fear of failure and concern about the consequences for his family and friends of such failure was one of the factors which had driven Hall to 'venture less in another place'. The loss of opportunity to lecture, as he had done in Edinburgh, was his only regret. Nevertheless, publication was seen as the avenue for presenting what was described as a 'new, useful and *practical* plan'. A manuscript on diagnosis had already been prepared although Dr Spens of the Edinburgh Infirmary had clearly advised further development of the work before submission for publication.[70]

Financial considerations had also played a part in the change of plans. His father was urged to keep strictly to the original allowance and not to deprive his siblings for his benefit. A ready willingness to borrow if need be, was expressed although this was at odds with later views about debt. A horror of it had led to Hall leaving all the financial arrangements to his wife. Perhaps also the money collecting from patients during his Clerkship had put him off. He did occasionally check that the family was not in debt. In the early days of his marriage money had been lost through ill-advised loans to others, poor financial advice and bad investments. From that time on he had conservatively adhered to 'the English funds, "the amiable Three per Cents" ....'[71]

In a second letter to his father from Edinburgh further details about future plans were disclosed. At least some months were to be spent in London and then, if possible, as was the way with many American and other European young doctors, travel on the Continent visiting clinics in France, Germany and Austria was to follow. After this he opined, 'I think I ought to look about me and begin to think of fixing my tent.'[72] A research programme was outlined, during which for several years little money would be earned from medical practice. Much time was to be devoted to general reading, especially to developing the specific topic of Diagnosis for which part of each day would be spent gaining clinical experience in the local infirmary. This research and the subsequent publications would, Hall anticipated, 'give me a reputation that would promote my views of practice'. Only gradually would a practice be built up.

Of the locations for embarking on a career, Bristol was contemplated.[73] It was still a fashionable health resort and an expanding industrial city which Hall originally thought could offer a desirable situation. But, as Neve has shown,[74] its scientific activities were at the time never prolonged or particularly successful. What scientific culture did exist was predominantly associated with the Anglican bourgeoisie rather than with marginal men. Nevertheless W. B. Carpenter a physiologist who reviewed for the prestigious *British and Foreign Medical Review,* was the son of a Bristol Unitarian Minister who ran an educational Academy in the City. His son gave a series of popular lectures. Such a conservative city could have been at odds with Hall's political leanings which, though not marked, inclined towards the Whigs.[75] Furthermore, Bristol had begun to decline around the turn of the century and non-establishment religious groups were growing. A sense of threat, which would not have helped Hall, was felt from the city's Methodist groups and from disturbances in mining communities not far from Bristol.[76] And so the idea of going westward was abandoned.

Following his departure from Edinburgh and a short visit to friends, Hall accompanied a wealthy Yorkshire gentleman Dr Harrison, to Paris. In Paris, as well as visiting the medical schools he toured art galleries and was enchanted by the Louvre which still held many treasures of sculpture and paintings seized by Napoleon but since returned to their countries of origin. The medical schools of Berlin and Göttingen which were opened up for a brief interval by Napoleon's reverses at Moscow were also visited. According to Charlotte Hall, in November 1814 her husband walked the 600 miles from Paris to Göttingen, a somewhat longer and lonelier journey than his Highland hike. He crossed territory over which the armies of Napoleon and the Allies had marched and carried a cocked pistol for fear of wolves.[77] At Göttingen he met the physiologist and anatomist Johann Friedrich Blumenbach, Professor of Medicine, the third edition of whose major work *Institutiones Physiologicae,* originally published in 1787, was translated and published in English in 1817.[78]

Nottingham Period

Having returned from the Continent in 1815, Marshall Hall still ruled out Bristol. Instead he eventually took up residence near home, in Bridgewater,[79] staying for only six months before moving into Nottingham itself in February 1817 and remaining there until August 1826. The time between his return from the Continent and his move to Bridgewater had been devoted to new research and to writing up some of his Edinburgh work which was published. Two general medical articles appeared in the *Edinburgh Medical and Surgical Journal*[80] to which Hall had also contributed four papers on diagnosis in 1816.[81] A further article was published in 1817[82] the year which saw the first edition of *On Diagnosis.* The Edinburgh publications brought election as a Fellow of the Royal Society of Edinburgh in January 1819, although Hall never published in the Society's *Transactions.* His election as a Corresponding Member of the Manchester Literary and Philosophical Society had taken place two years earlier following submission to the Society of a chemical paper on the relations between zinc and water.[83] A similar paper involving iron and water had already been submitted to the *Quarterly Journal* of the Royal Institution in London. During his early years Hall submitted several papers, on a range of topics, to this prestigious journal.[84]

Despite the scaled-down career beginning, to which Hall had become reconciled, he nevertheless intended to promote himself by catching the public eye and emphasizing his presence in Nottingham. A residence in the fashionable High Pavement area was sought out[85] and active membership of the community was pursued. The building up of a successful practice was also on his agenda. Involvement in a wide range of societies followed. Many of these societies were philanthropic or religious, had been established by evangelicals and were already supported by members of his own family. The enthusiasm which evangelicals showed for setting up societies to convert and reform their contemporaries was one of their most enduring legacies to the Victorians. From it, says Bradley, 'developed that passion for doing and organizing which still characterizes the English today'.[86] This passion was an aspect of Hall's own evangelical background which left its mark on him. Throughout his life he took up a number of causes with greater or lesser commitment and on more than one occasion proposed the setting up of a fund, organization or society to further the aims of a particular pressure group.

One of the advantages which London, like Edinburgh, would have provided for Hall was access to good libraries both general and specialist. He must have felt this deprivation in Nottingham and so joined the Nottingham Subscription Library and by April 1816 was on its committee.[87] This Library was one of the many which proliferated in the late eighteenth century, largely in response to the emerging taste for fiction. These libraries, and Hall would have benefited from this, circulated books and provided reference services to paying members. The Nottingham Library, with a list of more than 200 subscribers had, as was common with public institutions, a member of the aristocracy as one of its proprietors. In this case it was the Duke of Newcastle. Subscriptions were one pound and five shillings (£1.25) for 'gentlemen residing in the town of Nottingham, or within four miles'. Strangers paid an annual subscription of one guinea (£1.05). Women appear not to have been members in their own right. Shares in the Library of 3,000 volumes stood at ten guineas[88] and in 1821 the library moved to the more spacious accommodation of Bromley House[89] where it remains one of the city's most elegant small libraries.

A more professionally specialist library which Hall joined was the Nottingham Medical Book Society, associated with the Nottingham Infirmary.[90] This Society was the predecessor of the Nottingham Medico-Chirurgical Society, founded in 1828 two years after Hall's departure from the city. At this time provincial doctors were becoming politicized and forming special interest groups to facilitate exchange of professional views.[91] They sought to promote career opportunities, to attack the élitism and influence of the London medical world and also to safeguard standards in the profession against what was seen as the growing strength of commercial apothecaries and an array of quacks. Hall's brother-in-law John Higginbottom was President of the Medico-Chirurgical Society and Hall was the first, and, according to Rolleston 'the most distinguished'[92] to be named on the list of Honorary Life Members of the new Society. Membership of local scientific societies was one way in which the social position of the general practitioner could be raised.[93] Such membership enabled him to share in the cultural life of the community, offered a prestigious social image and, perhaps most important, gave access to fellow lay members who were potential patients. In May 1854, when in retirement, Hall addressed the Nottingham Medico-Chirurgical Society on the chief subject of his life's work – the physiology and pathology of the spinal system.[94]

Of the Nottingham philanthropic societies which Hall joined, the Nottingham Benevolent Society existed 'to extend relief to such cases of Distress as cannot be assisted or sufficiently relieved by the Parochial Establishments of the County'.[95] To achieve this objective, monthly visitors were appointed to personally inspect recommended cases of hardship. These might include destitute women in childbirth, widows and orphans. The Society encouraged industry, cleanliness and honesty among its recipients and exhorted those members acting as visitors to beware of the 'Imposition of idle and dissolute persons'.

A Protestant, inter-denominational, lay organization to which Hall also belonged was the Nottingham Auxiliary Religious Tract Society[96] of which the parent association was the London Tract Society. This latter was the first real Bible Society, founded in 1804 by members of the Clapham Sect, a breakaway group within the Anglican Church who had proposed the idea of a Bible Society to the Religious Tract Society in London. The British and Foreign Bible Society was mainly concerned with making vernacular translations of the Scriptures available to peoples of all races at a price they could afford to pay. During the time of Hall's membership the Society offered financial assistance to Bible Societies in other countries.

The Tract Society, half of whose twelve committee members were from the Established Church and half from 'the different Denominations of Dissenters', elected Hall on to its committee in January 1819.[97] Members of the Society were called Friends of Religion. The Society urged its members to contribute to the production of tracts which might 'promote Moral and Religious Improvement'. In this period of active production and dissemination of short tracts on political or religious issues or for propaganda purposes, members were reminded that,

> ... a Tract may reach those whom a Sermon cannot, and may be the Means of saving those, towards whom no other Means can be used; that one Tract given to an ignorant or dissolute Person may be the instrument of his conversation; and that he who converteth a Sinner from the error of his way, shall save a Soul from death, and shall hide a multitude of sins.[98]

Of Hall's professional work in Nottingham we are told by his wife, who had not at that period even met him, that 'his practice was almost entirely among the county families and his fees, like those of London physicians, were guineas'.[99] Hall's 'territory' included the 'Dukeries', which held the Seats of the Dukes of Portland and of Newcastle and of the Earl of Manvers. It also embraced neighbouring county towns such as Derby and Leicester. Sir Arthur Keith attributed Hall's lucrative practice among the county families of the Midlands to his having been recommended to the wealthy family of the Earl of Manvers by Dr Matthew Baillie.[100] Baillie had been Physician Extraordinary to George III and a Fellow of the Royal College of Physicians. He had been presented with a copy of Hall's work on diagnosis and when visited by the author, Baillie had thought him too young to have produced the work.[101]

Whatever initial arrangements for study time Hall had adhered to in Nottingham, professional duties came to occupy much of it especially since his patients were widely dispersed in the countryside with its winding roads and bad surfaces. For these visits Hall rode on horseback, the usual mode of transport for country physicians, since a horse could dart across fields and use other short cuts. At the end of the eighteenth century a horse cost 12–18 guineas (£12.60–£18.90) and was expected to work for about six years.[102] Hall had three or four horses, often using different animals for day and night time visiting, before he resorted to a gig, which was quite expensive but which allowed reading during the journeys. He loved horses, rode them fast and was often present when they were fed. Asked why his horses never fell, Hall replied, 'I never give them time to fall.'[103] A large portrait of one of his favourite horses was painted for him.[104]

Hall did nevertheless find some time to pursue research and set up a laboratory in the High Pavement residence. Dr Wilkins, Archdeacon of Nottingham, was a neighbour in the early days. He recalled how Hall had shunned social activities preferring, if he did not have to visit patients, to spend his time studying in his laboratory or library. Wilkins remembers Hall as stout, of middling stature and possessed of a ready smile.[105] It was thought that Hall did not take regular meals but helped himself from an ever present side table of food. It looks as if the stoutness in his late twenties and early thirties was due to a habit of constant nibbling.

Despite the fact that as a physician Hall would tend to have wealthier patients, he did throughout his life show concern for the under-privileged including the indigent poor in Nottingham, for whom the philanthropic groups with which he was associated made some provision. Thus more modest individuals were also treated and even visited for little or no charge. Wilkins confirms Hall's generosity towards numerous poor people seen at Wilkins' special request. One of these, Edward Morrill died suddenly, early one morning in January 1820. An inquest on his death was carried out at the 'Flaming Sword' public house before Jonathan Dunn Gent, one of the Nottingham coroners. Morrill had been troubled with a violent cough and a painfully sore leg which prevented him from sleeping. In the habit of indulging in self-medication he had on the night before his death taken sleeping tablets. At 2 a.m. an episode of convulsions was followed by his death four hours later. It was thought he had died of poisoning but Hall's brother-in-law, who carried out the autopsy declared that Morrill had 'died in a fit of apoplexy and not otherwise'.[106] Hall's genuine concern for the under-privileged was evident on a number of occasions. And later, his sympathy for the Negroes of America whose unfavourable conditions and treatment as slaves, or as Free men, he witnessed during a tour of the United States in 1853–4, was expressed in his writing.

One of the ways in which a nineteenth-century physician could enhance his reputation and build up his practice was by obtaining an honorary post in the local infirmary especially if, as in London or Edinburgh, that infirmary was also a teaching hospital. In 1825, at the age of 35, Marshall Hall applied for the post of honorary assistant physician to the Nottingham General Hospital. Qualified to apply on account of his Edinburgh training and through having resided in Nottingham for longer than the minimum period of one year[107] the application was confidently and successfully put forward. Sources vary as to the date of his appointment, his wife stating 12 October 1825[108] and Rolleston erroneously giving the date and duration as 1826–7.[109] In a brief biography of Samuel Hall, Felkin believed that his brother Marshall had 'received a sound training in the general hospital at Nottingham'.[110] It certainly must have put more than a few new patients his way.

Removal to London

Less than a year after his appointment to the Nottingham General Hospital Hall had left the city. While his own patients regretted his departure some of the older apothecaries and surgeons, who as medical practitioners would have had to confer with Hall when he was called in as a consultant physician, felt no such sorrow. As his wife admitted, some of these people 'manifested great bitterness, and even refused to meet in consultation a physician so much their junior'.[111] Hall had already begun to antagonize certain individuals. Family and friends had tried to dissuade him from undertaking such a risky venture as moving to London. Unable to face their concern or reproach, according to his wife, no one was told that he was actually leaving. In August 1826 a visit to London was followed by him simply staying there. Thus, in what seems an irresponsible manner, Hall walked out on the job, almost as he had done from Edinburgh. On this occasion he had not even informed the General Hospital who, after a week's absence enquired about his duties there. Higginbottom wrote and was instructed to sell furniture, horses, gig and so on and to pay all bills.[112] Hall's resignation from the Hospital followed.

Once in London a sum of money was deposited with the firm of Sir C. Forbes, to defray expenses during the early years. This fund was never drawn on and the first year's practice income was £800. By 1833 his income had risen to £2,200 and was £4,000 towards the end of his career.[113] Some former Nottingham patients continued to consult Hall especially since they were in London during the season. Soon new patients were attracted and he was quickly able to acquire the real status symbol of a successful physician, a carriage with a pair of horses together with a coachman and footman.

A spacious house at 15 Keppell Street, Russell Square near the British Museum was Hall's first London home.[114] This location was the site of the University of London, already being planned and where Hall was very interested in gaining a post. The house was shared with a Nottingham friend, Mr William Burnside who had moved with his family to London, to form a partnership with the well known publisher Mr Seeley, publisher of a number of Hall's books. Mrs Burnside took on all the household duties and Hall, a stranger in London, suffered no loneliness even when in the following year his father died.

Following his marriage to Charlotte Green of Leicester in November 1829, Hall moved to 14 Manchester Square where their only child, Marshall, was born in 1831. In 1850 the family moved to 38 Grosvenor Square from where he retired, in ill health, in 1852. There is nowhere any mention of lost infants nor expressions of regret that for some medical reason the couple were unable to have more children in an era when an only child, except where illness and death had decimated a family, was not common. It seems likely, especially from Charlotte Hall's details of her husband's lifestyle with its evenings devoted to work until 11 o'clock or beyond, followed perhaps by more reading in bed, probably in his own room, that Hall was simply too preoccupied with his scientific and medical work to devote himself to the upbringing of more children.[115] In addition, Hall was almost 40 years old when he married. He may later have drawn support from the French physician Louis who also had only one child and who, writing to a former American student in 1840 said of his family, '... et j'espère qu'elle n'augmentera'.[116]

Although Marshall Hall unwillingly became deeply involved in the institutional politics of the Royal Society and was associated with a range of reform activities relating to public health and the abolition of slavery and was a founder member of the British Medical Association, he had no real taste or at least no talent for overt politicking. Greater success within the world of science and medicine might have followed if he had. And although closely associated with Thomas Wakley, founder and editor of the *Lancet* who was also a Member of Parliament, Hall had no time for national politics. His inclinations would probably have aligned him with the Whigs. Loyalty and patriotism in the highest degree were aspects of Hall's personality as, *faute de mieux* it seems, was an ardent admiration of the constitution. This admiration of the constitution resulted from 'observing, and contrasting with its influence, the state of things elsewhere'.[117]

Interestingly the monarchy and its ceremonies were indirectly responsible for two mishaps which befell Hall. The first concerns the loss of a manuscript section of his early work, *A Critical and Experimental Essay on the Circulation of the Blood* which was being transported by stage coach on the evening of the Coronation of William IV[118] from Hall's publishers, Seeley and Burnside to their printer at Thames Ditton. The manuscript, Hall's only copy, was stolen from the pocket of the coach by someone in the crowd which had gathered to see the illuminations. The second was his inability, on account of the disruption caused by the Coronation of Queen Victoria in 1837,[119] to obtain a turtle for demonstrating some of his work on the nervous system to a visiting American medical friend.[120]

Just as he revealed no allegiance to a political party so, but more surprisingly, Hall joined no denominational religious group, despite his membership of societies with a marked religious affiliation. Throughout his life Hall remained a committed Christian. He believed in a simple mode of worship. For him the heart of his religion lay in his personal relationship with God and he saw no need to join any particular group or acquire any special label. His wife is quite definite in her affirmation that Hall never joined the Wesleyans[121] even though he retained many of the attributes of his early Methodist upbringing. Like the Methodists he was a great believer in organizing pressure groups for mutual encouragement and for achieving action.[122] It would be interesting to know if the views of Wesley, who had been interested in medical cures including the therapeutic use of electric shocks, had turned Hall against Wesleyism through Wesley's belief that insanity is typically caused by diabolical possession.[123] This for Hall was irrational and unscientific.

As a child, reading had been one of Hall's great delights. He always had one of the juvenile classics such as *Robinson Crusoe, The Arabian Nights* or *Pilgrim's Progress* beneath his pillow so that he could start the day with a good read. Although as a student he was an early riser, according to his wife he later came to love a lie-in when perhaps he returned to his earlier fiction reading. He had been obliged by his father to learn by heart many portions of Scripture and throughout life was never without a Bible. He was what was referred to as a 'Bible person'.

Hall was convinced that he suffered for his steadfast religious views. In 1839 University College London was accused by him of religious discrimination over his application for a Professorship.[124] On another occasion when he had been subpoenaed as a specialist witness in a notorious murder case he had to step down because, on religious grounds, he refused to take the oath and, according to his wife, thereby risked imprisonment.[125] A legacy from a relative is said to have been forfeited for the same reason.[126] Like many Quakers, Hall addressed his wife as 'thee' and 'thou' although his son was always 'you' as were other professional contacts including Quakers such as Thomas Hodgkin, who were part of Hall's professional and personal network in London. Despite the absence of reference to Quakerism in his wife's *Memoirs* it is possible that Hall espoused some of their beliefs and practices. In Nottingham, the Nottingham Benevolent Society to which Hall and his brother Samuel had been subscribers, although interdenominational at the time, had been founded in 1776 by members of the Society of Friends.

Charlotte Hall compared her husband's eminent powers of reasoning on scientific subjects with his child-like religious faith and obedience. Hall's almost unquestioning acceptance of the Scriptures included his belief that matters of faith were beyond the realm of intellectual enquiry. His published work contains numerous, brief but unabashed references to Christian beliefs and to their influence on his life and behaviour. In the regular, affectionate correspondence with his young son, during his boarding school days at Ventnor on the Isle of Wight, his period at a crammer at Ramsgate and then at Eton, Hall almost never failed to make some brief religious reference as a vehicle for inculcating his own religious views and values. Nevertheless his son was cautioned to be discreet about his religion and in one letter of June 1846 was warned, 'I forgot to tell you not to *dispute* on religious subjects' and, perhaps wishing to spare his son the pain his own scientific confrontations had already and would continue to cause him had added 'or indeed any other (subject)'. The father also proffered the advice, which he himself was not always able to follow that he should, 'State modestly your opinion if you like but the less the better; ....'[127] Two days later he returned to the subject affirming that, 'There is no religion in disputing; disputing is the sign of no religion, ...' adding 'generally speaking'.[128]

Impressions of Hall

Of Hall's character, the most personal portrait is that presented by his wife Charlotte in her *Memoirs of Marshall Hall.* She herself was reasonably well connected as the daughter of a landowner whose parents and grandparents included a Member of Parliament and a physician to the father of George III. Not surprisingly her *Memoirs* are a eulogy from a widow who seems to have had a very happy marriage with Marshall Hall. Nevertheless she is not an unreliable source being an educated woman who was his constant companion when he was not actually engaged on his professional duties, and someone who had some acquaintance with aspects of his work. She wrote out for publication certain of his scientific papers. Even she acknowledged the difficulty of deciphering his handwriting which was done at great speed. In one of his Royal Society papers she failed to decipher his spelling of 'capillaries' and misspelt the word throughout.[129] She also talked of his fastidiousness regarding personal habits, which at the end of his life enabled a family servant who had supposed herself unable to nurse a sick person, to administer to him with great care.[130] This fastidiousness extended to almost everything about him save one thing – the condition of his papers, which were always in disarray. Any attempt by others to tidy them up was most unwelcome.[131]

Following her husband's death, Charlotte Hall approached members of his family, former student companions and some of his professional colleagues, seeking reminiscences and impressions of her husband. These individuals were of course more likely to seek to console her with kind rather than with critical words. Nevertheless, Hall appears as a considerate, trustworthy, energetic, generous and even sometimes, in spite of an overall seriousness, a friendly and happy individual. A person of great integrity he could though, even on his wife's admission, speak his mind frankly. He was clearly outspoken. Described as being of below medium height, he was in fact short. His wife recalls her mother-in-law's account of the severe childhood illness which Marshall Hall suffered at the age of four and which the mother felt had retarded his growth whilst not affecting his lively behaviour.[132] It would seem, therefore, that he was shorter than normal for his family. As is not uncommon with children of small stature, he had been bullied at school. But one day, his wife recalls, and John Tillotson has illustrated the episode,[133] Hall decided to tackle his chief tyrannizer. Although this bully was twice his size Hall fought and beat him before the whole school, ***Plate I.*** The incident gained for him a reputation for spirit and pluck but it was also said of him, 'Though he is but a little one, it is best to leave him alone', which suggests that he could be pugnacious in his own right.[134] Charlotte Hall wonders whether this incident of the fight contained the germ of the independent spirit which was characteristic of Hall in adult life.

Plate I: Marshall Hall as a boy facing a school bully

Tillotson, J., *Our Untitled Nobility,* London, n.d., by kind permission of the British Library, London.

There seems not to have been any serious illness until 1830 when Hall was 40 years old and developed severe jaundice which took some time and a Continental tour to clear up. There remained a slight enlargement and a tenderness of the liver for which he took a daily digestive pill based on his own trial and error personal medication. He finally settled on a preparation which was also dispensed for his patients and he submitted the formula to the Pharmacopoeia.[135]

A life-long friend was Dr George Webster, who had been a younger fellow student at Edinburgh and who was the first President of the South East London British Medical Association, one of the numerous local medical associations which were the precursors of the British Medical Association. Webster spoke of Hall's unceasing ardour and enthusiasm for his professional work and for his scientific investigations.[136] James Paget made a less flattering comment about Hall's personality when in making a pun he said,

> I have made the acquaintance of Dr Marshall Hall lately. Budd introduced me to him, as able to tell him what Muller said of his reflex functions. This I have done, and he has sent me copies of his original paper in the Phil. Trans. and his Lecturers on the Nervous System. He is certainly a sharp fellow, but I should think rather monomaniac on the *reflections*.[137]

But Webster observed that Hall had nothing of the pompous or solemn manner associated with physicians half a century earlier.

Webster also provided an anecdote concerning Hall's professional manner. A rich elderly lady living in the country had been advised to contact Marshall Hall, then newly arrived in London. Entering her room with his frequently noted quick step he had asked some plain straightforward questions, given a few precise instructions to the nurse and retired with the lady's regular medical attendant or general practitioner. The woman, clearly used to ordering her physicians about, had been indignantly surprised by his unceremonious conduct. She doubted his skill but was convinced by her subsequent improvement that he was at least as successful as the 'old boys' she had been used to.[138] Sir Edward Bulwer-Lytton, a Whig MP who spoke in defence of the 1831 reform Bill and for the abolition of Negro apprenticeships, included in his novel *The Caxtons. A Family Picture* a favourable portrait of Hall who is actually named and presented as kind, able, and just the physician to turn to in a dire emergency.[139]

A tremendous capacity for hard work was an observation of Wakley who published some of Hall's lectures[140] and who once described Hall as the most conscientious man he had ever met. A workaholic would have been his contemporary label. In referring to his own industry, young people were advised that they too should always have a subject. In his wife's opinion he could make use of the scraps of time which most people threw away or spent on trivial pursuits.[141] Since his experimental research was carried out at the end of a busy day with patients and much of the writing immediately followed the experiments or was done in his carriage en route to visits, it is clear that he can have had little time to pursue his own non-professional interests. Concentration and writing were not marred by the chatting of people around him, provided he did not have to join in and he positively loved music in the background. Professional calls which interrupted his evening work never irritated him; he quickly settled back to it.

Hall had a life-long admiration of France, its institutions especially those relating to science and medicine and its language which he spoke and wrote well. Unlike German works, French sources presented no difficulties. He had travelled each year to the Continent and was proud to include among his admirers Flourens and Louis to whom he had dedicated some of his works.[142] Louis was once entertained in London. Towards the end of his career Hall contributed further articles on the nervous system and on artificial respiration to the *Comptes Rendus.*[143] His tendency to sprinkle his lectures with French expressions and quotations provoked the irritation in 1836 of a reviewer for the *British and Foreign Medical Review,* in his analysis of a collection of medical lectures, syllabuses, addresses and hospital reports. Among these was Hall's 'Syllabus of Lectures on the Diseases of the Nervous System; to be delivered in the Aldersgate School of Medicine'.[144] The reviewer had admonished Hall for using terms such as *Méthode Numérique* which had a perfectly well understood and established English counterpart, that is, *Numerical Method.* He was further accused of the misleading use of terms such as *cliquetis* to describe a particular type of breathing when in fact the adjective rattle or ronchus, itself hardly English, was what was meant. Finally, in the reviewer's opinion Hall's unnecessary partiality for the French idiom was reflected in his reference to *the* anatomy, *the* pathology, and *the* morbid anatomy which was certainly not idiomatic English.[145]

Hall's works, some of which were translated into foreign languages,[146] brought him recognition abroad through honorary membership or foreign associate status of a number of prestigious scientific societies.[147] For instance, he was elected Foreign Associate of the Royal Academy of Medicine in Paris 22 July 1845.[148] In 1850 Flourens made a verbal report to the Academy of Sciences in Paris, of *Synopsis of the Diastaltic Nervous System,* an outline of Hall's first Croonian Lecture delivered at the Royal College of Physicians that year.[149] In the following year Hall presented a paper to the Academy on his latest experiments on the nervous system.[150] In September 1855, following the death of Fodera, which created a vacancy, and three months after his departure from Paris following the completion of *Aperçu,* Hall learned that he had been nominated Corresponding Member of the French Academy of Sciences in the Medical and Chirurgical section. His election was noted in several publications including the *Comptes Rendus*[151] and *Gentleman's Magazine.*[152] The voting figures, announced by Claude Bernard, showed that Marshall Hall had gained 39 out of 41 votes, in a contest in which his four opponents came from Vienna, Edinburgh, Turin and Heidelberg. The two remaining votes were cast for Riberi of Turin.[153] Understandably Hall's wife contrasted this recognition by France's premier scientific institution with his treatment at the hands of the Royal Society.[154]

Relations with his Family

That Hall was regarded by his wife as a loving and affectionate husband, can be seen from the letters to her when on holiday with their son, during periods of childhood convalescence. It was with some reluctance that she later reproduced one of his final notes scribbled during his terminal illness when unable to speak. The note read,

> Brighton, May 30th, 1857
>
> Dear, dear, dear Charlotte – I love thee more dearly, more tenderly, more fondly than ever. I can sometimes scarcely look at thee and watch thy care of me without tears. And my prayers in the night are for thee – Ever, ever thine, entirely thine.
>
> *MARSHALL HALL*[155]

As a father, he was devoted to his son, whose health in childhood and whose future he constantly worried about. But the son was driven hard from an early age and could have found his father's regime hard and oppressive. He faced high parental expectations and ambitions in terms of learning and developing an interest and expertise in a wide range of scientific subjects as well as philosophy, art and languages, the subjects lacking in Hall's own early curriculum. On one occasion the child, not yet nine, was in Brighton recovering from a cold and had received a *History of England* albeit 'full of illustrations'. The work was abridged from Hume and Smollett and the boy was informed that his mother would tell him about the authors. At the same time he learned that he was expected to have mastered his Latin.[156]

Young Marshall knew he had to grow up to be well informed 'for it is *education* which makes the best difference between one man and another'.[157] In his father's rather extreme opinion, a man without science and literature is lower in the scale of creation than some animals. Earlier the same year he had been assured that '*truth* and *virtue* and *science* are the most beautiful things in the world' and a fortnight later, in the religious part of the next letter he was informed that 'all history, and all philosophy, and all knowledge is nothing compared with the knowledge of Jesus Christ ....'[158] The child was frequently reminded that he should 'Play when at play; work when at work' since this was the way to be both clever and active.[159] Hall almost certainly did not adequately attend to his own maxim even though he loved playing bowls with his friend Dr Webster at Dulwich if he could snatch a couple of hours recreation. His son clearly did not enjoy too many idle hours either.

During the boy's school days Hall's letters were frequently mini-lectures and the topics he raised ranged from astronomy to volcanoes. He even wrote to Professor J. D. Forbes of Edinburgh, in November 1846 on his son's behalf concerning his interest in geology.[160] Davy's work on metal oxides, gunpowder, oxygen, the functioning of a stocking machine and the process of bleaching, as well as the life cycle of animals and the unforgettable architecture of Athens and Rome were the subjects of these letters. They also contained frequent reference to the boy's progress in languages. At one time the lad was studying the classics as well as learning French, German and Italian and Hall revealed an attitude of determination even about the acquisition of linguistic competence. His 14-year-old son was urged on with,

> I hope to hear you speak the German fluently when we come. This will give me real delight, and your dear mamma too. Think of this, and lose no opportunity of practising and speaking. Never mind a few blunders. Dash at a phrase on all occasions, and you will soon speak correctly as well as easily. Do the same with Italian and French with your masters. – Your most affectionate papa,
>
> *MARSHALL HALL*[161]

But it was the classical languages, absent from his own early schooling, about which Hall cared most. Having sent his son a copy of Wordsworth's *Athens and Attica* and promising him a copy of the same author's *Greece,* Hall reflected,

> What a *colouring* a knowledge of the classics diffuses over these works! What a delightful, or rather delicious thing is the possession of such knowledge! It sheds its lustre over a man's life, occupations, reading, conversation, correspondence – everything! and fits him for the highest society.[162]

Hall always demanded much of himself in terms of languages. He was self-taught in Latin in order to cope with his Edinburgh medical examinations and he studied German at the age of 47 and Hebrew intermittently over the years.[163]

There was no shirking from providing moral guidance to his son and once when he needed comforting about some unhappy incident at school his father warned him that in the world are people who are selfish and who show little consideration for others. On the other hand concern was expressed at what he once detected as over indulgence towards his son by a friend. It was hoped that his friend did not always give in to him '... for this would not be good for you or any boy. *Boys* should experience opposition, for, as men they will meet with plenty.'[164] It is not quite true, as Charlotte Hall claimed, that their son gave them no trouble. In 1852 the son had been sent to America in advance of his parent's visit to the United States and Hall wrote to one of his medical contacts there about his unspecified anxiety and 'affliction' concerning some 'evil'. He begged his contact to speak plainly to the son as if he were the parent.[165]

Later, there was no hesitation in delineating the character and qualities seen as desirable for a future daughter-in-law. His own wife was charmingly proposed as a good model, but Hall unequivocally believed that women, like children, should be seen and not heard. His son was told,

> Display in a man is disagreeable; in a woman it is odious. Seek for some meek and gentle and kindly creature, without pretension and without display; and *be* the worthy object of her admiration for your virtues, acquirements, and well-earned distinction. It is an odious thing to see, as I have seen it, the wife everything in a house and the husband nothing. It is unnatural and monstrous.[166]

According to the *Lancet* biography in 1850,[167] Hall had hoped that his son would follow him into the medical profession by beginning his studies at Harvey's alma mater of Caius College, Cambridge and then moving on to Hunter's school of training at St George's Hospital in London.[168] The son did enter Caius College but did not graduate and seems to have been unsettled about future career plans.

Hall had successfully persuaded his brother John to enter the medical profession but he did not succeed with his son of whose eventual profession he knew nothing. On his own retirement, Hall had written to his son, 'May you, my dear boy, live to attain my age, amidst some useful and worthy career, and be thus equally blessed ....'[169] His son's lack of enthusiasm for medicine was probably a disappointment and his own interest in and encouragement to another young doctor, Mr C. Hunter who helped with some breathing experiments for one of his later works on resuscitation, was probably a substitute for not being able to guide his own son.[170] As Hall himself had contemplated doing on leaving Edinburgh and while looking around for a good practice, his son joined the army, and was a captain with the First Royal East Middlesex militia. He then changed his career plans and entered Lincoln's Inn in November 1859 two years after Hall's death and when he was almost 29 years of age. He was called to the bar three years later, practised in Wiltshire and became a J.P.[171]

Sporting and outdoor activities were his son's passions but he retained his scientific interest throughout life, was a member of the Scientific Club, a near neighbour of the Savile Club in Savile Row, and was made a Fellow of both the Geographical and the Chemical Societies. As a keen mountaineer he was a member of the Alpine Club. He was also an expert yachtsman and later in life undertook marine dredging work.[172] His other enthusiasms were combined in rock climbing and geology and in 1894, two years before his own death in April 1896, he wrote to Sir Archibald Geikie (1835–1924) on a number of issues.[173] Permission was sought for examining Geikie's Hebridean rock specimens and slides and Geikie was told of a recent article in the Alpine Club *Journal*[174] with views and maps which could be of use should the author bring out a new edition of what was flatteringly described as Geikie's 'most delightful book on Scottish scenery'.[175] In spite of being professionally a barrister Hall seems to have been more enthusiastic about pursuing his interests than in making money for on his death[176] he left very little money to his wife.[177]

Notes

1. Marshall Hall's wife recorded that the family claimed descent from John Fitzwilliam who had been at the Court of William the Conqueror. Robert Hall brought up his family in a residence which was later called Basford Hall and which during the present century became a Welfare Centre for the National Union of Mineworkers. Although he was not a target of Luddite action around 1815, Robert Hall's extensive cotton mill and works caught fire and were destroyed in 1820.
2. According to one story related after Hall's death by his widow in her memoirs of her husband, Hall had early applied some of his acquired knowledge on science and technology to the solution of theological problems. Visiting his father's book-keeper (nowadays accountant), who was also a respected Methodist preacher, Hall asked him if Hell was under the sea. If it were, he had been thinking, then holes bored at the bottom of the sea would let sea water through which would

'put the fire out'. Hall, C., *Memoirs,* 1861, p.6.

3. A number of Hall's letters to his family are reproduced in his wife's memoirs. In one letter in 1812 following his graduation, he told his father that through the experiments he was planning 'I expect to be Chief Physician to his Majesty ....' Hall, C., *Memoirs,* p.29.
4. *Ibid.,* p.30.
5. Hall, M., 'On the Combination of Oxygen. In a letter from the Author', *Journ. Nat. Phil. Chem. and Arts,* 27, 1810, pp.13–217.
6. In one of his early works, Henry, T., *Experiments and Observations,* 1773, he was described on the title page simply as 'Apothecary'.
7. Henry, W., *Tribute,* 1819. William Henry explains that his father did have ambitions to go to Oxford prior to entering the Church but his parents could not afford the costs. Thus he became an apprentice to an 'eminent apothecary of Wrexham'. On completion of his apprenticeship he did reach Oxford as assistant to the City's leading apothecary but later moved to Manchester where Percival was one of his professional colleagues and friends.
8. *Ibid.* The College of Arts and Science was one of several, providing primarily for the education of intending non-conformist ministers. Charlton, H.B., *Manchester University,* 1951, pp.13–14.
9. Marshall Hall gives this information in a note appended to his *Letter to Prince Albert,* 1851, p.11. Hall was appealing for patronage for his brother Samuel who, despite numerous earlier lucrative technological successes, which Hall cites, *ibid.,* p.12, had now fallen on hard times and was living in London's East End rather than near his brother in the smart Grosvenor area.
10. Henry Thomas's son William gave the title of one of his father's publications on this subject of bleaching and dyeing, 'Considerations relative to the Nature of Wool, Silk, and Cotton as Objects of the Art of Dyeing; on the various Preparations and Mordants requisite for these different substances; and on the Nature and Properties of Colouring Matter – Together with some Observations on the Theory of Dyeing in general, and particularly the Turkey-Red', Henry, *Tribute,* p.16.
11. Samuel Parkes maintained, with the support of Professor Copland, that 'oxymuriatic bleaching was employed at Aberdeen in preparing goods for sale, many months prior to any such application of it at Manchester, or at any other place in Great Britain', apart from Watt who was actually the first person in Great Britain 'who introduced *science* into the bleaching process'. Parkes did allow that others including Henry may independently have developed the same process. Parkes, S., *Reply to Dr Henry's Letter,* 1816. William Henry also submitted a letter, and accompanying correspondence, to *Annals of Phil.,* 6, 1815, pp.421–4 in which he asks that the Rev. A. Rees should in the forthcoming reprint of his *Cyclopaedia,* acknowledge the contribution to the practice of using oxymuriatic acid for

bleaching, by his father Thomas Henry, as well as Mr Watt of Birmingham. The two men exchanged views on the issue.

12. Charlotte Hall's claim that 'Success crowned Mr Hall's persevering efforts, and to his genius this country owes the present mode of bleaching....' was clearly an exaggeration. Hall, C., *Memoirs,* p.2.
13. Liebig, Justus von, *Letters on Chemistry,* 1843, pp.27–8. Hall reproduced a substantial amount of this quotation in his *Letter to Prince Albert,* note b, p.11.
14. Like his wife, Hall exaggerated his father's role, when he wrote to his son in 1847, 'I much wish you to understand the former and present processes of bleaching; your grandpapa first introduced the latter.' Hall, C., *Memoirs,* p.271.
15. *Dict. Nat. Biog., 24,* London, 1890, p.80. See also: (i) Felkin, W., *History of Machine-Wrought Hosiery,* 1867, pp.302–7, for details of the patents granted to Marshall Hall's father and to his brother Samuel. Felkin himself was associated with the Nottingham hosiery and lace industry and belonged to a family of Dissenters. (ii) [Felkin, W.], *Centenary Edition,* 1967.
16. Hall, *Letter to Prince Albert,* p.5.
17. The machine for 'gassing' lace had been displayed in 'the department No. 96, Class VI', at the Great Exhibition of 1851. Apparently Queen Victoria and Prince Albert had noticed it but in Hall's opinion '... to have been seen to effect, it ought to have been in operation'. *Ibid.,* p.11.
18. It was this misfortune which had prompted Hall's letter to the Duke of Sussex.
19. *Dict. Nat. Biog.,* 24, pp.85–6.
20. Hall, C. *Memoirs,* p.420.
21. *Ibid.,* p.23. His first letter from Edinburgh in December 1809 had been to Samuel explaining how careful he was being with his money but pointing out that he did nevertheless need further financial help.
22. According to his son, the father joined '... a congregation of Protestant Dissenters'. Henry, *Tribute,* pp.34–5.
23. Hall, C., *Memoirs,* p.4.
24. *Ibid.,* p.29.
25. *Ibid.,* p.7.
26. Peterson, M. J., *Medical Profession,* 1978, p.53.
27. Henry, W., 'Experiments on Carbonated Hydrogenous Gas; with a View to determine whether Carbon be a Simple or Compound Substance', *Phil. Trans.,* 87, Part II, 1897, pp.401–15. 'Experiments on the Quantity of Gas Absorbed by Water, at Different Temperatures and under Different Pressures', *ibid.,* 93, Part I, 1803, pp.29–43. 'Appendix to Mr William Henry's Paper, on the Quantity of Gases Absorbed by Water at Different Temperatures and under Different Pressures', *ibid.,* pp.274–6. 'On the Action of Finely Divided Platinum on Gaseous Mixtures, and its Application to their Analysis',

ibid., 114, Part II, pp.266–89.

28. Watson, R., *Chemical Essays,* 1800. This is almost certainly the edition – the last – which Hall would have consulted. Richard Watson read Mathematics at Cambridge and, following his appointment as Professor of Chemistry set about teaching himself the subject and survived the shattering of his workshop by an explosion. He abandoned chemistry for divinity – on which he published extensively – gaining, through some skilful footwork the qualifying D.D. just in time for his appointment to the Regius Chair of Divinity at Cambridge on 31 October 1771, following his predecessor's death less than a month earlier. Watson described himself as a Christian Whig. He became Bishop of Llandaff. *Dict. Nat. Biog.,* 60, 1899, pp.24–7. The Preface to the final volume of *Chemical Essays* is particularly directed to young people. The reference to the deleterious influence of the 'Manners of the Age' on 'a young man of seventeen' would certainly have appealed to young Marshall Hall (who was 15 years old at the time) and its message of morality on sex, drinking, excessive spending etc. was clearly taken to heart by Hall. It is less likely that Hall would have approved of Watson having burnt all his chemistry books on completion of *Chemical Essays* so that he 'could wean myself from an attachment to [chemistry]'.
29. Hall, C., *Memoirs,* p.8.
30. Loudon, I., *Medical Care,* 1986, pp.19–21.
31. Peterson, *Medical Profession,* p.8.
32. Schunck, E., 'Dr R. Angus Smith. A Memoir', reprinted from the *Proc. Manch. Lit. and Phil. Soc.,* 1885, p.2.
33. Morrell, J. B., 'The University of Edinburgh in the Late Eighteenth Century: Its Scientific Eminence and Academic Structure', *Isis,* 62, 1971, pp.158–71.
34. Emblen, D. L., *Roget,* 1970, p.18.
35. Morrell, 'University of Edinburgh', p.167. Morrell indicates that living costs for the 'abstemious or parsimonious' – probably a good description of Hall – could be as low as £15–£20 per session.
36. Hall, M., *On Diagnosis,* 1817. Hall's course of unofficial lectures, given in 1813 on 'The Principles of Diagnosis' were for a group of about 50 including his friends such as Robert Grant and Robert Lee. Sir Humphry Davy's brother, also a medical student, attended these lectures.
37. MS *List of Students Attending Lectures in the University of Edinburgh from 1790,* University of Edinburgh, p.246. MS *Lists of Students, General and Medical,* 1811, 1812, 7, University of Edinburgh. MS *ibid.,* 1812, 1813, 8, University of Edinburgh. It seems that Hall continued to attend lectures after his graduation in 1812.
38. MS *Medical Matriculation Lists, 1810–1811,* 6, University of Edinburgh.
39. Hall, C., *Memoirs,* p.10. Andrew Fyfe provided Charlotte Hall with

details about her husband, following his death.

40. *Ibid.,* p.11.
41. Class size varied considerably. Those for anatomy or chemistry which were popular, vocational and compulsory for students intending to graduate, attracted large audiences. Joseph Black illustrated his lectures with demonstrations and had classes of about 200. His pupil and successor Thomas Charles Hope attracted, in the 1820s, classes of more than 500. He became very rich. Morrell compares the Scottish situation where those who attracted small audiences could be distressingly poor, with the situation in Oxford and Cambridge where Professors received fixed and adequate salaries irrespective of class size. Morrell, 'University of Edinburgh', p.161.
42. According to Emblen, on average only 22 students out of a class of four to five hundred were awarded the MD degree, Emblen, *Roget,* p.31.
43. Hall, C., *Memoirs,* p.10.
44. *Ibid.,* p.12.
45. Hall, M., *De Febribus,* 1812.
46. Ackerknecht, E. H., *History of Medicine,* 1982, p.172
47. *Edin. Med. and Surg. Journ.,* 8, 1812, p.525. It would have been his research for this dissertation which enabled Hall to write the review of Thomas Sutton's work, Sutton, T., *Delirium Tremens,* 1813, in *Edin. Med. and Surg. Journ.,* 9, 1813, p.352. The review is unattributed but in the *Dictionary of National Biography* Hall is strongly tipped as the reviewer.
48. Gray, J., *Royal Medical Society,* 1952. As Morrell has pointed out, these societies mitigated to some extent the impersonality of big lectures. Morrell, 'University of Edinburgh', p.170.
49. Power, Sir D'Arcy (ed.), *British Medical Societies,* 1939, p.12.
50. *Laws of the Medical Society,* 1823, pp.13–15. [In a collection of booklets on the Edinburgh Medical Society, entitled: *List of Members, Laws, ...,* 1820–1827].
51. *Members of the Medical Society,* 1823, p.57.
52. The minutes of the Society's meeting 8 December 1809 reads: 'Messrs. Cheyne, Minoch, Hall, McCullock, White and Kemble signed the Laws. M. Holland.' MS *Minutes of the Royal Medical Society,* 1805–11, University of Edinburgh.
53. MS *Records of the Royal Medical Society Commencing in February 1811, Dissertations, 1811–1812,* 66, University of Edinburgh, pp.193–209. Hall opened this dissertation with: 'In my paper on Nosology ....' which suggests that he had already presented a paper to the Society.
54. *Ibid.,* 67, 1811–13, pp.339–49. This dissertation was included in *Dissertations by Eminent Members of the Royal Medical Society,* 1892, pp.84–94. Hall thanked his fellow students 'Messrs. Bigsby, Fyfe, Wright, Sandberg and Davy, who have all had the kindness to make some of these experiments for me....'

55. MS *Records of the Royal Medical Society, Dissertations 1811–1813,* 67, University of Edinburgh, pp. 417–27
56. Gray, *History Medical Society,* p.62, p.114.
57. *Members of the Medical Society,* p.103.
58. Hall described his election as his 'crowning honour'. Hall, C., *Memoirs,* p.15. *Catalogue of the Library of the Royal Medical Society,* 1896, p.50, lists 18 items of Hall's works.
59. Hale-White, Sir William, *Great Doctors,* 1935, p.86.
60. MS Managers' Meeting 15 Jan, 1811, *Minute book of Royal Infirmary of Edinburgh,* 7, Ser.1, 1 Jan. 1801–6 Dec. 1813, Royal Infirmary Edinburgh.
61. In a letter from Marshall Hall to his father. (Hall, C., *Memoirs,* p.27.) In January 1751, a decade after the opening of the first part of the Royal Infirmary, the salaries of the two physicians, Doctors David Clerk and Colin Drummond, were 30 pounds per annum. Clerks' salaries are not given or that of the Infirmary Treasurer for whom the first item in his job description was that 'The treasurer shall take the oath *de fideli* before he acts'. The second item was that 'He shall receive and pay out the money of the corporation.' *History of Royal Infirmary,* 1778, p.20, p.55.
62. *Ibid.,* pp.65–9.
63. Risse, G. B., *Hospital Life,* 1986, p.69.
64. *Ibid.,* pp.69–70. See also *History of Royal Infirmary,* p.68.
65. MS Managers' Extraordinary Meeting 11 August 1812, *Minute Book of Royal Infirmary Edinburgh,* 7, Royal Infirmary Edinburgh, 1812.
66. Hall, C., *Memoirs,* pp.41–2.
67. MS Minutes Managers' Meeting 7 February 1814, *Minute Book of Royal Infirmary Edinburgh,* 8, Royal Infirmary Edinburgh, 1814.
68. *Ibid.,* 14 February 1814.
69. Hall, C., *Memoirs,* pp.30–1. In a letter to his father.
70. *Ibid.,* p.32.
71. *Ibid.,* p.223.
72. *Ibid.,* p.32.
73. *Ibid.,* p.33.
74. Neve, M., 'Science in a Commercial City: Bristol 1820–60' in Inkster, I. and Morrell, J. (eds), *Metropolis and Province,* 1983, pp.179–204.
75. George Webster said of him, 'He was a Liberal in the best sense of the word, in all his feelings, opinions, and practice.' Hall, C., *Memoirs,* p.455. His wife reproduced a letter written after Hall's death by his former Nottingham neighbour and friend, Archdeacon Dr Wilkins. He recalled a conversation, following Hall's return from the New World, and relating to Hall's politics. Wilkins remembered Hall as an Ultra Whig but Hall pointed out '... I went to America "a Whig and something more"', but I returned a Whig indeed still, but very much less of one!' *ibid.,* p.60.

76. Neve, 'Bristol', p.185.
77. Hall, C., *Memoirs*, pp.38–40.
78. The first edition of Blumenbach's work was: Blumenbachii, [D]. J. F., *Institutiones Physiologicae*, 1787. The second edition in English was translated from the third Latin edition by John Elliotson, MD who like Hall had been a President of the Royal Medical Society of Edinburgh. Blumenbach, J. F., *Institutions of Physiology*, 1817. There was a subsequent English translation by Elliotson, of the fourth Latin edition, in 1828. In his work on human comparative anatomy, Blumenbach built up the most complete contemporary collection of skulls. He divided mankind broadly into five races – the caucasians (whites) the mongolians (yellow) the ethiopians (black) the americans (red) and the malaysians.
79. According to his wife he had tried to settle in Nottingham but the presence of four already well established physicians caused him to move to Bridgewater. Hall, C., *Memoirs*, p.40.
80. Hall, M., 'Cases of Painful Subcutaneous Tubercle', *Edin. Med. and Surg. Journ.*, 11, 1815, pp.466–9; 'Case of the Effects of Tobacco', *ibid.*, 12, 1816, pp.11–13.
81. Hall, M., 'Contributions to Diagnosis', *ibid.*, 12, 1816, pp.423–8; *ibid.*, 13, 1817, pp.63–8; pp.189–95; pp.303–8.
82. Hall, M., 'Contribution to Diagnosis', *ibid.*, 13, 1817, pp.63–8. Hall's work on Diagnosis is dealt with in greater detail in Chapter 2.
83. Smith, R. Angus, 'A Centenary of Science in Manchester', *Mem. Lit. and Phil. Soc. Manch.*, 3rd series, 20, 1883, p.441. The Society had elected 90 Corresponding Members between 1781–1881.
84. Hall, M.,'Suggestion of a New Principle for the Register Thermometer', *Journ. Sci. and Arts*, 4, 1818, pp.43–6. The title of this journal changed to *Quarterly Journal of Literature, Science and the Arts* in 1820. In the present work it will be cited as *Quart. Journ.*. 'Experiments on Stoving Cotton Goods with Sulphur' by Mr and Mr S. Hall, communicated with a note by Marshall Hall, M.D.', *ibid.*, 4, 1818, pp.196–9. 'Description of an Aerometer, for Making the Necessary Corrections in Pneumatic Experiments, for Reducing the Volumes of the Gases to a Given Standard', *ibid.*, 5, 1818, pp.52–5. 'Experiments and Observations Relative to Vision', *ibid.*, 5, 1818, pp.249–57. 'On a Mode of Preserving Some Vegetable Remedies', *ibid.*, 5, 1818, pp.366–7. 'On the Spontaneous Combustion of Cotton Goods Which have Been Imbued with Linseed Oil', *ibid.*, 5, 1818, pp.367–8. 'A Memoir on the Combined Agencies of Oxygen Gas and of Water, in the Oxidation of Iron', *ibid.*, 7, 1819, pp.55–63. 'On the Mode of Producing Cold, Proposed by M. Gay Lussac', *ibid.*, 7, 1819, pp.383–4. 'Some Observations on the Physiology of Speech', *ibid.*, 19, 1825, pp.8–16. 'On the Concealed Agency of Carbonic Acid in Determining the Decomposition of Water by the Contact of Iron', *ibid.*, n.s. 6, 1828, pp.262–8.

85. *Glover's Nottingham Directory,* 1825, p.77, p.110. Hall is attributed with an 'FRS' but 'Edinburgh' has been omitted. The Directory also lists the profession and abode of his brother Samuel. *Ibid.,* p.110. There were 10 physicians in Nottingham, listed between 20 plumbers and 4 portrait painters.
86. Bradley, I., *Call to Seriousness,* 1976, p.135.
87. Russell, J., *Nottingham Subscription Library,* 1916, pp.22–8. Hall's name inadvertently appears on the 'List of Trustees, 1858', the year following his death.
88. *Ibid.*
89. *Nottingham Journal,* 30 June 1821.
90. Rolleston, Sir Humphry, *Nottingham Medico-Chirurgical Society,* 1928, p.5. Rolleston also provides a biography of Marshall Hall, *ibid.,* pp.15–19.
91. And Charles Hastings started a quarterly journal in 1828 'with the view of raising the tone of provincial medical practice and supplying a medium for the interchange of opinion and the diffusion of knowledge ....', Little, E.M., *British Medical Association,* [1932], p.12.
92. Rolleston, *Nottingham Medico-Chirurgical Society,* p.15. His nephew Marshall Hall Higginbottom was President in 1856. *Ibid.,* Appendix A, p.34.
93. Durey, M., 'Medical Elites, the General Practitioner and the Patient Power in Britain during the Cholera Epidemic of 1831–2', in Inkster & Morrell, *Metropolis and Province,* pp.259–60.
94. Hall, C., *Memoirs,* p.338. Also Rolleston, *Nottingham Medico-Chirurgical Society,* p.15.
95. *Nottingham Journal,* 10 February 1821. Mr Henry Smith who helped Hall over many years with his experimental work was also a member of the Society.
96. *Ibid.,* 6 February 1819.
97. *Ibid.*
98. *Ibid.*
99. Hall, C., *Memoirs,* p.62.
100. Keith, A., *Menders of the Maimed,* 1919, p.81. Keith gives the spelling Manners for Manvers as the family to whom Matthew Baillie recommended Marshall Hall. This is one of several errors in Keith's account. *Ibid.,* pp.78–90.
101. According to Baillie, the book '... would have done credit to the greyest-headed philosopher in our profession'. Briscoe, J. P. (ed.), *Nottinghamshire Facts and Fictions,* 1877, p.67. Tillotson, J., *Untitled Nobility,* n.d., p.214.
102. Loudon, *Medical Care,* p.105; pp.118–25.
103. Hall, C., *Memoirs,* p.58.
104. *Ibid.*
105. *Ibid,* p.57.
106. *Nottingham Journal,* 22 January, 1820.

107. 'No person shall be qualified to become a candidate for the office of physician to the institution unless he be a graduate of the Universities of Oxford or Cambridge, or of Edinburgh, Glasgow or Dublin ... the candidate shall have resided and practised one year in Nottingham.' *Statutes and Directions for the General Hospital, Near Nottingham,* 2nd edn., 1821, pp.17–18. The first edition, *Statutes and Rules, for the General Hospital, Near Nottingham,* 1781, did not include these qualifying requirements.
108. Hall, C., *Memoirs,* p.55. Other sources give the same date. Jacob, F. H., *History of the General Hospital Near Nottingham,* 1951, p.105. Hale-White, *Great Doctors,* p.86.
109. Rolleston, *Nottingham Medico-Chirurgical Society,* p.15.
110. Felkin, W., *History of Machine-Wrought Hosiery,* p.301.
111. Hall, C., *Memoirs,* p.63.
112. *Ibid.,* p.64.
113. *Ibid.,* pp.120–1. Hall certainly fulfilled the adage that while Oxford taught men how to spend a thousand a year, Edinburgh taught them how to make a thousand a year. Nevertheless, Hall always insisted that he was never primarily interested in making money.
114. *Ibid.,* pp.66–8.
115. And perhaps Charlotte was content with her dependent, sexually undemanding, delicate angel role of tender comforter. Was she Marshall Hall's 'Angel'? See Taylor, G. Rattray, *Angel-Makers,* 1958.
116. MS Letter P. C. A. Louis to H. I. Bowditch, dated 1 Oct. 1840. *B MS Misc.* 2 Folders, Folder 2. Francis A. Countway Library of the History of Medicine, Boston, Massachusetts.
117. Hall, C., *Memoirs,* p.246.
118. *Ibid.,* p.76. The Coronation of William IV was on 8 September 1831.
119. Victoria's Coronation was on 28 June 1837.
120. In a letter to the visiting American, one of the Boston medical contacts who was part of Hall's overseas network which he greatly cherished, he explained, 'I have been to inquire about a Turtle but the hurry of the Coronation has prevented me from having one at my disposal.' But Hall assured him, 'I will certainly arrange to repeat my principal experiments before your departure.' MS Letter dated 6 July 1838, *Warren Papers,* 19, 1838–40, Massachusetts Historical Society, Boston. In a subsequent letter he acknowledged that he had 'actually failed in my endeavours to procure a Turtle'. MS Letter dated 18 July 1838, *ibid.* With James Jackson, John C. Warren established the Massachusetts General Hospital in Boston.
121. 'Marshall Hall himself never joined the Wesleyans.' Hall, C., *Memoirs,* p.3 in a note to a section of the text in which she records the correspondence between Hall's father and John Wesley.
122. See Bradley, *Call to Seriousness,* note 86 *Supra.* Among the 'causes' Hall took up were his appeals for the setting up of an institute for the

study of physiology, for the abandonment of open railway carriages on the grounds of their damage to health especially in winter, for a new City road and rail system combined with a new system of sewage management, and for the exposure of cruelty in military flogging due to its serious physiological damage. Hall cited the case of Private Frederick White of the 7th Hussars who died of his flogging injuries sustained on 15 June 1846 in Hounslow Barracks where he received 150 lashes.

123. Porter, R., *Disease, Medicine and Society,* 1987, p.27.
124. In his letter to University College, the Council was accused of persecuting him for his religious views. Hall expressed his dismay since, as he claimed, the College was 'founded upon a principle of freedom and toleration as to religion'. Hall hinted darkly at the damage done to him by a 'painful affair', in which his own involvement had demanded of him the response of a Christian. MS Letter Marshall Hall to the Council of University College 23 July 1839. *College Correspondence* 4604. Library of University College London.
125. Hall, C., *Memoirs,* p.247.
126. *Ibid.* Hall had referred to this instance in his letter to the Council of University College in 1839 when he alleged that for his adherence to his religious principles 'I have foregone large sums of money.' MS Letter Marshall Hall to the Council of University College, *Ibid.*
127. Hall, C., *Memoirs,* p.270.
128. *Ibid.,* p.271.
129. Her transcription of the paper from Hall's notes had clearly caused difficulty, because there were long gaps containing additions in Hall's writing. MS 'On the Anatomy and Physiology of the Minute and Capillary Vessels by Marshall Hall'. Communicated by J. G. Children, Sec. RS., Read 28 April, 1831, *Archives 1831, AP 15,* 32, 6, Royal Society, London. In 1837 the *Lancet* printer rejected Hall's MS of his lectures due to be published in the journal. His amanuensis, Charlotte Hall came to the rescue and recopied the whole series. She excused her husband on the grounds that his own version had been made while travelling in his carriage. Hall, C., *Memoirs,* p.222. On the other hand, his application for the Chair at University College was almost a model of beautiful calligraphy!
130. Hall, C., *Memoirs,* p.413.
131. *Ibid.,* p.245.
132. *Ibid.,* p.5.
133. Tillotson, *Untitled Nobility,* facing p.206. The caption beneath the illustration reads 'Marshall Hall knew that there was only one way of dealing with this autocrat – he must be thrashed.'
134. *Ibid.*
135. Hall, C., *Memoirs,* p.274. Hall's self-medication pill consisted principally of Barbadoes aloes, extract of liquorice, soft soap and

treacle in equal parts, dissolved in pure water and then inspissated (thickened) with gentle heat until the mixture reached the appropriate consistency for the pills. The dosage was 3–8 grains as required, to be taken with dinner.

136. Webster reckoned that, next to Hall's wife, he himself knew Marshall Hall better and more intimately than anyone else. He provided Charlotte Hall with a brief sketch which she included in her *Memoirs,* pp.442–58.
137. Paget, S. (ed.), *Memoirs of Sir James Paget,* 1901, p.93.
138. Hall, C., *Memoirs,* pp.445–6.
139. The narrator has discovered that his Uncle Roland has been taken very ill. He continues, 'In another moment I had singled out, like an inspiration, from a long file of those ministrants of our Trivia, the cab of the lightest shape and with the strongest horse and was on my way, not to my mother's, but to Dr M- H-, Manchester Square, whom I knew as the medical adviser of the Trevanions. Fortunately, that kind and able physician was at home, and he promised to be with the sufferer before I myself could join him.' The uncle recovered. Bulwer Lytton, Bart., Sir E., *The Caxtons,* 3 vols, Part 8, Chapter 1, 1849, p.42. Charlotte Hall quoted from a letter of Bulwer Lytton who had met Marshall Hall and formed a favourable opinion of him. Hall, C., *Memoirs,* pp.135–6.
140. For example, Hall, M. 'Lectures on the Theory and Practice of Medicine', *Lancet,* vol. i, 1837–8, pp.33–43; 649–57; 681–8; 729–35; 761–8; 793–800; 840–8; 873–80; 921–9; *ibid.,* vol. ii, 1837–8, pp.1–8; 33–42; 65–72; 97–105; 129–36; 177–83; 209–14; 241–7; 273–8; 321–8; 353–60; 385–90; 417–20; 465–72; 498–504; 529–35; 561–7; 593–8; 641–6; 673–8; 705–11; 737–43; 818–23; 849–57. These lectures were described as being 'Now in Course of Delivery at the Theatre of Anatomy and Medicine, Webb Street, Southwark.'
141. Hall, C., *Memoirs,* p.218.
142. For example: Hall, M., *Aperçu du Système Spinal,* 1855, was dedicated 'à M. Flourens .... Temoigne d'estime et d'admiration.' *Idem, Diseases and Derangements of the Nervous System,* 1841, was dedicated 'To P.Ch.A.Louis, M.D. etc. who has accomplished for medicine what BACON projected for the Natural Sciences in General ....'
143. Hall, M., 'Comparaison Entre les Éffets Tétanoides des États Électrogéniques, et ceux de la Strychnine de la Narcotine etc.', *Comptes Rendus,* 24, 1847, p.1054. 'De l'État de l'Irritabilité Musculaire dans les Paralysies Cérébrales et Spinales', *ibid.,* 33, 1851, pp.80–4. 'Épilepsie Traitée par la Trachéotemie', *ibid.,* 35, 1852, pp.571–3. 'La Physiologie de l'Épilepsie et de l'Apoplexie d'Origine Inorganique', *ibid.,* pp.781–6. 'Des Éffets de l'Acetate de Strychnine, *ibid.,* 37, 1853, pp.289–91. 'Note Sur le Mal de Mer', *ibid.,*

pp.600–1. 'Sur la Physiologie des Paralysies', *ibid.*, 39, 1854, pp.1090–3. 'Règles pour le Traitement de l'Asphyxie', (Extract), *ibid.*, 43, 1856, pp.569–71. 'Méthode de Traitement de l'Apnée (ou Asphyxie)', *ibid.*, 44, 1857, p.595.

The *Comptes Rendus* also published a number of letters or extracts from letters, from Marshall Hall to Flourens, under the following headings: 'Sur la Division du Système Nerveux en Système Cerebral, Système Spinal et Système Ganglionnaire', *ibid.*, 24, 1847, pp.619–22; 'Recherches Expérimentales sur le Système Nerveux', *ibid.*, 32, 1851, pp.633–4; 'Note sur la Théorie de l'Épilepsie et de la Convulsion Générale', *ibid.*, pp.832–4. 'De la Position la Plus Favourable à Donner aux Individus Asphyxies sur lesquels on Tente la Respiration Artificielle', *ibid.*, 41, 1855, p.949.

In addition there appeared a brief review of Hall's published Croonian Lectures under:'Tableau Synoptique Présentant la Théorie des Affections Nerveuses qui se Montrent sous Forme de Paroxysmes et plus', *ibid.*, 33, 1851, p.879.

144. 'Article IX', *Brit. For. Med. Rev.*, 1, 1836. p.205. According to the irritated reviewer, 'That the dedication of the lecture (to M. Louis,) should be in *French*, may be regarded merely as an injudicious whim; but, that three or four pages of the lecture should be in the French language .... For three-fourths of the hearers ... [they] were as little edifying as if they had been read in High Dutch.' The Aldersgate School was one of the Medical Schools in which Hall did part-time lecturing.
145. *Ibid.*
146. Hall's works were translated into at least French, German, Dutch and Italian. Hale-White, *Great Doctors*, p.93; *Dict. Nat. Biog.*, 8, p.965. There were also American editions of a number of Hall's works.
147. The Royal Society biographical entry on Marshall Hall for 1855 reads: Hall, Marshall, M.D., F.R.S.E., Coll. Reg. Med. Socius, Instit. Imp. Sc. Paris, Corresp. Acad. Sc. Reg. Brux., Acad. Imp. Med. Paris, Adsocius Extraneus, Soc. Imp. et Reg. Med. Vienn., Reg. Havn. Brux. Rotterod., Socc. Med. Dresd., Massach., Philad. Socius., 38 Grosvenor St.
148. 'Index Biographique des Membres, Associés et des Correspondants de l'Académie de Médecine de 1820 à 1970', *Catalogue de la Bibliothèque de l'Académie de Médecine*, 1971. Hall was informed of his election by Louis who dedicated the following work to Hall, Louis, P. C. A., *Recherches sur les Éffets de la Saignée*, 1835.
149. Under 'Correspondance', *Comptes Rendus*, 31, 1850, p.590.
150. The presentation of this paper was reported in 'Medical News', *Lancet*, vol.i, 1851, p.591.
151. *Comptes Rendus*, 41, 1855, p.93. Marshall Hall's expression of thanks was published in the journal, *ibid.*, p.1083.
152. *Gentleman's Magazine*, 45, 1856, p.56.

153. *Comptes Rendus,* 41, 1855, p.972.
154. Hall, C., *Memoirs.,* p.357.
155. *Ibid.,* p.411.
156. *Ibid.,* pp.259–60.
157. *Ibid.,* p.258.
158. *Ibid.,* p.259.
159. *Ibid.,* p.260.
160. There are three letters in this correspondence. Hall points out that his son is 16 years old and that his enthusiasm for geology was stimulated by Forbes's writing. He reminds Forbes that they share the same University and Royal Societies so that he hopes Forbes will pardon him for asking for help for his son. Forbes is most generous with his help. MS Two letters Marshall Hall to J. D. Forbes, 6 Nov., 14 Nov., 1846; MS Letter J. D. Forbes to Marshall Hall, 9 Nov. 1846. *Forbes Letter Box,* University Library, St Andrews, Scotland.
161. Hall, C., *Memoirs,* p.263.
162. *Ibid.*
163. *Ibid.,* pp.237–8. Charlotte Hall joined her husband in the German lessons. Although he could not devote enough time to become proficient in the language, he quickly acquired sufficient knowledge to serve his purpose, which included translating references to his own work.
164. Hall, C., *Memoirs,* p.265.
165. The letter was to Dr George Shattuck another eminent medical Bostonian and former student of Louis. It referred to Mr Smith a 'young medical' with a BA degree from Oxford who was also in America at the same time but with whom Hall did not wish his son to join up. Hall declared that his son needed someone to '[hold] up the mirror before him.' Hall implied that the 'evil' concerned his son's spending, for he added as a postscript. 'I do not know much of Mr Smith. I have told my son that if 300 pounds a year is not enough he is to return home.' MS Letter M. Hall to Dr George Shattuck, *Shattuck Papers,* 21a, 1848–54, The Francis A. Countway Library of Medicine, Boston.
166. Hall, C., *Memoirs,* p.272.
167. 'Biographical Sketch of Marshall Hall, M.D., F.R.S.' *Lancet,* vol. ii, 20 July, 1850, pp.120–8. The sketch was accompanied by a very recent Daguerreotype portrait taken by Mr Mayall in the Strand.
168. *Ibid.,* p.127.
169. Hall, C., *Memoirs,* p.280. In a letter to his son then already travelling in the United States of America. It provides details of the parents' travel arrangements to America.
170. Charles Hunter contributed Appendix IV 'On the Marshall Hall Method in Drowning and other forms of Apnoea or Suspended Respiration' to Hall, C., *Memoirs,* pp.502–13.
171. Venn, J. A., *Alumni Cantabrigienses,* Part 2, 3, 1947, p.202. Foster, J.,

Men-at-the-Bar, 1885, p.197.

172. In correspondence covering about fifteen months during 1870–1, Hall's son discusses his dredging and geological collecting work in the Channel and the Bay of Biscay and off Lisbon, for the British Museum and other institutions. He had designed some special aquaria and it seems he would have liked to make a new career out of this collecting at which he was evidently an informed and skilful practitioner. He was clearly in deep financial difficulties and was looking for economies including finding cheaper accommodation in Brighton. But throughout he retains a sense of humour and uses the kind of colourful language – describing a potential dredging assistant as being 'full of spunk', complaining about the weather in 'that confounded Bay of Biscay' and exclaiming 'My Crikey' about some potential hazard concerning the tanks and his specimens – of which his father would certainly have disapproved.

MS Letters, Marshall Hall to Mr H. Lee, FGS, dated:

14 April 1870, Cleveland Terrace, Hyde Park, London;
15 April 1870, *ibid.*;
23 May 1870, Yacht Novna;
30 July 1870, *ibid.*;
27 Aug. 1870, Grosvenor Crescent, Scarborough;
9 Sept. 1870, *ibid.*;
21 Sept. 1870, Cleveland Terrace;
24 Sept. 1870, *ibid.*;
26 Sept. 1870, *ibid.*;
24 Oct. 1870, Brighton;
16 Nov. 1870, *ibid.*;
14 June 1870, *ibid.*;

Documents 5382 – 1 to 12, Wellcome Institute for the History of Medicine, London.

173. MS Letter, Marshall Hall to Sir Archibald Geikie, dated 3 May 1894, *MS General 524,* University of Edinburgh.

174. *The Alpine Journal* was described, in its subtitle, as a 'record of mountain adventure and scientific observation.'

175. Geikie, A., *The Scenery of Scotland,* 1865. A second edition was published in 1887 and a third in 1901. H. H. Thomas, in the *Dict. Nat. Biog.,* 1939, described this work as 'charming'.

176. Marshall Hall (the younger) died 14 April 1896, two years younger than his father had been. His executors were his wife Mary Eliza, her brother John Edward Mammatt and his solicitor Charles S. Pemberton. *Will of Marshall Hall,* Lord Chancellor's Department, Somerset House, London.

177. Probate was granted 27 May 1896. Effects £685. *Certificate 482, ibid.*

2

Hall's Nottingham Period: 1817–26

It is clear that Hall had intended Nottingham to be a stepping stone to London. According to George Webster, Hall had 'always looked forward to London as the world of genius and science'. A solid foundation was to be laid in Nottingham with as many publications as possible and in the most prestigious journals. An effective medical training and practice depended on a sound theoretical base and while Hall longed to make a significant breakthrough in terms of presenting a major new principle in experimental medicine or physiology, in terms of credibility among fellow professionals and among patients, any new theory had to inform and enrich practice by suggesting new methods of treatment, new ways of trying to cure people and to save lives.

Some of Hall's early papers reveal that he had maintained his interest in chemistry. But he seems to have heeded the advice of Dr Balcombe of York, the parent of a fellow student at Edinburgh, who had observed, 'I never knew a great chemist make a good physician.' Hall moved away from pure chemistry to a study of the processes associated with his father's industry, as well as developing an interest in pharmacy. Thus between 1818 and 1819 he published in the *Journal* of the Royal Institution a paper on the Register Thermometer and details of his own design for an aerometer for simplifying the corrections in pneumatic experiments where it is necessary to reduce the volumes of gases to a given standard.[1] These were followed by papers on vision and on the oxidation of iron and a comment on Gay Lussac's method of producing intense cold.[2] Artificial refrigeration was a development of the early nineteenth century, although William Cullen is credited with having produced refrigeration in the University of Glasgow in 1748, by the evaporation of ethyl ether into a partial vacuum. There were also short articles on an improved method of preserving vegetable remedies which did not quickly deteriorate and which could be dispensed throughout the year until the next crop, and on the mishap of spontaneous combustion of cotton which was either intentionally or accidentally imbued with linseed oil.[3] Many cotton mills had suffered these fires and Hall was offering advice to owners and insurers. He continued to submit medical papers to the *Edinburgh Medical and Surgical Journal* but he had already turned his attention to the London scene. His association with the Medico-Chirurgical Society, which he served as office bearer once he was living in London, began with publications in its *Transactions.*

Hall's ten years in Nottingham are also marked by significant work on three major, medical themes which he rightly judged to be in need of careful scrutiny and revision. Hence they were potentially fertile ground for the major new contribution, the breakthrough he was so anxious to make and be recognized for. The three themes were the diagnosis of diseases, the disorders of women and the therapy of blood-letting. He certainly presented a fresh look at the rationale and practices associated with these themes. And though his Rule offering guidance on the appropriate use of blood-letting as a therapy must have saved numerous lives, Hall's Nottingham work did not blaze the trail he hoped for.

Hall's Work on Diagnosis

Of the early, more extended works published during the Nottingham period, *On Diagnosis* is the most significant.[4] This is the enthusiastic work of an ambitious young man setting forth what he saw as new ideas on the practice of diagnosis, presented against the background of a belief in the need for a closer relation between theory and practice in medicine. It reveals an attempt to trim and impose some new structure and organization, based on the criterion of user usefulness, on what was rightly perceived as the overburdened, existing jungle of disease identification and classification. This itself represented an advance on the old fevers and plagues categorization of disorders which nevertheless contained useful contributions on individual diseases. Hall certainly carried out some drastic pruning on existing schemes. Useful tables of symptoms were compiled which other young physicians could 'read back', arrive at a named disease condition, and then attempt to treat it appropriately. Adequate tribute was not paid to William Cullen (1710–90) former Professor of the Practice of Physic at Edinburgh, whose significant work – taken up by his own pupil John Brown (1735–88) and founded on his belief that the basic phenomenon of life and disease was a nervous force[5] – clearly contributed some invaluable ideas. In the event, *On Diagnosis,* although favourably commented on by Matthew Baillie, nephew of William Hunter, was not a forward-looking work since it lacked the significant new insights on organ and tissue pathology in disease definition, such as had been developing apace in Paris.

Definitions of Disease and Diagnosis

Hall's work opens with the premise, with which one would hardly quarrel, that 'The Diagnosis of diseases constitutes the foundation of the Practice of Medicine, ...'[6] Cullen had called this skill 'The art of Discerning and Distinguishing diseases.'[7] Hall's own justification for his work was that,

> Accuracy in Diagnosis is essential to the safe and successful treatment of diseases, because this can only be ensured by judiciously adapting the remedies to each individual case, and by cautiously avoiding those erroneous and dangerous modes of practice, to which a faulty Diagnosis must necessarily give origin.[8]

Definitions of Disease

A crucial aspect of the diagnosis issue was, as it has always been, the definition of disease and the disease status of certain conditions. Before diseases can be classified they must be defined as a group and individually named. Does a disease consist of the symptoms on which Hall spent so much time and by which it is recognized that something is wrong or is there a deeper essence which is properly 'the disease'? Despite his early emphasis on diagnosis, Hall nowhere overtly defines either 'disease' or 'diagnosis'. Medical dictionaries of the mid-eighteenth century did not always include the term disease although the *Edinburgh Medical and Physical Dictionary* provided the definition of George Fordyce,[9] a pupil of Cullen, who had also moved to London to teach. According to Fordyce,

> A disease is such an alteration of the chemical properties of the fluids or solids, or of their organization, ... as produces an inability or difficulty of performing the functions ... or pain, or praeternatural evacuation.[10]

A publication of Hall's own time, *The Edinburgh Medical and Physical Dictionary* published in 1807, acknowledged that the term disease had been variously defined by physicians, every founder of a new system of classifying them having given a different definition. Hall's wife claimed that the term diagnosis was little heard of applied to the study which it designates. For those 'of my own sex who are not Greek scholars' she explained, it referred to the accurate observation of the minutest symptoms, and the classification of those occurring in each disease; that is 'to the detection and distinction of diseases'.[11]

Cullen had already criticized Boerhaave (1668–1738) and his pupil Gaubius (1705–80) for producing definitions of disease which embodied notions of their cause. For Cullen a disease, nosologically speaking, is evident only from certain external appearances – its symptoms. From this it follows that diseases are distinguished 'only by their difference in these external appearances'. Cullen nevertheless highlighted in the opening to his *Nosology*,[12] the major problems in applying this principle,

> ... there are diseases quite different in their nature and proximate cause, and yet so similar in external appearance and symptoms, as to be with the greatest difficulty distinguished from one another, ... But as diseases, different in their nature, require not only different, but even contrary remedies, it is of the utmost consequence to the medical practitioner to be able, with certainty, to distinguish any disease from every other.[13]

Elisha Bartlett, one of many American physicians who had trained in France in the early nineteenth century when Paris had overtaken Edinburgh as the destination for American medical students studying abroad, reflected on the influence of the Paris school on his own generation. In his *Essay on the Philosophy of Medical Science* he explained that the locality of a particular disease 'may have reference to an entire organ, or to the anatomical elements, or tissues, which enter into that organ'. As exemplars he referred to inflammation of the internal lining membrane of the heart constituting one disease, whereas 'the inflammation of the external lining membrane constitutes another'.[14] In spite of his life-long Francophile attitude, Hall's own intellectual development in medicine remained in many ways very British. Had he been able, like Bartlett, to add study in Paris to his Edinburgh medical education he would probably have been a ready and enthusiastic disciple of the Paris medical school.

Present-day definitions of disease still reveal a certain lack of consensus. For instance, the *Encyclopaedia Britannica*, with echoes of the nineteenth century functional definition, explains that 'Disease commonly is considered to be a departure from the normal physiological state of living organisms sufficient to produce overt signs, or symptoms.' But whereas earlier figures, among them Cullen, listed a separate Order Vesaniae, which included mental disorders such as melancholia and mania, there are among contemporary practitioners individuals including Szasz, who like Bichat of the Paris school, always demand the existence of underlying physical abnormality in for example an organ, before accepting that disease is present.

Reznek and others, including Susan Sontag, see the concept of disease as value-laden and hence vulnerable to the changing whims of society.[15] Hall's own writings on disease reveal something of a normativist ideology most clearly evident in his work on women's diseases. Here are described a package of disorders under the generic term the Mimoses, almost none of which would be recognizable today.

Diagnosis

Diagnosis is what physicians have always done as a major part of their craft. It is both an active process and the conclusion reached by that process and is normally the forerunner to treatment which itself should be based on the diagnosis and the prognosis. Cullen had emphasized the significance of diagnosis in the medical school's curriculum. His opening remark to the Practice of Physic course was that, 'we endeavour to give instructions for *discerning, distinguishing, preventing,* and *curing* diseases as they occur in particular persons'.[16] As a physician Hall was obviously concerned with noting untoward signals or symptoms relevant to patients' health and with producing a name, or label which would identify the condition as a particular disease requiring a particular kind of treatment. But he wanted to go further by questioning contemporary classifications or nosologies of disease and proposing his own new system. Already as a student in Edinburgh, Hall had speculated that the practice of disease diagnosis, being in a state of confusion, offered opportunities for someone prepared to work at the bedside, systematically collect data and draw up an improved nosology of diseases.

And so he had embarked on work for publication while still a student. *On Diagnosis,* which embodies four papers already published in the *Edinburgh Medical and Surgical Journal* during 1816 and 1817,[17] is primarily yet another physician's guide. But although it does contain accounts of the natural history of diseases, it was intended to be more than a book of disease descriptions such as had been available for centuries. This publication, it was claimed, would guide the young doctor towards more accurate diagnosis. It was dedicated to the Managers of the Edinburgh Royal Infirmary and to the physicians James Hamilton and Thomas Spens who had all provided Hall with 'the most favourable situations for the study of diagnosis, which these kingdoms afford'.[18] Perhaps this dedication was intended as reparation for his untimely departure from the Infirmary.

George Webster, writing in 1850 and reminiscing about his own student days in Edinburgh noted that he had first encountered Marshall Hall as a Clinical Clerk in the Infirmary. According to Webster, much of the credit for the success of the Edinburgh system of practical teaching at the bedside, depended on the conduct and capacity of these clerks both in diagnosing the diseases and in accurately recording all new symptoms as they arose.[19] Hall himself gave a course of lectures in Edinburgh on the 'Principles of Diagnosis',[20] which would have been unscheduled and voluntary, to an audience of about 50 including his personal friends and other interested medical students. Among Hall's audience were Robert Lee, who experienced opposition from within the Royal Society, and Robert Grant who was appointed Professor of Zoology and Comparative Anatomy at University College London and held the post for 40 years.

Without doubt, Hall's period of Clerkship at the Edinburgh Royal Infirmary, which had been joined to the University in the late eighteenth century, provided him with the invaluable clinical experience of studying diseases at the bedside. It was routine at the Edinburgh Infirmary for the physicians and surgeons to visit their patients daily rather than twice weekly, the norm in other hospitals, and to be accompanied by the Medical or Surgical Clerks. It was also the practice in Edinburgh for the clinical Professors of the University who were attached to the clinical wards in the Infirmary to give regular lectures on the most important cases, and for the clinical clerks to supervise the work of students.[21] The clerks overburdened though they were with clerical and medical or surgical tasks within the hospital nevertheless had the invaluable experience, unmatched elsewhere, of working closely with physicians, surgeons and professors from whom they gained experience and skills of immense value to say nothing of joining a prestigious career and social network. Bedside teaching thus complemented that in the lecture theatre and, as Maulitz notes, 'interlocking directorates insured a continuum between the interests of hospital and [medical] school, if not necessarily of the patients'.[22] These facilities existed in Edinburgh long before such practices were introduced in London.

Early Classifications of Disease

There was nothing unusual about Hall's interest in the classification of diseases. Sauvages (1706–67), Linnaeus (1707–78) and others had already searched for and devised classifications, ways of ordering the objects in their domain of study.[23] For Linnaeus it was plants; for Hall it would be diseases. Linnaeus' classification reflected his belief in a natural order of plants which existed independently of any classifier and which, in a sense, Linnaeus 'discovered'. Linnaeus had not seen the need to construct a system based on some criterion such as convenience or usefulness, and Reznek refers to the Linnean type of classification as Taxonomic Realism.[24] Sydenham had already tried to apply Taxonomic Realism to diseases, but they proved then, and have continued to be, less biddable than plants.

The Edinburgh Medical and Physical Dictionary, while making passing reference to other classifications or nosologies, presents in great detail the nosology of William Cullen describing it as, though not unimpeachable, highly approved and 'generally adopted in this country'.[25] Although he paid tribute to ancient medical authors, Cullen considered that they had received undue veneration for their work on diseases,

> But, in my opinion, a foolish and superstitious veneration of antiquity, or a certain affectation of learning, has, in this matter, procured for the works of the ancient physician, an unmerited degree of estimation.[26]

According to Cullen, it was not until Sydenham's work that there were any full or accurately written histories of diseases. Sydenham, (1624–89) had been well aware of the need for a classificatory system, a methodical nosology, like the botanical systems then in use. But he had been clever enough to abandon the effort to satisfy this need. It was left to Linnaeus to pick up the challenge. Sauvages who established a close rapport with Linnaeus, had started to work on the classification problem in the 1730s but did not produce his major systematic nosology until 1762 when he announced the existence of 2,400 species of disease divided into 10 classes, 40 orders and 295 genera. In the words of the contemporary French translator from the then latest Latin edition of Sauvages,

> L'ouvrage dont j'ai l'honneur de vous offrir la traduction, comprend deux mille quatre cents éspèces de maladies observées jusqu'ici. Que d'ennemis à la fois déchâinés contre le genre humain![27]

In the meantime Linnaeus and Vogel, Professor at Göttingen, had produced their own versions. But, in Cullen's view they had stuck too closely to the same track as Sauvages and hence had offered little advance in the subject.[28]

Cullen's Nosology

Cullen himself had begun to compile a nosology on his appointment to the Chair of the Practice of Physic in Edinburgh in 1766, after several years of clinical lecturing in the Royal Infirmary. He acknowledged that the task of distinguishing diseases was difficult but believed that it could be accomplished with a properly constructed nosological method.

> ... at present the distinction of diseases lies in so much confusion, that Nosology must certainly be difficult; and some carry it so far as to assert that the attempt is impossible. But certainly that is going a great deal too far.[29]

Earlier nosologies were overburdened both in terms of their structure with their too numerous genera, orders and classes and in the actual number of diseases listed.

While earlier system builders had based their nosologies on an analogy between the natural order of plants and what they perceived as a natural order of diseases, that is on a belief that diseases are recognizable, discrete entities amenable to labelling and grouping, Cullen displayed something of what Reznek calls 'Taxonomic Skepticism'.[30] He did believe in the existence of disease entities which he called species. But in his opinion too many had been identified and the whole system building of genera, orders and classes was entirely due to 'an effort of the human mind', rather than being based on any natural order. In his opinion,

> Now nature has made nothing but species; the structure of genera is an effort of the human mind, which, till the species are well known and understood, must be fallacious and uncertain; and indeed, in constituting genera, unless we have perpetual reference to the species, all our labour will prove futile and vain.[31]

Cullen therefore showed both fewer genera and fewer species on the grounds that earlier writers had listed separately genera which should have been combined. He had moved towards Hall's idea of consumer convenience,

> In constructing my system, I have enumerated fewer genera than are to be found in the systems of others; a circumstance which I conceive will be exceedingly convenient for students, who are apt to be confounded by a multiplicity of things.[32]

Further, only genera considered to be idiopathic or primary were listed and some conditions which are so trifling that they do not deserve the attention of physicians or because they are congenital, 'born with the person' and 'elude the art of physic' were omitted. This group included instances of monstrosity and hermaphroditism.[33]

It was also commonly held that those diseases cured by the same remedies were also varieties because in the end the similarities between disorders really consist in their common causal agent and remedies cure maladies only in so far as they remove their proximate cause. One illustration of this line of reasoning, was the similar nature of the group of inflammatory diseases which, it was claimed, are cured by blood-letting. Another was the intermittent fevers cured by Peruvian bark. But Cullen thought this doctrine, although sometimes useful was in practice fallacious. Experienced physicians know that diseases differing in any degree from one another, as do those of separate species, are less often cured by the same medicine than is imagined. And showing awareness of drug resistance, Cullen added that in any case over-use of a medicine leads to its becoming ineffectual.

Even though he attempted to distinguish species of diseases from their varieties, like his predecessors Cullen was overwhelmed by the task and acknowledged that his effort was by no means perfect. He was 'leaving to others, who may hereafter prosecute the same subject the discovery of a more certain discrimination'.[34] It was Cullen's scheme that Hall was taught in Edinburgh; it was this nosology that Hall was seeking to improve.

Hall's Nosology

Criterion of Usefulness

For the construction of his classification of diseases and of the tables in the practical guidance section in *On Diagnosis,* Hall had searched for a new principle on which to base his classification, one which would, he hoped, through its popularity replace those of predecessors. The chosen principle was that of usefulness. Practical utility to a physician, it was believed, must have 'naturally occurred to every person of experience in medicine' as a proper basis for classifying diseases. In fact, 'This arrangement has no other claim to attention than that of its utility; and the principle on which it is conducted is too practical not to be eminently useful.'[35]

The first step involved clustering diseases on the principle of similarity of symptoms. Such diseases most resemble one another in appearance and as a result are most likely to be confused with each other. This kind of classification, based as it is on analogy between clusters of diseases, points out to the young practitioner what a particular disease may be and thus contracts the sphere of his enquiries and focuses his attention on some crucial points. The task of drawing the distinction between diseases within a cluster needs to be especially carefully and accurately carried out; it is based on the contrast between them. The scheme as a whole was compared to the mariner's chart which,

> ... exhibits a view of the rocks and sand-banks, the difficulties and dangers of his voyage, and suggests the prudence required and the means to be adopted in overcoming the one and in avoiding the other.[36]

One result of this utilitarian principle of classification with its 'Diagnostic view of diseases', was that pregnancy was placed with tumours and diseases of the abdomen 'not because pregnancy is a disease but because it requires to be distinguished from certain diseases'.[37] At the same time a particular disease could appear in several parts of the classification, if in its different forms or stages, it resembles other, distinct diseases. Thus,

> Epilepsy for instance, is arranged with diseases of the head and with hysteria, because these affections, although different in themselves, require to be distinguished from epilepsy in its different forms.[38]

Nature of Evidence in Making a Diagnosis

On Diagnosis is concerned with an investigation of the principles of diagnosis, part of which Cullen would have included under the course on the Institutes of Medicine. That is, it deals with the nature of evidence gathered and used in medicine. At a practical level the work provides guidance on making an accurate diagnosis which Cullen would have classified as an aspect of the Practice of Medicine. Hall's aim was to extend and impose a structure on this evidence, to render it more conclusive in identifying illnesses and also to render it capable of being transmitted between medical practitioners. It has always been the case that experienced and successful physicians can sometimes with only a 'coup d'oeil' of a patient identify the malady.[39] Such physicians do not indulge in any conscious and distinct process of analysis of the general impressions or the appearance and manner of the sick person. Furthermore, experienced clinicians may be almost unconscious of the sources of their judgement and discrimination. And yet the nature of experience in medicine consists to a large extent in acquiring the capacity for receiving and acting on these general impressions. But unfortunately, 'The inexperienced practitioner is incapable of receiving these general impressions. The experienced are, in general, incapable of explaining them.'[40]

Hall had highlighted a special feature of practical medicine – through experience it is caught, not taught. As such, the precious results of experience must necessarily die with their possessors. It was Hall's intention to provide, especially for young, inexperienced practitioners, an analysis and description of what a successful physician must be using as the basis of his diagnosis. He sought to render the process of diagnosis at least partly teachable.

On Written Sources of Evidence

As practitioners, medical men have an obligation and ample opportunity to study diseases and investigate remedies. But their success in these two areas of clinical practice depends on two major sources, reading and careful observation of patients carried out at the bedside. Hall was here combining in his philosophy of medicine, the library-based ethos of the Middle Ages together with the ideology of sick bed vigilance in the home, which characterized the following centuries. To these were added the more recent benefits of hospital experience and the available technology, where appropriate.

In his *Practical Observations and Suggestions in Medicine* published after many years of fitting his experimental, physiological investigations into a busy professional life, Hall insisted that it was essential for a physician to be well read in order to become an informed practitioner. And taking an exemplar from Cullen, Hall opined that a thorough knowledge of the nervous system for instance 'is at once a guide to the diagnosis, the pathology, and the treatment of its diseases'.[41] New knowledge could be as valuable as new instruments in the diagnosis and treatment of disease,

> A knowledge of the modes, forms, and principles of action in the true spinal system especially, has become in reference to diseases of the nervous system, in some degree what the stethoscope is in diseases of the thorax.[42]

At the same time Hall, like Cullen, held that the special knowledge most essential for physicians was physiology.[43] Hall did though warn that reading alone was not enough. Hands-on clinical observation was essential. But practical experience of any sort and on its own was also inadequate,

> When every member of our profession has a sound knowledge of physiology, derived from his own cautious observation, and not, at second and third hand, from books, medicine will take its just position, and quackery its departure. But whilst a medical person can be found to boast that he, forsooth, is a 'mere practical man', quackery will continue. One kind of mere empiricism is as good as another.[44]

Thus pursuit of knowledge in physiology should of course in addition to reading include experimental and clinical work for, 'The biblio-physiology of the day can issue in no good whatever.' Reading enabled one to keep up to date with scientific advances in diagnosis and treatment. Hall was of the opinion that, 'Disease was essentially the same in the days of Hippocrates and of Sydenham as in our own. But the science of medicine has made rapid advancement since the days of these writers.'[45] Failure to keep abreast of these advances exposes the practitioner 'to the mortification of being thought ignorant by those whose education has been later', as well as depriving patients of remedies which the physician might and should have known about if his reading was up to date.[46] And thus in its turn was the model of the 'mere man of practice' also firmly dismissed. Hall had little patience for what he termed,

> ... the folly and the presumption of those practitioners who indolently and ignorantly pretend that what they are pleased to term experience, is alone necessary to form the physician.[47]

On Observation, the Science of Medicine

Observation of Symptoms

Medicine has long been regarded as an art; Cullen had opened his lectures on the Institutions of Medicine with 'Medicine is the art of preventing and curing diseases.'[48] Hall though was more fascinated by its science. And the science of medicine was ultimately the science of observation, vital for identifying disease and monitoring the effect of treatment,

> ... the science of medicine is ultimately, and more than any other, the science of observation; from observation every new fact, both in the discrimination and in the treatment of diseases, must eventually flow. Conjecture and analogy may lead to experiment; but observation must be our guide in this experiment, by enabling us first to identify the disease, and, in the second place, to ascertain the real effects and virtues of the remedy.[49]

In his *Essay,* Bartlett had also propounded the primacy of observation in all the sciences including medicine and claimed that,

> Ever since the time of Francis Bacon, the language of philosophy has been almost uniform upon this subject. The world has been constantly told, that all science, except that of a purely speculative, or metaphysical character, depends upon observation.[50]

Bartlett, a disciple of the French physician Pierre Charles Alexandre Louis (1787–1872), lauded his mentor's 'Modern School of Medical Observation' with its numerical method approach, that is an emphasis on the statistical study and analysis of morbid phenomena. It rejected as an essential or legitimate element of science 'all a priori reasoning or speculation'.[51] Louis' school was one of the numerous new Societies created in post-revolutionary France.

While Bartlett regretted the marginal impact on Britain of Louis' school and its numerical method, he did allow that the '*British school of observation* may fairly enough be said to have commenced with Sydenham; ....'[52] This school displayed some of the best qualities of the British character such as sagacity, shrewdness and common sense. Its therapeutical resources were more various and extensive than those of the French school and if it did less than France to advance medicine as a science, Bartlett believed it accomplished more as 'a useful and beneficent art'.[53] Small comfort to Hall of course.

But Cullen had already warned that, applied to medicine the practice and usefulness of observation was not quite so straightforward as often claimed. It is extremely difficult and often requires knowledge of circumstances which are not themselves always obvious. The data gathered during observation are often as Hall had himself acknowledged highly variable and can be differently interpreted. And, medical judgements still today retain a degree of subjectivity, despite numerous technological aids which are intended to standardize evidence and iron out differences of opinion between physicians. Patients' desire to seek 'a second opinion' arises out of this element of subjectivity in making a diagnosis. As Hall admitted in his own day,

> The conclusions of different inquirers respecting the same point, have been noted as various and contradictory; the evidences on which medical knowledge is founded have therefore been deemed uncertain and equivocal.[54]

Cullen had emphasized the need for frequent repetitions of observations. In fact,

> If we consider this, with the prodigious number of subjects upon which observation is required, we shall readily perceive that the work of one man's life can go but a little way in this business; ....[55]

The significance of this need for repetition was embodied in Louis' 'Method' and for one of his major studies he withdrew from Paris for ten months to devote himself full-time to handling and interpreting his mountain of research data. Neither Hall nor any other contemporary British medical researcher could look forward to such conditions.

Since evidence on which to base a diagnosis was gained through clinical observation, the process involved noting details of the countenance of the patient, the general attitude, the general surface and appearance of the body together with the condition of the hands and feet of as many patients as possible. In the case of each symptom it was important, as Cullen had already emphasized, to note the variety and modifications possible. It is rarely the case that any symptom is truly pathognomic, that is unique to a particular disease, but the kind and character of the symptoms are frequently so. For example, the form of dyspnoea (difficulty in breathing) found in pneumonia is quite different from that in asthma and other diseases in which it is a symptom. To draw up a classification of the particular forms of a symptom and associate each with a specific disease was associated with attempts to establish a system of Pathognomics (a system depending on skill in judging symptoms of diseases) which Cullen had envisaged and which Hall attempted.

Hall had devoted much of what would otherwise have been his leisure time during his Edinburgh days, to the careful and accurate collection and arrangement of the symptoms manifested by a large number of patients. In addition to recording patients' observable symptoms, other details relating to age, occupation, lifestyle, changes in symptoms, and reaction to therapies were also noted. The progress of the patient had to be carefully monitored and recorded until recovery or dissolution took place.

Although he never adopted the statistical, clinical research methodology associated with Louis in Paris and later with his pupil William Farr in England, Hall's approach was nevertheless a quantitative one. Louis' handling of medical evidence required that diseases be viewed as specific entities for which it was possible to draw up a natural order and a set of value-free descriptions. He had tried to determine, by statistical methods, the main symptoms of the two diseases tuberculosis and typhoid on which he worked. Farr demonstrated Louis' influence on his own ideas and practice in that he too attempted to enunciate a law of disease, based on his medical, statistical work, which included the compilation of sickness tables. Hall's physiological researches were later publicized in Farr's short-lived journal, *British Annals of Medicine, Pharmacy, Vital Statistics and General Science.*[56]

Observations Based on Autopsies

The Monro medical dynasty in the University of Edinburgh had effected a closer link between medicine and surgery through the union of the hospital and the medical school. And to some extent lectures were based on dissection. Despite these developments and Cullen's assertion that 'It is, I think, now agreed, that the dissection of morbid bodies is one of the best means of improving us in the distinction of diseases',[57] the discipline of pathological anatomy did not flourish in Britain as it had under Xavier Bichat (1771–1802) in Paris around the turn of the century. Bichat had refined Morgagni's earlier concept of organ pathology. Thus lesions were seen as the seat of certain diseases not just in organs but in particular tissues of these organs. And, in contrast to Hall's symptoms concept of disease, the Paris medical school was convinced that lesions or other pathological changes were the reliable indicators of disease.[58] But both recognized an altered physiology as symptomatic of disease and over the years Hall came increasingly to regret his early neglect of pathological evidence. Nevertheless, as a part-time lecturer in St Thomas's medical school in the 1840s he was teaching Pathological Anatomy. The French clinical approach in its turn came to be criticized by German physiologists whose experimental approach to medical research was in fact much more in tune with Hall's own ideology.

The careful autopsy reports of Bichat at the Hôtel-Dieu, had played an important part in enabling him to move forward from the symptoms approach to diagnosis, with which Hall was still grappling, to his tissue-oriented pathological anatomy as the basis of disease description and classification. Hall himself supported the practice of carrying out autopsies wherever possible. He had been influenced by his period as Clerk for James Hamilton who had already in 1795 propounded the opinion that,

> A view of the morbid appearances in dead bodies, enables the medical practitioner to ascertain the causes of some diseases, and the consequences of others; and, therefore can be useful only when connected with the symptoms which preceded death, and when applied to explain such symptoms.[59]

While from the 1790s the Surgeons' Clerks at Edinburgh carried out the actual dissection, Hall as a Physicians' Clerk would, along with any interested, registered student have been able to witness such autopsies. Although for anatomy classes at this time, well before the Anatomy Act of 1832, it was difficult to obtain corpses apart from those of hanged criminals, it was certainly the practice in the Edinburgh Infirmary to dissect where possible the bodies of expired patients.

The sick in the Infirmary were usually of a lower class. That is they were poorer than those being treated in private practice where families would generally not surrender them for dissection in the event of death. And so permission to carry out autopsies in the hospital was not always necessary particularly if there was no one who could afford the funeral. But the hospital statutes, to protect the hospital's reputation, required permission. This was sometimes withheld and even Cullen was once refused permission to dissect a patient who died of typhus.[60] The Infirmary later became less dependent on the provisions of the Anatomy Act of 1832 which legalized (for teaching purposes) the dissection of the bodies of those who died in the workhouse. As Richardson has shown in her compassionate study, the marginalized poor were never safe from the anatomist's knife.[61]

Despite the regular practice of carrying out post-mortems at Edinburgh and the care taken by the Managers not to offend relatives, the actual numbers of autopsies were small in any one year, so low was the mortality rate among the Infirmary patients. The autopsies were often rushed affairs – whether because the dissection of not yet rigid bodies was easier and more instructive or, as Christison has suggested, because there could be a race between relatives and students, the former to carry off the body intact, the latter to dissect it – is not made clear. Christison tells how, as a student he once restrained his friend William Cullen (great-nephew of Professor Cullen), when he was dissecting a fresh cadaver from which the blood gushed forth and coagulated on the table 'exactly like living blood'.[62] Autopsy reports, although produced, were not as detailed as those of Bichat. Risse is of the opinion that only secondary importance was ascribed to pathological findings 'which were sought merely to illustrate preconceived diagnostic notions rather than help shape new ones'. He believed that 'No serious efforts were made to link the pathology to the patient's clinical manifestations or to use the lesions as criteria for redefining certain diseases.'[63]

On the issue of dissecting dead patients, Hall believed that,

> It is certainly the duty of the physician to seize every opportunity of confirming or of correcting his opinions and his Diagnosis by an inspection of the bodies of those patients whom his care and attention have not been able to restore.[64]

He knew that Cullen had already placed some importance on pathological anatomy and on clinico-pathological correlations. He was also aware that Cullen had commented on the salutary but educational impact of dissection, with the observation that 'If every patient that dies were opened, as ours has been, it would but too often discover the frivolity of our conjectures and practice.' Cullen had also pointed out the all too frequent, but face-saving corollary of not dissecting with, 'It is not improperly said that the earth hides the faults of the physician.'[65]

The carrying out of autopsies was not exactly an attractive aspect of a physician's work. Hall described it as 'disagreeable' and 'repugnant'. But as a student he had been an ardent dissector devoting many hours to it. It would seem that as a young man he had not identified in any way with his dissection corpses. Perhaps he lacked imagination or, as yet, compassion for the poor. In his view there was a direct relationship between the number of post-mortem examinations carried out and a physician's assiduity. The physician had a duty to carry out the dissections and, as he somewhat pompously claimed,

> A contrary conduct is altogether inexcusable, as it can only be prompted by a love of ease and an indisposition to exertion. Indeed the author has long considered the number of bodies a physician may inspect, as the just measure of his industry and desire for knowledge.[66]

Hall, however, did show some sensitivity regarding the feelings of relatives when asked if a body could be dissected. But surprisingly for such a devout Christian there is no stated appreciation of the possible causes of reluctance on the part of relatives such as a fear that the deceased, if dissected, might be denied resurrection, 'life in the world beyond the grave'. He seems not to have been troubled about this possibility. Hall believed that the mind/soul did not survive physical death so that dissection was not an assault upon it. On the contrary he would have justified the practice not only on grounds of utility to anatomical and medical studies but also on the grounds that it revealed the hand of God in the design of the human body. The Edinburgh Infirmary relatives were held to have a duty, for the good of others, to allow the dissections.

> It is painful to make the request, and it is painful to visit the house of mourning and meet eyes bathed in tears; .... It is no less incumbent on the friends of the deceased to further an object so replete with general good.[67]

His pursuit of science could override the religious scruples of others.

In his early years, Hall had not been entirely convinced about the prime role of pathological evidence in diagnosis. Just as the importance of distinguishing between diseases *per se* and symptoms of disease was stressed, so also was the caution against confusing the pathological effects of disease, which dissection might reveal, with the disease itself. This is especially important where the state of disease shortly before death is particularly active in inducing morbid appearances. Hall did regret, though, that despite Bichat's work, pathology in Britain was in a very inadequate state. As a pathologist, Bichat would not have approved of Hall's emphasis on symptoms. In his view, 'Several autopsies will give you more light than twenty years of observation of symptoms.'[68] And Farr in *British Annals of Medicine* had informed readers at the outset that, 'Pathological observations, which point out the alterations in the physiological state, and which constitute disease occupy a prominent position in the present volume ....'[69]

Hall admitted that he scarcely made any speculations about pathological conditions associated for instance with the condition in young women which he called Mimoses. He gave as his reason,

> ... for this part of medicine, notwithstanding the ingenuity of some late theorists, seems scarcely to have advanced from the state of conjecture and uncertainty described by Celsus, whose words are still, in every sense, but too admissible.[70]

But he agreed that Farr was contributing some excellent observations relevant to improving the diagnostic part of medicine.

By the time a second edition of *On Diagnosis* was published in 1834,[71] Hall's regret at not having pursued pathology was even more clearly implied in his remark that, 'it is to the study of morbid anatomy that we are principally indebted for the recent progress ... in medical science'.[72] Even so he still clung to the belief – or was it hope? – that at least for the practising physician, the primacy of diagnosis based on symptoms was all important for data gathering and sound clinical judgement,

> The progress of medicine as a science – might we not say, as an abstract science? – may be considered as greatly dependent on that of our knowledge of morbid anatomy; but the advancement of physic, as a practical art, is intimately linked with our knowledge of the history, symptoms, and the effects of remedies – with the diagnosis of the disease in the living patient.[73]

Ackerknecht might have described Hall, as he did Pinel, as essentially 'a man of the eighteenth century' for being still primarily interested in classifying diseases on the basis of symptoms. Hall himself must have realised that in continuing in this tradition he had not backed a winner in terms of making his reputation. Soon there were other technological advances such as thoracic auscultation developed by Laennec (1781–1826) which extended the range of observations to be made and which improved diagnostic work on living patients. These new techniques were never dismissed by Hall. As an experimental physiologist of acknowledged, manipulative dexterity in the development and use of apparatus, it is most likely that Hall would have used the available technology, for example the stethoscope. Physicians would still have to make their own diagnosis, but with the help of Hall's programme the task could be more systematic, more consistent and hence more reliable.

Tables of Diagnosis

On Diagnosis provides detailed descriptions of the symptoms of different anatomical regions and systems in the body such as 'The stomach, The uterine system', comparing the parts and systems in both health and in disease, thus enabling at least an inexperienced physician to recognize untoward symptoms. An overall scheme of types of diseases is followed by a description of the symptoms of individual disease conditions, for example, 'Disorders of the digestive organs and of Nervous Affections', or named individual diseases such as 'diabetes' together with a number of illustrative case histories. An enumeration of cases of suspended animation and of sudden death is included in an appendix which also contains cases of drowning and still births.[74] Hall returned to work on these conditions at the end of his life, when he devised a method of resuscitation which was successful for some instances of suspended animation.[75]

Hall summarized his overall classification in eight detailed 'Tables of Diagnosis'. Two tables each are devoted to the disease categories of Fevers and Acute Cutaneous Affections and one each to disorders or diseases of the Digestive Organs, Head, Thorax and Abdomen. Within each major category is listed the cluster of relevant diseases or different stages or forms of a disorder. For example the category of 'Acute Cutaneous Affections' contains such diseases as Scarlatina, Rubeola, Urticaria, Febrilis, Herpes, Erysipelas together with Eczema which is shown as having two forms – Solare and Rubrum. Each table has its horizontal axis divided into sections giving details where relevant, on symptoms relating to the Countenance, the Attitude, the General Surface, the Functions of the Head, the Respiration, the Circulation, the Chest, the Stomach, the Intestines, the Urinary System, and the Uterine System. These are recorded against the appropriate diseases listed vertically.

Not surprisingly the entries under 'The Countenance'[76] are numerous; evidence here is readily available to the unaided eye. The details which the physician is guided to check are: The Colour, Tumidity/Shrinking, Fulness/Emaciation, Muscular Action, Particular Features, the Expression, The Tongue and The Breath. By contrast there are only two entries relating to the inaccessible uterine system on which the physician can reliably look for evidence of morbidity. These are 'The Catamenia' and 'Local Uneasiness'. A physician using Hall's tables could locate the general site of the untoward symptoms and, by noting the nature of the symptoms could identify the specific disease.

It was the subjective nature of the conclusions of individual practitioners and the possibilities of observer variation and error, due to ignorance, inexperience, carelessness or unskilled handling of patients which had concerned Hall. This general concern continued to provide the main impetus for the development of medical technology during the nineteenth but especially the twentieth centuries. In principle, as Reiser notes, the instruments of medical technology are intended to sharpen the doctor's powers of observation.[77] But contemporary research has shown that while these tools can standardize the ways of gathering data on patients so that to a large extent practitioners receive similar sensory data, there still exists the problem of variation in clinical judgement due to the use of different criteria for interpreting the data.

Those who show concern about the increasing intrusion of medical technology, emphasize its deleterious effects on the patient–doctor relationship. They underline the danger of distancing the doctor from the whole range of non-measurable data on which Hall relied and which remain invaluable, for making sound clinical judgements, but which are not available to machines. As in Hall's day this information can only be obtained by practitioners using their own senses and developing skilful techniques of questioning and observation. While Hall did not directly record patients' own verbal contributions and Cullen had cast doubts on the value of patients' accounts in history taking of clinical cases, Hall did make his patients describe their own symptoms and feelings, and provide such details as individual lifestyle, occupation and standards of nutrition. Furthermore, as revealed by his Tables and by his coloured illustrations in work on the diseases of women, he himself paid especial attention to facial and other details, including nail colour and skin condition.

Following receipt of a copy of Hall's work, Matthew Baillie had commented, 'The object of it is most important, the execution of it is very able, and I have read it with great advantage.' Despite Baillie's encouraging words and despite Hall's pruning and attempt at systematizing the evidence for diagnosis, there is no record of Hall's work making a great impact. The Paris Clinical School had, through the work of other individuals such as Corvisart (1755–1821), developed further the practice of combining the results of physical examination with those obtained from autopsy, as the basis of clinical medicine. And it was only a couple of years after the publication of Hall's work, that Corvisart's pupil Laennec had advanced medical technology through his auscultation techniques.

Due partly to his extreme Royalist sympathies, Laennec did not enjoy great popularity in his own day and was overshadowed by François Broussais (1772–1838) a liberal who had served Napoleon. Broussais condemned the artificial creation of disease units which was still the basis of the work of systematists such as Marshall Hall but, as a pathologist he in his turn simply created another system of disease classification based on lesions. Broussais came to concentrate on diseases of the gut and his therapeutic methods were reduced to dietary measures and blood-letting with leeches.[78] In 1833 France imported no fewer than 42 million leeches. In spite of his immense popularity, Broussais' influence was soon undermined by the work on diagnosis and on blood-letting of his antagonist Louis. For his part, in the decade following the first edition of his work on diagnosis, Hall sought to apply its principles. He used them in attempting to distinguish between the different forms of Mimoses which afflict young women and to providing rules for blood-letting, still one of the most commonly used therapies in medicine.

Hall's Work on the Disorders of Women and Children

Women and children have always provided a steady flow of patients for medical practitioners, and early in his career Hall enjoyed a certain reputation for his skill in handling women's disorders. His publications on these diseases spanned more than a decade, 1818 to 1831 and followed hard on the heels of *On Diagnosis.* Some of them can be regarded as exemplar material of his principles of disease diagnosis and classification and his recommendations on treatment. Reflecting on one of the major problems already encountered during his work on diagnosis, Hall explained,

> Our systems of nosology have, I am persuaded, greatly erred, in attempting to separate diseases from each other, and describe them as distinct, when they far more frequently occur in conjunction; ....[79]

Hall's own system building concerning women's diseases, involved the grouping of some disorders and the jettisoning of some long-established nomenclature which had its root in the medieval humoral picture of disease. For instance, he abandoned terms such as dyspepsis and hypochondriasis as being useless.[80]

Range of Hall's Publications on Women's Disorders

Hall's first publication, on a collection of disorders mainly, but not exclusively, affecting young women appeared in 1818.[81] Each disorder was given the generic name of Mimosis (imitator). More than once Hall was criticized for introducing new, and it was alleged unnecessary terms into medical literature. One term which did not survive was this label of 'Mimoses'. The name had already been dropped from the title of the second edition in 1820. The new title, *A Descriptive, Diagnostic, and Practical Essay on Disorders of the Digestive Organs,* emphasized malfunction of the digestive organs and of the general health as the underlying causes of diseases which he had grouped together.

Towards the end of his life Hall explained that it was the high number of fatalities among puerperal cases, which had drawn him towards his study of women's diseases. He had observed that the symptoms commonly included diffuse abdominal pain combined with disorders of the head region. The cases were diagnosed as peritonitis and following accepted practice were uniformly bled.[82] Hall questioned both the diagnosis and the remedy in his first work on disorders relating to the childbearing years. This was *Cases of a Serious Morbid Affection ... and of a Similar Affection unconnected with the Puerperal State,* also published in 1820.[83] In questioning the diagnosis of peritonitis he substituted in many instances one of constipation which definitely did not require bleeding. His amended recommendation must have reduced the incidence of death among puerperal women. To this collection of cases was later added an essay on a most distressing condition in which puerperal women could develop destructive inflammation of the eye leading frequently to breakdown of the eyeball and permanent loss of sight in the affected eye.[84]

These separate publications were then collected to form the comprehensive *Commentaries on Some of the More Important of the Diseases of Females,*[85] in which the conditions were grouped around the three main episodes of youth, childbearing and the middle and later years. Three years later, the material was yet again slightly rearranged, but this time he dealt only with the earlier and later phases of a woman's life.[86] It was claimed that observations on the disorders associated with pregnancy had so expanded it was necessary to reserve them for a distinct and separate work. This never appeared.

Interestingly, his former chief James Hamilton, Professor of Midwifery at the Edinburgh Infirmary, brought out in 1836 *Practical Observations on Various Subjects relating to Midwifery* when, despite his advanced age and half a century of professional work, he was still in practice.[87] This work dealt with the disorders Hall had omitted and a second edition was going through the press in late 1839[88] when Hamilton died and Hall had moved to pastures new with his work on blood-letting and later, the nervous system. The advertisement to the first edition of Hamilton's work, refers to Hamilton's own rules, first of noting only the opinions of those who had deservedly acquired public estimation and second of stating his objections to those opinions, with the courtesy and respect due to such professional brethren. He made several favourable references to the work of 'his old pupil Dr Gooch of London' with whom Hall had conducted a dispute, over a paper on Hydrencephaloid Affection of children, read at the Medico-Chirurgical Society in 1828.[89]

Although Hamilton dealt with uterine haemorrhage during pregnancy and delivery, he made no reference to Hall's pertinent research on the effects of loss of blood, which contains cases of women in just these conditions. On the other hand he cites the work of Robert Lee, another former pupil, freely acknowledging Lee's disagreement with one of his practical techniques for stemming uterine bleeding. But then, Hall had walked out on his Infirmary Clerkship, no doubt causing considerable inconvenience to Hamilton, whereas Gooch and Lee had not.

As with other contemporary works on women, Hall's publications are written by a man for other men – fellow physicians. Nowhere do they contain the voices of women telling the story of their lives, or of their perceptions of their illnesses or of the problems of childbearing and child-rearing from their own point of view. Nor do women have a voice in commenting on the efficacy or otherwise of the therapies and regimens recommended for them and their children. These works do though, while linking the prevailing stereotype of female constitution with certain disorders reveal, in Hall's own perception of women, some interesting deviations from the prevailing notions.

Perception of Women – Femininity and Illness

Hall's perception of women is in important respects completely in tune with that of the age, namely women in their place, women as weak and modest, prone to nervousness and in need of man's protection. On the other hand there is ambivalence in his view and, as with his opinion that childlessness is not attributable exclusively to women, there sometimes emerges a more enlightened portrait of women. Women are the victims not only of their biological characters and qualities such as an innately delicate constitution, but also of certain unfortunate practices in the way of life imposed on them. Two further specific disabling constitutional qualities contribute to the prevailing linked perceptions of femininity and frailty. First, women possess a greater development of the capillary circulation – clearly a deduction based on female menstruation – and second, 'a greater susceptibility of the nervous system than [is] observed in the male sex'.[90]

The capillary development predisposes women to circulatory problems manifested by changes in skin colour which then produces symptoms such as the pallor of *Mimosis Decolour.* Other circulatory phenomena, including oedema and haemorrhage characterize some of the other forms of the Mimoses. Belief in the second constitutional infirmity, the susceptibility of their nervous systems has, in linking women with higher levels of mental illness than men been more damaging for them. It is the basis of the long-held and continuing view that women are, as Hall unconsciously and condescending, put it,

> ... far more sensitive and susceptible than the male, and extremely liable to those distressing affections, which, for want of some better term, have been denominated nervous ....[91]

Hall had in mind for example, *Mimosis Urgens* typically called hysteria. When Rudolph Hess during the early days of his captivity in the 1940s attempted suicide because he believed he was going mad he explained his actions with, 'I know that I have recently been behaving like a woman.' He believed he had a nervous illness with a feminine form.

The major problem with nervous disorders, and Cullen had already built this into his nosology, was that they can elicit their effects in any part of the body, including head, heart, side, spine and breasts, 'indeed, ... almost every part of the system'. This gives rise to the wider range of female disorders and, painful and distressing though these are, it is precisely these conditions in particular which are apt to be confounded with more serious conditions of an inflammatory and organic nature. In cases of false diagnosis nervous disorders may provoke the use of inappropriate and potentially dangerous remedies.

Throughout his work Hall thus sought to identify the idiopathic or primary symptoms peculiar to each disorder he described and to distinguish these from the highly variable and taxonomically unreliable collection of accompanying symptoms which can often mask the telltale signals of disease. He also tried, following his principle of utility, to group disorders such as the five described as the Mimoses. They were seen as related by a common underlying cause and by being amenable to similar therapies.

Skill in obstetrics established the reputations of surgeons like

Smellie and Hunter[92] in the eighteenth century. Despite the fact that it was respectable to write on obstetrics, as opposed to midwifery, writing about the general disorders of women and children has never been considered a prestigious, fashionable and career-wise activity. Indeed Hall's final work in this genre, *Eupædia, or Letters to a Mother on the Watchful Care of her Infant* (1831) was published anonymously.[93] Not surprisingly there was at the time something of a dearth of literature in the area of women's disorders and, according to Hall, several female diseases 'appeared not to have received the consideration from physicians which is due to them'. But Hall was named as the author of new annotations to the ninth edition in 1835 of the highly regarded *A Treatise on the Diseases of Children; with Directions for the Management of Infants* by the then deceased Michael Underwood.[94] Hall used this work to advertise his own earlier publications.

Disorders Affecting Mainly Young Women

The five disorders named the Mimoses consisted of *Mimosis Acuta, Mimosis Chronica* (more usually called dyspepsis or hypochondriasis) and *Mimosis Decolour,* often called the green sickness and supposedly characterized by chlorosis or anaemia plus some other morbid affections in which the complexion and skin usually suffer. To these were added *Mimosis Inquieta* and *Mimosis Urgens,* the familiar condition of hysteria, long thought to be an exclusively female disorder. The prime cause of all these disorders was most often 'sedentariness' and 'confinement' and Hall claimed that it was his work in industrial Nottingham which had particularly drawn his attention to them.

Hall was one of the first to attempt a more rational and systematic compilation and classification of women's disorders. The seat of each disorder in his cluster called the Mimoses, was looked for in a localized change in specific organs rather than it being attributed exclusively to some general, diffuse condition within the body. At the same time attention had to be paid to the effects of sympathy between organs, for example that between the digestive system and the uterus, when interpreting symptoms of disease in women. 'Perhaps no organ is subjected to the influence of the condition of the intestinal canal, in a more marked manner than the uterus.' Furthermore, the state of the alimentary canal, probably through its effect on the uterus could, it was believed, also affect the mammae. Tumours have been induced in them which have at first been mistaken for carcinoma. But these tumours have been 'removed' by restoring the functions of the organs primarily affected, that is the digestive tract.

Plate II: Young female with symptoms of one of the *Mimoses*

The original coloured plate shows dark rings encircling eyelids and lips of a healthy hue indicating chronic general disorder of health.

Hall, M., *Commentaries. In Three Parts,* 1827. Wellcome Institute Library, London.

Plate III: Young female with symptoms of one of the *Mimoses*

The original coloured plate shows a yellowish complexion, pale lips and puffiness of the face and eyelids indicating chlorosis (anaemia).

Hall, M., *Commentaries. In Three Parts,* 1827. Wellcome Institute Library, London.

Applying his principles of diagnosis, it was claimed that much could be inferred about the condition of the uterus from observation of external features such as the appearance of the tongue, complexion, fingers and nails. A set of four hand-coloured portraits commissioned by Hall, highlighted details of just these features in young women.[95] The face has long been appreciated as a reliable provider of invaluable evidence to the good observer – a window to the state of health of the body. And the French psychiatrist Esquirol came to place great importance on physiognomy. With his portraits Hall was, in a manner unusual for the time, demonstrating to young physicians the diagnostic importance of colour as in the complexion, beneath the eyes and on the lips, in illnesses such as those described in his earliest works on women. The facial features of the portraits present as accurately as possible the listed details of each disease.

It is interesting that Hall later describes the countenance of puerperal women as highly characteristic with a unique expression.[96] But, unlike Charles Bell with his medical iconography based on males, sometimes suffering, he makes no attempt to produce illustrations of women in such times of ghastly agony. Instead the portraits he includes, **Plates II and III,** are idealized representations of young, womanly beauty, one of the preoccupations of the time. And two of the portraits, with their slim necks and smooth faces surrounded by nun-like white caps suggest, albeit gently, and despite the illness being depicted, both saintly virtue and a degree of sexual attractiveness. They have something of the contemporary Crazy Jane face, a touching image of vulnerability, or the 'pretty features' Showalter refers to.[97] They are in fact somewhat erotic. Hall, at the time he produced these portraits was still only in his thirties and unmarried and he was often alone, working far into the night. It is possible that they represented for him the Victorian equivalent of pin-ups and what J. B. Priestley might have described as Hall's 'buried life'.[98] His wife in her *Memoirs,* simply refers to them as one would to professional visual aids.[99]

Hall's Role Models – Hamilton and Abernethy

In the subtitle of the work on the Mimoses, Hall claims to be trying to prosecute the views of Dr Hamilton, father of his former chief in Edinburgh, and of Mr Abernethy, whose works he had studied during his early years. Alexander Hamilton, who had also been Professor of Midwifery in the University of Edinburgh, had in 1780 published for midwives a non-technical *Treatise on the Management of Female Complaints* which, according to his son James, had been extensively circulated among private families in Edinburgh. Subsequent editions were deliberately modified and rendered more widely acceptable to both professionals and non-professionals, that is midwives, patients and their medical practitioners.[100] It sold well. James Hamilton continued to up-date his late father's work and added practical material on the management of children during early infancy. In addition to his post in the Edinburgh Infirmary, James Hamilton had, in his early years, been Assistant Physician to the first Edinburgh General Lying-in Hospital, founded only in November 1793. The fund-raising for establishing and developing the hospital had been disappointing and the directors had been obliged to abandon their original plans of setting aside some of the accommodation especially for unmarried mothers, 'except in very particular circumstances of distress'.

No doubt Hamilton envisaged the possibility of establishing the hospital as a teaching centre for obstetrics, thus enhancing his own chances of becoming a worthy Scottish successor to the renowned obstetricians William Smellie (1697–1763) and William Hunter (1718–83). In order to publicize the existence of the hospital and its financial difficulties, as well as to encourage its use as a teaching institution, Hamilton had published in 1795 a number of interesting maternity cases from the hospital. Some of these were of such a nature 'that they may not occur oftener than once or twice in the course of a very extensive practice' but which might on the other hand be encountered by young practitioners just setting out on their careers.[101] Working alongside James Hamilton and influenced by the writings of Hamilton's father, it is hardly surprising that Hall developed an interest in women's diseases or that he should try to apply his own system of diagnosis to them.

John Abernethy had been senior surgeon to St Bartholemew's Hospital in London and was involved in the foundation of the medical school attached to the hospital. Lecturing on anatomy and physiology at the College of Surgeons, he held office there until 1817 and retired from St Bartholemew's a decade later, after 40 years' service to the Hospital. Both Hamilton and Abernethy belonged to the fraternity of medical men who produced 'advice books' on what might be called home medicine, which encouraged an attitude of self-help, 'be your own physician', in the maintenance of health and prevention of disease.

Advice Books – Codes of Health

These vernacular works, have a long history rooted in the philosophy of preventive hygiene, one of the five Institutes of Classical Greek medical education. They were, as Ginnie Smith has indicated, ostensibly written for the lay individual.[102] Two-thirds of them were produced by medical men, physicians producing more costly, eye-catching publications than surgeons. They were for long considered as an antiquated source of old wives' tales, a view possibly enhanced by what would seem their effect of undercutting the interests of the medical profession. However, according to Smith and also to Hamilton's comment about his father's work, these advice books attracted a fairly broad, mixed lay and professional readership. Willich's *Lectures on Diet and Regimen,* of which the second edition was published in 1799 and a third edition in 1800, was titled 'the most rational means of preserving health and prolonging life; chiefly for the use of Families'. This work was clearly catering mainly for lay people but it was of a standard to serve as a reference for physicians 'on some of the most recent developments of European physiology'.[103]

The *Dictionary of National Biography* (*DNB*)refers to Abernethy's surgical[104] and physiological works published between 1793 and 1830, in which he makes frequent references to the variety of diseases, both direct and indirect, which may result from disorders of the digestive organs, and which may often be corrected, therefore, by attention to diet and bowel action. No reference is made to Abernethy's less popular, prestigious but more often pocket-sized, advice books. These small works, continued to be reissued after Abernethy's death in 1831.[105]

They advocated what was referred to as Abernethy's Code of Health and Longevity, based on the three main principles of diet and regimen, appropriate purgatives and finally vigorous exercise in the open air. Abernethy quoted John Paris' support of digging as an excellent source of exercise. They also provided numerous dietetic tables showing the comparative digestibility of different kinds of food. Abernethy had emphasized the reciprocity between the mind and the body which could both act as agents of disease and of distress. In a forward to a posthumous publication, the reader was informed that,

> The amount of suffering which is inflicted on the body through the agency of the mind, is only equalled by the retributive misery reflected on the mind through the medium of the body. Hitherto the reciprocity of sympathies between the intellectual and material portions of our nature, have not been sufficiently attended to in the investigation and management of diseases. This error has been avoided in the ensuing treatise, ...[106]

In his personal lifestyle and in his writing Hall was clearly influenced by Abernethy's philosophy, which was of course being recycled from older sources, that 'everyone's health (is) in his own keeping'. The principles embodied in the Code permeate Hall's writings on women's diseases although he is not writing for a popular audience. His rather dry volumes, where slim enough, are designed for young physicians' pockets rather than for those of mothers. On occasions, Hall wrote almost as if the principles were actually of his creation. There was insistent emphasis on the importance of everyone – men, women and children – maintaining a high standard of general health as the best means of warding off disease of any kind. In conducting his own life, Hall tried to follow the Code at least with regard to the modest size of his meals and his brisk style of walking, often commented on. He reached only 67 but even when proper attention is paid to its three major points, the Abernethian Code allowed for failure to achieve longevity in those cases, when '*gradual decay* or *unforeseen casualties* happen, which no precaution can guard against'.[107] The Code could not fail.

Nevertheless, Hall violated in a number of ways the health maxims, mottos and proverbs handed down from antiquity and used to embellish a later edition of Abernethy's Code, in order to 'send into the world an "old friend with a new face"'. Hall's habit of studying late into the night ignored both the advice to 'Avoid night studies, and preserve your health and intellect' as well as the reminder that 'One hour's sleep before midnight is worth two hours after.' Also in being a nibbler, Hall was forgetting the dire warning that 'often and little eating, maketh a man fat';[108] he was living with the consequences. He had also been somewhat tardy in following the recommendation that males should marry between 25 and 36; Hall was almost 40.

References to the need for preserving one's general health at all stages in life have a modern ring and resonate well with aspects of contemporary alternative medicine. Hall was an enthusiastic advocate of preventive measures to preserve good health both against day-to-day disorders as well as against the more insidious and more dangerous long-term conditions of both sexes, for 'the noblest part of the art of physic is the prevention of disease'. This was the key to protection against consumption and cancer, 'the respective banes of youth and middle age',

> It is highly probable, that if the health and tone of the general system were duly maintained, and if the first derangements of action and function in the various organs were assiduously watched and checked, these tuberculous and scirrhous formations might not occur.[109]

In any case, and echoing Cullen's reminder, Hall cautioned that such prevention, 'is most essential, inasmuch as those diseases do not, in the present state of our knowledge, admit of cure'.

Hand in hand with this philosophy of preventive medicine was stressed the need for early detection of and prompt action upon any untoward symptoms. 'Let the first, the slightest deviation from health in the system at large, and in particular organs, be promptly and perseveringly corrected.' Details for check-up procedures were provided,

> ... let the complexion, the tongue, the state of the bowels, and the uterine discharges, be strictly noticed; and let every deviation from their healthy condition, however slight, excite due attention: we shall then probably prevent the incursion of maladies which do not admit of remedy.[110]

And both practitioners and patients, mainly the women in the household, had a role in this early detection, and self-monitoring. Medical colleagues were expected to gather information directly from patients and were encouraged to train them, especially women, to keep a strict eye on their own and their family's complexion, bowel condition and uterine discharges, in order to make useful reports to their doctors. It was acknowledged that 'All this will require watchful and steady inquiry or observation on the part of parents or of the patients themselves.' More responsibilities for women.

The strategies for maintaining a good standard of general health, rule out the over-indulgence of the rich; the work of Sadler on the Law of Population and its relevance to the incidence of sterility among the wealthy was cited.[111] Hall could never resist a dig about 'high-living' which he criticized on numerous occasions. The rich, in his opinion, reaped what they deserved in terms of their infertility but, significantly, women alone were not held responsible for their childlessness,

> No fact is so calculated to demonstrate the silent and hidden yet powerful influence of an impaired tone of the general system over the functions of individual organs, as the history of the operation of the causes and preventives of sterility .... The fecundity of the human race is diminished by the indolent and luxurious modes of life prevalent among the rich, whilst it is augmented by the laborious habits and spare diet of the poor.[112]

Saddler's data suggested that the proportionate infecundity of the two classes, rich and poor, was in general six to one, which tended to lead to the dying out of indolent families and the multiplication of the poor. Hard physical work never induced impotence but mental workaholics in any age seem less fecund. Hall made no reference to his own meagre procreative activities. He probably considered his more advanced age at marriage reason enough.

Lifestyle, Environmental, Gender, Class and Educational Influences

In terms of female lifestyles, of which Hall was highly critical since they were held to predispose women to disease and could have long-term deleterious effects, Hall included 'the baneful but prevalent habits of sedentariness and inactivity in which young persons pass their early years'.[113] And with special reference to young females, the lack of fresh air and exercise in their daily regime whether at school or college or at home was deplored. Such environments were blamed for frequently 'laying the foundation of derangement of the general system and entailing pallor, feebleness and indisposition on after life'. So, instead of having their health invigorated by exposure to the open air, and by a regular plan of active exercises females,

> ... are enfeebled and disordered by a system of sedentary studies, pursued in warm and close apartments; an occasional walk, in fine weather, being taken merely as a sort of apology for the total neglect of what alone deserves the name of exercise.[114]

Hall thought he could draw some comfort for the future for both sexes from recent slight changes in pedagogical practice in which, due to the influence of Pestalozzi, physical education was being introduced into some educational programmes,

> Happily, indeed, for the rising generation, the culture of bodily health, activity and vigour, as well as of the mental faculties, is becoming an object of increased attention.[115]

Pestalozzi's direct effect was short lived, and in spite of the beneficial influence of his successors in incorporating sound programmes of physical education into all schools, Hall might be disappointed to find, a century and a half later, how apposite his comments about lack of enthusiasm for exercise are for many contemporary female students during their teenage years.

These unsatisfactory lifestyles applied in the main to middle-class, privileged women who did have some control over their lives. This was probably why Hall was so scornful of them. One patient, the indolent widow of an Indian Army colonel, was in Hall's opinion severely in need of exercise. The treatment prescribed was a daily walk from her home near Hyde Park to the Serpentine into which she had to dip her finger.[116] Minimalist hydrotherapy of a sort. But he was mindful of the seriously unfavourable but compulsorily imposed factory conditions of confinement and airlessness which affected the more numerous working classes of both sexes.

As a physician living and working in an increasingly industrialized city like Nottingham, Hall was aware of the link between manufacturing industries and certain diseases. Those individuals – men, women and children of all ages – working in the city's cotton, stocking and lace manufacturing had a particularly sedentary life imposed upon them 'from morning till night'. They did in fact work in conditions which deprived them of pure air and even of gentle exercise. And, environmental conditions apart, malnutrition was probably an even greater problem for many of these factory workers. The effects of the disadvantageous factory conditions were compounded by the poor diet of the workers, who were rendered particularly vulnerable. Like the rich, these workers needed exercise but unlike them they also needed a more nutritious and varied diet which would promote peristalsis and avoid those conditions associated with constipation, still held to be at the root of many disorders, especially in women.

Thus a major consequence of women's constitution whether in the pampered rich or the factory-imprisoned poor, was a 'loaded' condition of the bowels, that is constipation. This, in its turn, especially when prolonged, led to a host of other disorders including an icterode hue a yellowish, jaundiced appearance in the complexion which was more often *Mimosis décolor* but which could suggest a more sinister disorder of the liver. This basic condition of constipation was exacerbated by the habit of young females to delay 'in yielding to the first solicitations of nature to evacuate the bowels'.[117] Women were strongly urged to abandon this harmful habit especially since, it was supposed, the greater volume of the abdomen and pelvis of females already predisposed them to constipation.

Hall originally claimed that he had scarcely ever seen in young men and boys the disorders which he attributed to lack of exercise in females. At this stage so firmly did he associate femininity with weak, illness-disposing qualities that he only reluctantly admitted that 'the delicate and sedentary of the male sex', are also liable to the same disorders as women. Specific reference was made to 'literary persons of a delicate mode of life ... and tailors ....'[118] Already by 1827, having gained more experience and expertise through his practice, he had added a special 'Postscript' to his.Mimoses Section observing that,

> ... there is no one of the disorders of which it treats, which does not, mutatis mutandis, occur in the male, as well as in the female sex.
>
> I have observed even chlorosis and hysteria, in their most marked forms, in the young and delicate of the male sex.[119]

It still remained the case, that to have this set of disorders was to be 'like a woman'. It is interesting that at this later date, 'clergymen and other ministers' were added as frequent victims of the disorder.

Lack of exercise was considered to lead to a general feebleness of the frame and of the mind as well as interfering with growth and development. As someone whose own growth and development was, according to his mother, impeded by a childhood illness, Hall not unnaturally paid especial attention to factors affecting growth, especially since, in his view, 'not infrequently a defective development of the form during growth' could be among the causes of later disorders. A complex relationship between the rate of development and the general state of health was certainly recognized. Too rapid growth could lead to a condition of debility which in its turn could then impede growth and the proper development of the spine and maybe even of the main body cavities, the thorax and abdomen. In women these deficiencies could be responsible for serious problems at childbirth.

While genuine concern is expressed about the ill-effects of factory work on the health of women and girls working in them, Hall is nevertheless writing with the upper and increasingly numerous middle-class women mainly in mind. They were after all the majority of his well-paying patients. Thus, one course of treatment for constipation suggests starting the day by preceding the prescribed dose of medicine with tea or coffee in bed. The number of working-class women with a growing family who could have availed themselves of such luxuries as tea in bed 'on doctor's orders' must have been almost non-existent. On the whole, the large army of female domestic workers or those putting in up to sixty hours of work in the same factories as their husbands were, as J. B. Priestley put it, assumed to be as strong as horses.[120]

Another aspect of women's lifestyle and a target probably associated with Hall's narrow, very proper and Methodist upbringing had been the 'absurd ... defective system of clothing'[121] of females. He was obviously deploring the fact that contemporary women's fashion, which included décolleté dresses, needlessly and hazardously exposed parts of their bodies. In view of their weak constitution and poor circulation, especially to the skin (an echo here of the old humoral theory that women have less heat than men), Hall would in winter have had women looking like walking teddy bears with body surface swathed in flannel and the feet encased in lamb's-wool stockings.

And yet, about 20 years earlier, the essayist John Corry, a supporter of cool or cold regimens attacked those who advised flannel next to the skin and equated such practices with quackery, asking, 'Do these

Quacks wish to introduce the sweating sickness once so fatal to Englishmen ...?'[122] But then he may not have been including English women – and Hall was. Charlotte Hall referred to her husband's own sartorial indifference;[123] it is highly unlikely that even though she managed her husband's professional finances, she was allowed to deck herself out in the latest stylish confections.

Underlining the fixity of his conception of a weak and irretrievably flawed feminine constitution despite remedial measures, Hall acknowledged that even when attention is paid to their need for exercise and for more appropriate winter clothing, there would still exist 'causes enough why girls should suffer more in their health than boys'.[124] But, by way of contrast, Hall believed that not only did women's brains and bodies, including their procreative capabilities, not suffer from healthy exercise; neither did they suffer from education. By contrast, John Conolly's son-in-law Henry Maudsley believed, even in the second half of the nineteenth century, that education damaged girls' minds. Hall held that educated women could be of great assistance to their husbands and, as usual, was critical of rich indolent females. He had undoubtedly experienced the contempt which they could display to their medical advisers, including qualified physicians, especially when they were not feeling really ill. His own wife was an excellent model of a professional helpmate. But she had only one child and could be seen as serving Hall in the role of a ministering angel.[125] Predictably she did not complain and seems to have been proud to help him.

It is not known to what extent, if at all, Hall's aspirations on promoting good health were fulfilled even within his own practice. And, as he readily admitted, one would never know what evil had been averted. The chances of making progress with new ideas are always enhanced if they can be institutionalized. Hall certainly considered that he deserved the gratitude of women when he pleaded on grounds of their sensitivity, for positive discrimination in the form of special institutions, women's hospitals and clinics, which would be given over to handling the distressing conditions peculiar to them,

> I would beg to observe that, notwithstanding the numerous existing establishments for the care of the sick poor, there are still wanting medical institutions devoted to the distressing cases peculiar to females, to which they might resort with the confidence that the utmost attention would be paid to their sufferings, and to avoid wounding their feelings.[126]

A few years earlier, in 1814, Charles Mansfield Clarke surgeon and lecturer in Midwifery had put forward an urgent and professional reason for concern at the lack of such institutions. Clarke, had already noted that 'diseases of the sexual organs in females, ... are perhaps less generally known and understood by practitioners, than any other complaints to which the human body is subject'.[127] Women themselves were blamed for this ignorance, since out of a sense of delicacy they conceal their symptoms in the early stages and often make them known only 'when they have proceeded so far as to be beyond the reach of remedy'.[128] And by this time the initial disease may have become complicated with many symptoms not originally belonging to it.

Women were in a no-win situation; men demanded that they be modest but doctors then blamed this characteristic for what was at least in part due to their own professional shortcomings. And women were hardly likely to act in their own best interests when directly attacked in print by doctors themselves – albeit sometimes anonymously – for lack of 'Virtue and Delicacy' and for disregarding the 'Dangers and Immorality' of even allowing men-midwives to attend them in childbirth.[129] (See cartoon in **Plate IV.**)

Clarke pointed out that the corresponding major reason for medical students' defective knowledge of women's diseases was sheer lack of any real opportunity of observing them 'there being no hospital into which these complaints are expressly received'. At the same time, since many of these disorders are classified as chronic they are excluded from many public general hospitals. But, claimed Clarke harking back to the problem of women's modesty, students would still gain little even if these cases were admitted to the hospitals,

> ... for it could not be expected that a patient would submit to be exposed to the manual examination of a number of students; and without such examination, little good could arise, except in cases of disease in the external organs, which bear a very small proportion to those of the internal or remote parts.[130]

Plate IV: Cartoon of Man-Midwife

An Important Address to Wives and Mothers..., 1830.

By kind permission of the British Library, London.

Disorders of Women During their Childbearing Years

Of the disorders which affect women during their childbearing years the major affections occur during pregnancy, parturition and the puerperal state. Problems during pregnancy centre around the uterus and, through its enlargement, the increasing pressure it exerts on surrounding organs and tissues. Parturition is sometimes accompanied by hazardous haemorrhage which in itself can complicate disorders of the ensuing puerperal state. This is the most vulnerable stage when, in the nineteenth century, women were all too likely to develop the often fatal puerperal fever, one of the commonest causes of maternal mortality in childbirth.

Puerperal Fever

True puerperal fever, childbed fever, or puerperal disease as Hall called it, was still, even at the beginning of the present century one of the commonest causes of the high incidence of mortality in childbirth. Thackeray's wife is said to have become mad following puerperal fever after only four years of marriage. Unfortunately, and as Hall noted, in some of the more serious cases, puerperal disease supervenes insidiously and makes a slow and probably initially unheeded but fatal progress. New reliable guidance on early and accurate diagnosis and appropriate treatment would, if accepted by the medical fraternity, have been a major clinical contribution.

Through study of large numbers of puerperal women some of whose cases were published, Hall hoped to present the medical world with the means of making an accurate diagnosis irrespective of the host of variable and distracting symptoms which accompany the condition. His observations on women in the puerperal state had first been made in the Edinburgh Infirmary; to these were added those from the Nottingham Infirmary and from his own practice. Risse's tables of cases from the Edinburgh teaching ward show only one case of puerperal fever over the period 1771–99, during which time Cullen was in post. But if complications arose, and puerperal fever was a complication of childbirth, women in the midwifery ward could be transferred to the ordinary wards. These tables, therefore, are not giving a complete picture of the incidence of puerperal fever.

Both Edinburgh and Nottingham, like so many other cities in Britain at the time, were expanding with increasing industrialization. Overcrowding, together with the deteriorating and inadequate public health facilities to cope with the growing population, was accompanied by a higher incidence of disease and a higher death-rate among the poor. Puerperal women and their infants were one of the most vulnerable groups.

The puerperal condition was apt to be accompanied by a host of symptoms involving disorders of both the head and abdomen. The whole process of pregnancy and parturition was bound to be accompanied by a range of what medical men considered to be acceptable disorders and discomfort with which women must expect to cope, since given time and restoration of the general health they would clear up. These conditions were simply part of a woman's lot, to be endured. Of importance, was the identification of those symptoms definitely associated with the potentially serious condition of puerperal fever with its life-threatening inflammation and hence serious infection. This needed to be distinguished from less serious non-inflammatory abdominal conditions.

Puerperal fever is caused by infection with septicaemia (commonly called blood poisoning), or through pelvic or generalized peritonitis or abdominal inflammation, the terms already used by Hall. A complex picture emerges of symptoms which are rendered even more complicated by the fact that the severity of the illness depends on the virulence of the infecting organism. This variation has a significant effect on the likely outcome of the disease – recovery or death. From his first-hand hospital experience, Hall knew that puerperal fever was infectious and that epidemics could occur, especially in lying-in hospitals. They could be even greater death traps than a woman's home. James Hamilton had instituted at least one enlightened practice in his lying-in hospital in Edinburgh. Each patient used the same bedding throughout her stay in the hospital and on her discharge,

> ..., the bedding on which she had lain is removed into a drying-house fitted up for the purpose, and detached from the principal building; and there it remains for at least a fortnight before it be allotted to any other patient.[131]

Hamilton claimed that through these precautions, despite many indigent Edinburgh women entering in very poor shape, only one had died since the hospital was opened. And this death was avoidable having been caused by,

> ... the immoderate use of spiritous liquors immediately after delivery, which were privately conveyed to the unfortunate woman by the person who had the nearest interest in her recovery.[132]

Presumably the culprit had been her husband.

Although he never wrote on cholera, with which he had direct contact both in London and during sojourns on the Continent, Hall does not discuss the possible mechanism for the spread of diseases such as puerperal fever. No mention is made of the possibility of it being contagious even though the issue of contagion was provoking lively debate at the time. Neither is there speculation in detail on any form of miasma, that is direct infection by bad air, being responsible. There is only the simple statement that, 'The great questions of contagion, of the influence of the state of the atmosphere, of the pathology, and of the treatment, still remain, not satisfactorily determined.'[133] Hall did believe, though, that the state of the atmosphere played some important part but resorted to his reliance on 'the condition of the general health [as being] doubtless, their more immediate cause'.

The really major but initially disregarded breakthrough in identifying the cause of puerperal fever came 20 years after Hall's work, with the research of the Hungarian assistant obstetrician Ignaz Semmelweis in the Vienna General Hospital, in which Hall had longed to gain some experience if only he could obtain financial support. Through his statistical monitoring of two maternity clinics, one served by midwives the other by students, Semmelweis confirmed his own hypothesis that puerperal fever was produced primarily by contact of puerperal women during parturition, with the contaminated hands of doctors and medical students coming from the autopsy room. Puerperal women were, therefore, in the hands of such doctors, exposed to greater risks than in earlier periods when cared for exclusively by midwives.

Had Hall, particularly when he later became associated with Farr and his quantitative medical research, followed up one of his own hunches of 1827 and monitored his results, he would almost certainly have found a drop in the incidence of puerperal fever among his own patients. Maybe this was the breakthrough he missed. For, re-emphasizing his belief in prevention and in the association between filth (faeces) and disease, Hall speculated,

> A point not less important than the treatment of these affections, is their prevention. I believe no means would conduce so much to this purpose, as the invariable administration of copious warm water injections at some period before or during labour. The large intestines would thus be relieved of their load, and a great and fertile source of future disease would be removed.[134]

Properly carried out, such a procedure with its controlled expulsion of faeces would undoubtedly in some cases have removed the organisms *Escherichia coli* and *Clostridium welchie,* which are part of the flora of the lower bowel and of the faeces. These organisms, along with a range of Staphylococci and Streptococci which would have flourished on the skin and tissue of cadavers in the autopsy rooms, are among the micro-organisms which cause puerperal fever. In addition, with Hall's recommendation, the flow of some of the warm water over the attending physician's hands would have removed from them at least some of the dangerous micro-organisms involved and would have further reduced the incidence of puerperal infection. Hall had had a good hunch but failed to follow it up and monitor the results. Neither did he do any quantitative analysis of the case studies which he published, in order to propose a set of guidelines for early and accurate identification of puerperal fever.

Sexually Transmitted Diseases

There is throughout Hall's work on women, an absence of explicit reference to female sexual characteristics and behaviour and their relationship to sexually transmitted diseases. Hall had been brought up in a close-knit, highly conventional houschold with rigid Christian morals. Throughout his life he believed that the modesty of women should at all times be preserved. Hall was never a clubman and did not entertain or dine out; he led a very narrow and sheltered personal life. Perhaps at the outset he had not fully appreciated the sensitive territory into which a comprehensive coverage of women's disorders would take him.

While James Hamilton was quite prepared to acknowledge that 'An unmarried lady consulted the author respecting the time she should disappear for the purpose of being confined secretly',[135] Hall would never have publicized his willingness to offer such professional advice in similar circumstances. Similarly, in his additions to his father's work Hamilton added details on the 'Qualifications expected in a Hired Nurse' and stressed that she should have 'Good breasts, prominent nipples ...'[136] whereas Hall who generally referred to 'the mamma' rather than 'the mammae' would have used more covert language to refer to the need for a regular and copious supply of milk.

For a comprehensive coverage of women's disorders Hall would have needed to include the venereal diseases, rampant in the nineteenth century, which naturally afflicted both sexes and on which there was contemporary literature. Even though Hamilton had pointed out that women could be the innocent victims of their husband's indiscretions, the majority of female patients with venereal diseases would have been prostitutes. In his writing, Hall showed none of the social, ethical or even medical interest of contemporaries such as William Acton who had been Externe to the Female Venereal Hospital in Paris and who in 1841 published *A Complete Practical Treatise on Venereal Diseases.*[137] In the year of Hall's death, Acton published a work on *Prostitution, Considered in its Moral, Social and Sanitary Aspects, in London and Other Large Cities.*[138]

Hall was nevertheless aware of some of these gaps in his own work and on sensitive aspects of women's diseases involving 'the mammary and uterine systems' he makes uncharacteristically generous reference to the work of Sir Astley Cooper and Dr C. M. Clarke who have 'so admirably treated' these conditions 'that it is only necessary to refer my readers to the valuable and well-known works of those authors'.[139] One almost hears a sigh of relief that someone else had charted this territory and he could pass on to safer ground.

Hall's early sensitivity on sexual matters persisted if not increased. In *Diseases of the Nervous System,* 1841, there is a short section on the effects of 'Sexual Abuse or Excess'.[140] With an implied criticism that the morals of his beloved French were not always entirely admirable and with a touch of prudery, Hall explained that Lallemand had fully discussed the subject but, 'in a manner not suited to *our* modes of thinking'. Taking refuge in Latin to mention the unmentionable one is told that *venus solitaria* (masturbation) and *venus nimia* (excessive sexual indulgence), 'are the real source of much deplorable suffering'.[141]

Since only women who are depraved, and hence not deserving of help, were believed to indulge in such activities, dire warnings of what lies in store for the miscreant, 'the miserable patient', are directed at men. These individuals could be prey to a whole range of unpleasant symptoms including, in addition to general body malfunctioning, loss of mental energy, imbecility of mind, impotence, depression, and 'a peculiar sort of *spasmodic paralysis* of the legs, and even of the hands and arms'. Furthermore some forms of mania, are actually due to unacknowledged solitary vice. For this reason, 'the secret must be discovered' so that the physician can recommend the most appropriate remedy. This consisted of 'the most absolute and vigilant restraint' combined with what was clearly the precursor of this century's cold shower regimen for damping down youthful ardour,

> ... a system of very active exercises and recreations, regulated thoughts, well-chosen studies, early hours, the salt-water shower bath, – tepid at first, afterwards cold; the mildest nutritious diet, without stimulus.[142]

This section was neatly rounded off with Hall's triple formula of what was required successfully to cope with this 'most deplorable state of things', namely '... the diagnosis, the detection of the cause, and its removal'.

It is not clear how this or any other public health message was disseminated, apart from within the confines of the doctor's consulting room and by word of mouth within the population at large. Hall produced no case studies linking these sexual activities with the threatened consequences. But the prospect of such debilitating outcomes coupled with such remedial discomfort, surely deterred some. In the meantime Hall had added notes to the ninth edition of Michael Underwood's *Treatise on the Diseases of Children* and was less embarrassed to refer to the physical damage caused by sexual abuse in female children. It was Hall himself who extended Underwood's brief discussion of the nature and origin of discharges from the vagina in prepubescent girls, with details of a case[143] from Dr Percival's *Medical Ethics.*[144] This concerned Jane Hampson, a child of four who was treated at the Manchester Infirmary in 1791 for inflamed and painful 'female organs' which had developed trouble following the child's having slept two or three nights in the same bed as a fourteen year old boy. She complained that she had been hurt by him in the night.

Despite treatment, the child died. At the coroner's inquest, where Percival reported the mother's account of the child's charge of being hurt by the boy, the jury, who were informed that examination of the child's body revealed no evidence of internal disease, decided that her death 'was occasioned by external violence'. A charge of murder was returned against the youth with whom she had slept. A warrant was issued against him and the fact that he had absconded was taken as confirmation of his guilt 'when added to the circumstantial evidence alleged against him'.

However, there occurred other similar cases in which there was no reason to suspect external violence and yet others in which it was absolutely certain none had taken place. A few of these patients died. Percival became convinced that he had been wrong in the Hampson case. The coroner was informed of the reasons for his change of mind and he testified to the wrongful charge against the boy, who was prevailed upon to come out of hiding. When called to the bar at Lancaster the jury was informed by the judge that there was insufficient evidence to convict him and he was acquitted.[145]

Disorders of Women During their Middle and Later Years

In his altogether briefer treatment of female disorders affecting the middle and later years, Hall presents what is basically the present epidemiological picture – an increased risk of deterioration in the general health, together with an increased incidence of disease associated with the two climacteric or physiological episodes of this stage in a woman's life. These are the 'cessation of the catamenia' or the menopause, a term not in use until much later in the nineteenth century and the later 'decline of the vital powers in old age'. The menopause heralds increased risk of the serious diseases of apoplexy or strokes and of uterine and mammary cancer. Nevertheless, and again insisting on the benefits of maintaining a high standard of general health, Hall asserts,

> I am of opinion that, if the general health were perfectly good, the cessation of the catamenia would always be accomplished naturally, without either disturbance or danger to the patient.[146]

Observations of the symptoms accompanying and following the menopause are brief. The symptoms which Hall associated with the cessation of menstruation included flushes, attacks of vertigo, insomnia and weight gain. He also notes the more subtle but more variable symptoms of 'indistinct vision, such as a waving appearance when the eyes are open, or a sensation of sparks when they are closed'.[147] These often worrying symptoms had not been ignored by him nor dismissed as so often still happens, and they were accurately described. It was acknowledged that this sometimes troublesome period of a woman's life can be quite protracted and the irregularity of menstruation can lead to uncertainty about the diagnosis of menopause or the onset of more sinister conditions such as cancer, which can be attended by uterine bleeding. It is also the case that in some females both the uterus and the mammae become cancerous at this time.

Special treatment for this period include the recommendation that the hair should be cut very short or even shaved and the head washed frequently in cold water while at the same time the feet should be kept warm. Supper was to be avoided and the head was to be kept high in bed. The abstinence from supper may have been a method of weight control since, as Hall rightly observed 'For some time after the disappearance of the catamenia, the person is frequently observed to grow corpulent.'[148] The even later period of decline is still marked by the possibility of strokes leading to paralysis as well as by the continuing possibility of cancer. To these conditions are added the risk of bronchial troubles and intestinal disorders including enlargement of the liver.

The whole period, as Hall recognized, is clearly marked by what we should describe as aspects of ageing, including increased sluggishness of the large intestine, which exacerbates problems associated with constipation, and increased vulnerability to disease of a range of body organs. Nevertheless, and his theory fits in with present-day models of diseases such as cancer, with sometimes long incubation periods of up to 20 or 30 years, Hall believed that many of the diseases of this period, actually have their origins, unremarked, at a much earlier stage. For this reason there was insistence on the principle that successful recognition and treatment of the diseases of later years was dependent on familiarity with 'those of female youth and of the puerperal state'.[149]

Hall's Work on Blood-letting

Overview

Both loss of blood through haemorrhage and lack of blood from such conditions as chlorosis or anaemia figured prominently among the symptoms Hall had regularly to deal with as a physician, even if one considers only the large number of women he attended in childbirth, some of whom must have suffered complications such as haemorrhage, or the numbers of pale young women who visited him and were diagnosed as suffering from one form or another of anaemia.

In an early work on blood-letting Hall points to the similarity in the symptoms accompanying these two states of blood deficiency albeit from two such different causes. On the other hand, therapeutically induced loss of blood by venesection with a lancet, through cupping or from the application of blood-thirsty leeches to strategic areas of the skin, (where each animal actually ingests only a small quantity of blood) was still one of the most frequent physical methods of therapy Hall will have seen in use at the Edinburgh, and later the Nottingham, Infirmary.

This practice of deliberate and sometimes painful withdrawal of blood, which is still in use in plastic surgery for instance, has a long history. And the dual aspect of dreaded, uncontrollable haemorrhage on the one hand and the controlled widespread therapeutic blood-letting on the other has maintained the fascination of the practice. Blood-letting was described by classical authors who justified its use as a method of correcting humoral imbalance within the body, especially in cases of inflammation, which were often cases of infection. More modern explanations, including those of William Cullen in the late eighteenth century, saw blood-letting as a means of lessening tension.[150]

Such theories certainly fitted the familiar observation that removal of four to eight ounces of blood generally produced sudden and dramatic change in fevered patients. The pulse rate and blood temperature fell, pain diminished and the patient was suffused with a feeling of relaxation and faintness or sleepiness. Even if these physiological changes, bringing a sense of well-being, were only temporary, the blood-letting had certainly in many cases been effective in producing some relief. If these effects did not follow then further bleeding might be recommended. Samuel Pepys, recounting details of his health and ailments in his 45th year, described himself as having excess of natural heat and he seems to have undergone bleeding once or twice a year.

But there was a darker side to this practice. Patients sometimes died. Those patients entering the Edinburgh Infirmary were often diagnosed as suffering from typhus fever, an increasingly common condition in overcrowded Edinburgh and other industrial towns and cities, and blood-letting was considered an ideal therapy. But many of these patients were malnourished and so weak that they could not tolerate it and if treated might either succumb immediately or fall prey to other opportunistic diseases. Cullen had warned of the hazards of using blood-letting while James Gregory forbade it as a therapy for typhus patients.[151] The artist David Wilkie was, in July 1825, described as having fallen into the hands of Dr Benet in Paris for a malady upon his nerves. Benet ordered the same treatment as Wilkie had received in Cheltenham, that is, 'leeches to each of his feet: *they bled well*'.[152]

Wilkie did not respond favourably despite having 'remained in the dangerous society of the Paris doctors for more than a month ....' Following his departure from Paris it was 'very noticeable how steadily his powers of observation and appreciation awaken as he travels away from the doctors, and through the pure mountain air and wonderful scenery of the Alps'. Mollett clearly had a poor opinion of any form of blood-letting,

> It is painful, in reading the accounts of the last illnesses of David's father and mother and of his own treatment, to observe the extent to which they were each of them subjected to bleeding, blistering, and leeches.[153]

Following publication of his work on diagnosis, Hall published in 1820 details of a number of cases of a female indisposition on which he had collected notes during his early years in practice.[154] This condition generally followed parturition or miscarriage. But in some instances it was not connected with the puerperal state. Each case had though involved loss of blood through accidental haemorrhage or through therapeutically applied blood-letting procedures. The underlying cause of the indisposition was seen as either what was termed 'Irritation' due to a loaded state of the bowels – often constipation – or what was labelled 'Exhaustion'.

These conditions in their turn had in some cases provoked symptoms similar to, but not in fact, inflammation which commonly affected the abdomen or head. Whereas blood-letting was the longstanding preferred and recommended remedy for genuine inflammation, Hall disapproved of its use in cases of irritation and exhaustion. Indeed, he believed that the majority of cases of the indisposition which ended fatally, were due to 'a mistaken use of the lancet'. Applying his principles of diagnosis to this new enquiry, Hall saw an urgent need to distinguish reliably between cases of true inflammation, for which blood-letting was appropriate and those which were not and for which blood-letting was a mischievous measure. There was need for a means of accurate diagnosis, a procedure or set of rules, which could be recorded and reliably followed, especially by young physicians,

> And as blood-letting constitutes the most powerful of our remedies, and haemorrhage one of the most formidable of diseases, it became of the utmost moment to fix the rules and limits of the employment of the former, and to ascertain the most efficacious mode of restoration in the latter.[155]

Hall had contemplated an experimental enquiry on animals to gather data on which to base a theory of guidance on the appropriate therapeutic use of blood-letting. Such an enquiry was later undertaken and published, but in the meantime his notebooks contained a host of relevant observations including patients' own comments about their symptoms and feelings on the sequence of events which often followed loss of blood. A paper, read to the Medico-Chirurgical Society in 1824 detailed a number of cases, involving both men and women who had suffered either or both, accidental haemorrhage and therapeutic blood-letting.[156]

The physiological event following blood loss would most commonly be syncope or fainting from which the patient should generally recover, due to the body's reaction. In some cases, however, fainting could be succeeded either by the potentially dangerous overreaction to external stimuli or, even more seriously by exhaustion which in its turn could be followed by further deterioration and sinking and finally, dissolution.

The link between syncope and recovery, when it occurred, was described in terms of noting the changes in the complexion, body heat and pulse which attended recovery; absence of these indicators denoted impending dissolution. Where recovery was accompanied by over-reaction, more common in the young 'of robust constitution who have been subjected to repeated blood-letting'[157] there could be intolerance to light and even more frequently to sound. This intolerance could be so marked that complete darkness, muffled door knockers and the spreading of straw along the hard pavement beneath street windows to dull the sound of footsteps and carriage wheels, were required.[158]

Evidence of over-reaction, which was sometimes mistakenly interpreted as improving vigour, needed to be taken as a warning sign that deterioration rather than improvement could be setting in and that the last thing the patient needed was further bleeding. In fact the received wisdom of an inevitable state of well-being in the body's system following bleeding was questioned and external evidence of the effects of blood loss, from whatever cause, on these systems was looked for as part of the theory building plan. 'We are altogether in want of a series of observations on the effects of loss of blood on the internal organs.'[159]

Age was a factor in the rate and incidence of recovery from loss of blood, both young and old being vulnerable. Robustness of health also played a part as did the amount of blood lost. Not surprisingly, where blood-letting was used therapeutically, each episode was actually attended with increasing risk of trouble. Hall himself noted that the major body organs faltered in the order of the brain, the lungs, heart and finally the gut.

A further unpublished paper by Hall, 'On the Due Administration of Blood-letting' was read before the Medico-Chirurgical Society in 1828, the year in which Pierre Charles Alexandre Louis, the French physician with whom Hall established a life-long friendship, published a short paper in the *Archives Générales de Médecine* on the effects of blood-letting in some inflammatory diseases. Two years later, in *Researches Principally Relative to the Morbid and Curative Effects of Loss of Blood*,[160] the 1828 paper was extended to provide the forum in which Hall presented his rule to guide doctors in the safe use of therapeutic blood-letting. A further objective in the paper was to distinguish between irritation and inflammation. Numerous case histories from his own practice, corroborated by the observations of others, were cited as the source of Hall's principles for rules on blood-letting because, 'It is by the multitude of facts alone, that the propositions which have been stated can be established or corrected.'

A structure was proposed for recording the data to be gathered and, as with the tabulations in *On Diagnosis,* tables were drawn up. Details of the disease at the beginning of the blood-letting therapy along with the age and strength of the patient were noted. This was followed by an accurate record of the quantity of blood withdrawn at each blood-letting, the appearance of the blood and the effects of its abstraction on the disease. In fact, 'nothing may be left unattended ... which may throw additional light on the subject ....' Nevertheless there was to be some selection such that 'none but the most unequivocal cases should be registered'. Cases where the diagnosis was not perfectly clear, 'would only add their own obscurity to the investigation'.[161]

Hall's Rule for the Use of Blood-letting

Hall's research led him to propose a rule which consisted of 'placing the patient perfectly erect, making a moderate sized orifice into the vein, and allowing the blood to flow until incipient syncope occurs'.[162] The upright position was crucial as a marker for limiting the amount of blood drawn, since in the horizontal position syncope did not intervene before a potentially dangerous quantity of blood had been withdrawn and the patient was in danger of passing into the stage of sinking and then death. At most, with the patient in the upright position, convulsions which can follow haemorrhages, might occur. These usefully put an end to the blood-letting. In terms of appropriate quantities of blood to be withdrawn to incipient syncope, Hall pronounced in his 1836 publication, 'No man *can* say *a priori,* how much blood ought to be withdrawn in any case of disease: ....'[163] His rule on bleeding to incipient syncope, therefore, provided a criterion of guidance for each individual treated.

Hall's rule also provided him with his diagnostic tool for distinguishing between irritation and inflammation. Where there was uncertainty, the patient could be bled to syncope. In cases which were merely the less serious condition of irritation and where blood-letting is inappropriate, syncope occurs early and little blood is lost. By contrast where inflammation really is present, much blood flows, depending on the degree, duration and severity of the disease, before unconsciousness occurs. But, reasoned Hall, 'this is exactly what is required to subdue the disease, and is well supported by the actual condition of the system'.[164]

Hall's Experimental Investigations

Hall did undertake his promised experimental investigations on blood-letting which extended and confirmed his rule for its safe use as a therapy. The details and results were contained in a further paper read to the Medico-Chirurgical Society in April 1832 and subsequently published by the Society.[165] This research had involved work on seven dogs, mainly adult terriers but including a couple of mongrels one of which was described as being 'fat and of a dull heavy disposition'. The animals were of roughly comparable body weight and age being from fifteen months to two years old. They were subjected to a range of programmes of blood-letting in terms of the total number of bleedings during the experiment, the quantities of blood drawn and the intervals between successive bleedings. Other variables included the fact that an animal might be bled directly and deliberately to syncope and its position might be varied during the experiment, so that the head could be placed either higher or lower than the rest of the body.

Hall was looking for confirmatory reliable data, from work on animals, on the relation between blood loss and the physiological phases which may follow. Thus, as in humans he anticipated syncope, reaction, over-reaction, sinking and dissolution. He also sought further evidence on precisely which body systems were impaired at each stage, namely, brain and nervous system, respiratory system, the heart and circulatory system and the alimentary canal.

This work on dogs indicated that in order to induce the state of syncope the dog needed to be healthy and, as in humans, in a raised position and the blood speedily withdrawn. Failure to observe these conditions could actually lead to dissolution and another dead dog. Syncope indicates primarily effects on the brain and secondarily on the heart itself. It also involves the other body systems. Thus, the eyes and the dog's expression are affected and it pants or sighs indicating diminution in the action of the heart and arteries. There is also debility of the voluntary muscles which leads to relaxation of the bladder and rectal sphincters and escape of the contents. The influence of body position upon the action of the heart in syncope was of the greatest interest and as in humans occurred without variation; when the head was raised the action of the heart was subdued and when it was lowered, heart function was restored. The phenomena associated with syncope could be revived after they had disappeared, simply by raising the animal's head. Lowering the head brought immediate relief.

Syncope was relatively easy to bring about. Inducing a controllable state of sinking was much more difficult. Overshooting the mark could so easily lead to dissolution rather than the more gradual sinking, even though the dog's death might be delayed until during the night. A list of omens of a fatal outcome to the experiment included total loss of appetite over several hours, spasmodic twitching of the ears and deep thoracic breathing particularly if laboured. Towards the end, the animal was often restless and twitching before the limbs stiffened. And while breathing stopped and the anal sphincter relaxed freeing the rectal contents, the heart sometimes continued to beat, possibly for several minutes showing that as in syncope it was the brain which had ceased functioning first. While, even with extensive bleeding, the animal might just recover if left in the horizontal position the vertical state more frequently led to dissolution. This situation is not unknown in the human subject where 'sudden dissolution has been the unexpected result of effort or of the upright posture, ... by haemorrhagy or other loss of blood'.[166]

Combining his observations on patients with the results of these experimental investigation on dogs, Hall confidently repeated that therapeutically one could safely bleed to syncope provided the patient was not in a horizontal position and that when syncope was reached the blood-letting should cease and the patient be deliberately put in the horizontal position for recovery,

> I am alike persuaded that blood-letting ought never to be employed in the horizontal position. The quantity of blood taken, although apparently moderate may be beyond the powers of the system to bear. But if the erect position be adopted, the occurrence of syncope limits the flow of blood, and the change to the horizontal position affords a prompt refuge from danger.[167]

Paper on the Circulatory System Submitted to the Royal Society

While working on blood-letting Hall was turning his attention to a definitely related but less immediately utilitarian aspect of the blood system associated with disease and therapy. He embarked on a more purely scientific, experimental study of the circulatory system itself. He hoped this change of direction looking, for example, for the causes of phenomena such as bleeding, would gain him a Fellowship at the Royal Society. In April and May 1831 two unsuccessful papers 'On the Anatomy and Physiology of the Minute and Capillary Vessels' and 'On the Effect of Water Raised to Temperatures Moderately Higher than that of the Atmosphere, upon Batrachian Reptiles'[168] were read at the Royal Society. Hall's wife copied out these papers for him and in her *Memoirs* she recalls how he would sometimes call her to look at particularly splendid microscope preparations associated with the research. The toad lung provided one of these memorable spectacles. According to Charlotte Hall, it 'displays a most marvellous scene, which can never be effaced from the memory of those who have beheld it'.[169]

The rejected papers were assembled to form *A Critical and Experimental Essay on the Circulation of the Blood* published later that same year.[170] In this work the 'Introduction' consisted of an essay on Hall's principles for conducting physiological experiments.[171] It contained Hall's proposed code of ethics that is, a set of guidelines for physiological research. While 1832 had seen the publication in *Medical-Chirurgical Transactions* of Hall's experimental work on the dogs and his award of a Fellowship of the Royal Society, it was further marked by his first pronouncements on the nervous system at the Zoological Society. This marked his shift of interest to the nervous system. Louis in Paris was still to produce his major work on blood-letting. But for his part Hall had, according to D'Arcy Power, made a significant contribution to clinical practice, because he had helped to wean the profession 'from the pernicious habit of blood-letting, not because he disbelieved in its efficacy but because he endeavoured to lay down rules for its rational employment'.[172]

Louis on Blood-letting

By 1835 Louis reissued as the first of three chapters, together with a more detailed analysis of his research method, his 1828 paper on blood-letting which had appeared in the *Archives Générales de Médecine.* The later work, *Recherches sur les Éffets de la Saignée dans quelques Maladies Inflammatoires et sur l'Action de l'Émétique et des Vésicatoires dans la Pneumonie* was dedicated to Marshall Hall.[173] In the following year J. Jackson, Senior, probably in memory of his lost son so admired by Louis, with whom he had studied, had arranged a translation of the work into English. He himself wrote the Preface and a lengthy Appendix detailing his own findings, based on 34 cases of Pneumonitis, which had been treated through blood-letting and which were included in the detailed casebooks of the Massachusetts General Hospital where Jackson worked.[174]

One of the most surprising features of Louis's 1835 publication, is that while it is dedicated to Marshall Hall it makes absolutely no reference to or acknowledgement of Hall's work. Equally surprising is that Hall, generally never slow to put into the public arena even the slightest hint of plagiarism against him, is equally silent on Louis's lack of recognition. Louis's work is concerned with questioning, as Hall had already done, the efficacy of blood-letting as a therapeutic measure. Of the contemporary state of the art of therapeutics Louis believed that,

> In fine, it is by the results that the value of methods is to be appreciated; men have for ages devoted themselves to therapeutics, and the science is still in its infancy.[175]

For his new investigations in *Recherches* Louis had worked on the three inflammatory disease conditions of *pneumonitis* (pleuro-pneumonia), *Erysipelas* of the face and *Angina Tonsillaris* which were all currently treated by blood-letting. His data gathering on the first was as always most thorough and included in addition to age and related personal data, clinical details on pulse acceleration, appearance of sputum, effects on voice and lung phenomena, and the presence or absence of 'heats and sweats'. A record was made of the date of onset of the disease under scrutiny. This not always easy, as Louis acknowledged,[176] but for pneumonitis it was taken as the period when the patient experienced a marked febrile affection accompanied by pain on one side and rusty coloured sputa.

In addition the date of entry to the hospital was always entered in the case notes, along with details of first bleeding plus the programme of subsequent bleedings and the quantities withdrawn, generally 10–15 ounces at each episode, as well as the method of bleeding used – leeches or lancet. The date of onset of convalescence was also recorded. This was a less accurate indicator since it was marked by the patient's acceptance of light food and individual variation influenced this. The duration between onset of the disease itself and of convalescence, in the case of those who survived, was used to 'measure' the efficacy of blood-letting. Louis did record if other parallel therapies such as blistering, were used but he believed that 'vesication had no appreciable influence on the progress of the disease ....'[177] The cases on which Louis based his analysis of the efficacy of blood-letting were carefully selected in order to remove variables which would have invalidated any generalizations.

A mixed reception followed the original *Archives* article. Some individuals erroneously accused Louis of abandoning a time-honoured remedy. In cases of severe inflammation Louis, like Hall, in fact specifically advocated its retention. Others simply could not accept his research findings since they were at such variance with their own beliefs. And so his research technique was questioned. One unnamed critic – Louis refused to name him 'for fear of the imputation of revenge, and this I am far from entertaining' – had said of Louis's numerical method,

> By invoking the inflexibility of arithmetic, in order to escape the encroachments of the imagination, one commits an outrage upon

> good sense; as if it were possible to heap together flowers, houses and birds, and to derive from the medley, fish and fruits![178]

In other words, opined the despairing Louis, his method of grouping a number of cases of pneumonitis of equal severity and occurring under similar circumstances but with some small differences of detail, was seen as 'the same thing as the bringing together of flowers and houses!' But of course Louis was not combining items so disparate. Nevertheless, some physicians received his work with enthusiasm regarding his method as a valuable tool in research on therapeutics.[179]

Summarizing his results[180] Louis claimed that with regard to the three diseases studied, the influence of blood-letting was not more marked in the cases bled copiously and repeatedly, than in those bled only once and to a small amount. Neither, contrary to public opinion, does blood-letting immediately arrest cases of inflammation. Apparently contradictory cases are more often due either to a faulty diagnosis or to the fact that blood-letting was practised when the disease was nearly over anyway. Blood-letting was still recommended for cases of severe inflammation in important organs. It was only in urgent cases of imminent hazard and as a last resort that, without associating the technique with Marshall Hall, Louis would recommend trying to bleed to syncope. He had earlier in any case definitely cast doubt on the method 'used in England',

> En obtiendrait-on de plus grands résultats, si, comme c'est assez d'usage en Angleterre, on portait la première saignée des pneumoniques jusqu'à la syncope? Cette pratique mérite d'être éprouvée; mais son grand succès me semble douteux; ....[181]

Comparing local blood-letting treatment through leeches, with general blood-letting through venesection, Louis felt that if blood-letting had to be used he preferred the lancet to the leech.

Jackson on Blood-letting

Jackson began his Appendix to Louis' work with the warning to his fellow medical colleagues, that Louis' enquiry 'will, no doubt, surprise many, if not most, medical men. They certainly do not accord with my own previous impressions in various respects.'[182] Jackson had firmly believed that more copious bleeding and at an earlier stage than at La Charité rendered the practice more efficacious by mitigating the disease and shortening it. Confirmation of his own belief was sought by Jackson through examination of the records of the Massachusetts General Hospital where he worked and which, since it had become fully operational in 1824, had received annually about 300 medical patients. It was not possible to make comparisons of the three diseases investigated by Louis because *Erysipelas* of the face and *Angina Tonsillaris* were not as frequent in the Boston area as in Paris. Jackson therefore analysed in detail 34 cases of *Pneumonitis*, giving minute details on his selection criteria,

> Pneumonitis, a name which I use as equivalent to pleuro-pneumonia, is then the only one of the three inflammatory diseases, of which our records will furnish cases for comparison ....[183]

With a certain honesty tinged with regret, Jackson admitted that some 30 years earlier and based on George Fordyce's plan he had prepared tables, or 'blanks' as he called them, for recording data in his own practice. He had then intended to apply the numerical method of his mentor Louis. But 'the difficulties attending the plan in private practice discouraged me too soon.'[184]

Jackson did indeed find that the recovery period for patients in his hospital was shorter but acknowledged that a variety of factors could have influenced the outcome, including the fact that the bleeding was started earlier, that his hospital was kept uniformly warmer than French ones especially in winter and that Boston patients appeared not to share the French detestation of hospitals and actually got into hospital sooner than their Parisian counterparts. Jackson plotted data on just this issue of entry to hospital following onset of illness and period of convalescence. His results suggested that whether or not the patients had been bled was less important than whether they had entered hospital early or late.

Another factor which could have affected recovery rates was the American practice of using emetics and cathartics and also blistering quite freely as well as blood-letting in the treatment of inflammatory diseases. Money also played its part since patients paying their own way sought prompt discharge from the hospital whereas those enjoying a free bed delayed theirs as long as possible. Personality differences also affected outcomes. Thus those who were habitually careful about themselves did not fall prey to opportunistic secondary infection during their convalescence as did those who were careless and ignored, for instance, the perils of inclement winter weather.[185] Nevertheless, in his final remarks Jackson did not directly compare his own findings with those of Louis,

> We have added all the observations that we have of sufficient accuracy to be compared with his, which will be received for what they are worth. The whole are to be regarded as materials, to which others are solicited to make additions from time to time; that, at

> length, so many cases, impartially collected, may be brought together, as shall justify entire confidence in the inferences to be made from them. Ten hospitals, under the care of honest physicians, may settle the question discussed in this work within five years, so that our posterity will not for ages be able to make any material correction in the answers. Seasons and epidemics will vary no doubt; but the general laws will be found the same, and little else would remain for future ages than to settle the allowance to be made for disturbing forces.[186]

Louis himself had never pretended to answer all the questions relating to the efficacy or otherwise of blood-letting. He had, like Hall, posed important questions on the issue and had made a contribution to the methodology of data collection and analysis to inform its continuing but more circumscribed use as a clinical therapy.

Notes

1. Hall, M. 'Suggestion of a new Principle for the Register Thermometer', *Journ. Sci. and Arts,* 4, 1818, pp.43–6; 'Description of an Aerometer, for making the necessary Corrections in Pneumatic Experiments, for reducing the Volumes of the Gases to a given Standard', *ibid.,* 5., 1818, pp.52–5.
2. Hall, M., 'Experiments and Observations relative to Vision', *ibid.,* pp.249–57; 'A Memoir on the Combined Agencies of Oxygen Gas and of Water in the Oxidation of Iron', *ibid.,* 7, 1819, pp.55–63; 'On the mode of producing Cold, proposed by M. Gay Lussac', *ibid.,* pp.383–4. In this latter note Hall draws attention to the fact that he was interested in Gay Lussac's method, which appeared in an earlier number of the *Journ. Sci. and Arts,* because he had proposed the same method to the editor of *Nicholson's Journal,* who had inserted it in his publication six years earlier.
3. Hall, M., 'On a Mode of preserving some Vegetable Remedies', *Journ. Sci. and Arts,* 5, pp.366–7; 'On the Spontaneous Combustion of Cotton Goods, which have been imbued with Linseed Oil', *ibid.,* pp.367–8. There was a flourishing overseas market for preserved pharmaceuticals but there is no suggestion that either Hall or his brothers had, despite their work on preserving vegetable remedies, any entrepreneurial interest in it.
4. Hall, M., *On Diagnosis,* 1817. Hall's book received a lengthy and sympathetic review in *Edin. Med. and Surg. Journ.,* 14, 1818, pp.236–42.
5. Ackerknecht, E. H., *History of Medicine,* 1982, p.129.
6. Hall, *On Diagnosis,* p.1.
7. Thomson, J. (ed.), *Works of William Cullen,* 2 vols, 1827, 1, p.471. Hereafter in the notes this will be referred to as [Cullen], *Works.*
8. Hall, *On Diagnosis,* p.1.

9. George Fordyce was a favoured pupil of Cullen who continued to give him encouragement and guidance on his studies and even on his writing style, following his departure from Scotland for London. Thomson, J., *Life, Lectures and Writings, of William Cullen,* 1832, pp.125–9.
10. Morris, R., *et al.* (eds), Edinburgh Medical and Physical Dictionary, 2 vols, 1807, 1, under 'DIS.' (no pagination).
11. Hall, C., *Memoirs,* p.44.
12. Cullen's work was originally published in Latin, [Cullen, W. (ed.)], *Synopsis Nosologiae Methodicae,* 1769.
13. [Cullen, W.], *Nosology,* 1800, p.5.
14. Bartlett, E., *Philosophy of Medical Science,* 1844, p.134. This work was dedicated to P. C. A. Louis, his mentor during his student days in Paris.
15. Reznek, L., *Nature of Disease,* 1987, p.15f.
16. [Cullen], *Works,* 1 p.471.
17. Hall, M., 'Contributions to Diagnosis. Contribution First: Containing 1. The Effects of the Habit of giving Opiates on the Infantine Constitution. 2. Inflammatory Affection of the Chest in Infants. 3. Inflammation of the Large Intestines in Adults. 4. Notice on the Beneficial Effect of a Warm and Regulated Temperature in Conducting the Mercurial Course, and in the Cure of Syphilis.' *Edin. Med. and Surg. Journ.,* 12, 1816, pp.423–8; 'Contributions to Diagnosis. Contribution Second: Containing 1. An Instance of Permanent Spasmodic Contraction of the Muscles of one Side of the Face. 2. Remarks on the Character and Cure of Sycosis. 3. The Description of a Particular Ulcer, which appears on the Nose. 4. The Suggestion of a Mode of Preparing and Preserving Vegetable Remedies', already noted 3 *Supra, ibid.,* 13, 1817, pp.63–8; 'Reflections on the Arrangement of Cutaneous Diseases; with the Suggestion of a Practical and Diagnostic Mode of Classification', *ibid.,* pp.189–95: 'Contributions to Diagnosis. Contribution Third: On the Characteristic Symptoms of Hydrothorax', *ibid.,* pp.303–8.
18. Hall, *On Diagnosis,* Dedication, p.iii.
19. Hall, C., *Memoirs,* p.21.
20. *Ibid.,* p.22.
21. Risse, G. B., *Hospital Life,* 1986, pp.67–71.
22. Maulitz, R.C., *Morbid Appearances,* 1987, p.113.
23. See Sauvages, F. B. de [ed.], *Nosologie Méthodique,* 1772.
24. Reznek, *Nature of Disease,* p.25.
25. Morris, *Edinburgh Medical Dictionary,* under 'Nosology'.
26. [Cullen], *Nosology,* p.ii.
27. Sauvages, *Nosologie Méthodique,* pp.1–2.
28. [Cullen], *Nosology,* p.vi.
29. [Cullen], *Works,* 1, p.447.
30. Reznek, *Nature of Disease,* pp.25–33.
31. [Cullen], *Nosology,* p.vii. Nevertheless after his initial contempt for it – calling it jargon – Cullen applauded Linnaeus' 'Methodical' approach

with its applicability over space and time, '... it is impossible to convey any accurate knowledge of things from one country to another, or from one age to another, without more or less of this method; ....' [Cullen], *Works,* p.445.

32. [Cullen], *Nosology,* p.x.
33. *Ibid.,* pp.xv–xvii.
34. *Ibid.,* p.xvii.
35. Hall, *On Diagnosis,* p.13. See also p.viii and pp.16–17.
36. *Ibid.,* p.13.
37. *Ibid.,* p.16.
38. *Ibid.*
39. *Ibid.,* p.11.
40. *Ibid.,* p.ix.
41. Hall, M., *Practical Observations,* 1845; *Practical Observations. Second Series,* 1846. p.vi.
42. *Ibid.*
43. *Ibid.,* pp.v–vi. According to Cullen, 'Medicine is the art of preserving health and of curing diseases;.... The branch which treats of health has been named Physiology; ....' [Cullen], *Works,* 1, pp.3–4.
44. Hall, *Practical Observations. Second Series,* p.vi.
45. Hall, *On Diagnosis,* p.xix.
46. *Ibid.,* p.xx.
47. *Ibid.*
48. [Cullen], *Works,* 1, p.3.
49. Hall, *On Diagnosis,* p.vi.
50. Bartlett, *Philosophy of Medical Science,* p.11.
51. *Ibid.,* p.295.
52. *Ibid.,* p.301.
53. *Ibid.*
54. Hall, *On Diagnosis,* p.v.
55. [Cullen], *Works,* 1, p.420.
56. *Brit. Ann. Med.,* 1, 1837. This new, innovative journal did not survive beyond its first volume.
57. [Cullen], *Works,* 1, p.423.
58. Ackerknecht, *History of Medicine,* p.147.
59. Hamilton, J., *Seats and Causes of Diseases,* 1, 1795, p.ix. These were Hamilton's opening words to his Preface.
60. Risse, *Hospital Life,* p.262.
61. Richardson, R., *Death, Dissection and the Destitute,* 1989, pp.52–72.
62. [Christison, R.], *Life of Sir Robert Christison,* 2 vols, 1885–6, 1, pp.192–3.
63. Risse, *Hospital Life,* p.266.
64. Hall, *On Diagnosis,* p.xiv.
65. Thomson, Life of William Cullen, p.108.
66. Hall, *On Diagnosis,* p.xiv.
67. *Ibid.*

68. Ackerknecht, *History of Medicine,* p.147.
69. *Brit. Ann. Med.,* 1, p.v.
70. Hall, M., *Descriptive, Diagnostic and Practical Essay,* 1820, p.2. This edition was dedicated to Matthew Baillie.
71. Hall, M., *Principles of Diagnosis,* 2nd edn, two vols in one, 1834.
72. *Ibid.,* p.3. In 1837 a 'third edition' appeared incorporated in another compilation of earlier work, Hall, M., *Principles of the Theory and Practice of Medicine,* 1837. Subsequent reference will be to the first edition of *On Diagnosis* unless otherwise stated.
73. Hall, *Principles of Diagnosis,* 1, p.3.
74. Hall, *On Diagnosis,* pp.337–9.
75. Hall, M., *Prone and Postural Respiration in Drowning,* 1857.
76. See the folded 'Tables of Diagnosis' attached at the rear of *On Diagnosis* and capable of being pulled out to 3½ times normal page width.
77. Reiser, S. J., *Medicine and Technology,* 1981, pp.227–31.
78. Ackerknecht, *History of Medicine,* p.138 for a caricature of Broussais in which the translation of his therapeutic directions for the patient were, 'Ninety more leeches ... and continue a strict diet.'
79. Hall, M., *Commentaries. In Three Parts,* 1827, p.248.
80. *Dyspepsia* referred to a range of disorders of the digestive organs and of the bodily processes and states associated with the digestive system and could include a feeling of lowness of spirits. In a letter to Mrs Gaskell, Charlotte Bronte explained that headaches and dyspepsia were her worst ailments. By dyspepsia she almost certainly meant indigestion. At the time of Hall's work, *hypochondriasis* was sometimes held to have a strong resemblance to dyspepsia. It is now often regarded as being characterized by a morbid preoccupation with bodily health and can embrace patients' unfounded beliefs that they are suffering from serious bodily disease.
81. Hall, M., *On the Mimoses,* 1818. The work was dedicated to Matthew Baillie.
82. In a memorandum dictated to his wife during his final illness. Hall, C., *Memoirs,* pp.48–9.
83. Hall, M., *Cases of a Serious Morbid Affection,* 1820.
84. Hall, M., and Higginbottom, J., 'Cases of a Destructive Inflammation of the Eye and of Suppurative Inflammation of the Integuments, occurring in the Puerperal State and apparently from Constitutional Causes', *Med.-Chir. Trans.,* 13, 1827, pp.189–201.
85. See note 79 *Supra.*
86. Hall, M., *Diseases which are Constitutional,* 2nd edn, 1830.
87. Hamilton, J., *Observations Relating to Midwifery,* Parts I and II, 1836.
88. Hamilton, J., *Observations Relating to Midwifery,* 2nd edn, revised and enlarged, 1840.
89. The paper, with its reference to the difference of opinion with Gooch, had been published as, Hall, M., *On a Morbid Affection of*

Infancy, 1829.

90. Hall, *Commentaries. In Three Parts,* p.1.
91. *Ibid.,* p.2.
92. Both William Smellie (1697–1763), and William Hunter (1718–83) were born in Lanarkshire, Scotland. Smellie published *A Treatise on the Theory and Practice of Midwifery* (1752–64). This work was published in three volumes for which the title of the second and third volumes changed to *A Collection of Preternatural Cases and Observations in Midwifery.* Several editions of the work were subsequently published. Smellie was called 'Man-midwife' and was a friend of William Cullen. Hunter, who was Cullen's resident pupil (1737–40), later lived with Smellie in London. He became the leading obstetrician and was appointed physician extraordinary to Queen Charlotte in 1764, having been consulted by her during a pregnancy two years earlier. Hunter established the prestigious Great Windmill Street Medical School in 1770.
93. [Hall, M.], *Eupædia; or Letters to a Mother,* [1831]. In the Advertisement note at the front of the book Hall explains that, 'The term *Eupædia* ... means "the happiness of having good children". I have ventured to change its meaning from *good* to *healthy;* ....'
94. [Underwood, M.], *Diseases of Children,* 9th edn, with Notes by Marshall Hall, 1835.
95. These are part of a set of Plates (IV, V, VI and VII) by the artist H. Adlard and are included at the rear of Hall, *Commentaries. In Three Parts.* The portraits are accompanied by further plates showing details of the gums, tongue and nails in the afflictions shown in the portraits. A less engaging and smaller set of portraits by J. Stewart accompanies the 1830 publication, *Diseases which are Constitutional.*
96. *Commentaries. In Three Parts,* pp.179–81.
97. Showalter, E., *The Female Malady,* 1985, p.10f.
98. Priestley, J. B., *Victoria's Heyday,* 1972, p.109.
99. Hall, C., *Memoirs,* p.70. Charlotte Hall noted that Hall's work 'was illustrated by plates beautifully executed, showing peculiar conditions of the tongue, the complexion, the lips, the hands, the nails – their form, hue, brittleness, etc., – ....'
100. Hamilton, A., *Treatise on the Management of Female Complaints,* 7th edn revised, 1813.
101. Hamilton, J., *Selected Cases in Midwifery,* 1795, p.viii.
102. Smith, G., 'Prescribing the Rules of Health: Self-Help and Advice in the Late Eighteenth Century', in Porter, R. (ed.), *Patients and Practitioners,* 1985, pp.249–82.
103. *Ibid.,* p.278. Willich was a co-editor of the *Medical and Physical Journal,* 1799–1801.
104. [Abernethy, J.], *Surgical Works of John Abernethy,* 2 vols, 1811.
105. Even before Abernethy's death, a range of unnamed authors were publishing on the Abernethian Code: *The Abernethian Code of Health*

and Longevity, 1829. *Common Sense; or, the Abernethian Code of Health and Longevity*, 1830.

106. *Everyone his own Physician or the Abernethian Code of Health and Longevity*, 1836, p.3.
107. *Common Sense; or the Abernethian* Code, p.23.
108. *Dr Abernethy's Code of Health and Long Life*, [1857], p.19 '... an empty stomach is not, as used to be supposed, necessarily productive of injury; and the old, but (alas!) still much used adage, 'eat little and often', is as absurd in theory as it has long been found to be injurious in practice', was another warning against nibbling. *Every One His Own Physician*, 1837, p.59.
109. Hall, *Diseases which are Constitutional*, p.6.
110. *Ibid.*
111. Sadler, M. T., *The Law of Population*, Vol. 1, 1830. Sadler, an MP, was a political economist and social reformer with Methodist sympathies. His publication on population, an attempt to refute Malthus' view on population increase, provoked lively debate in the press.
112. Hall, *Diseases which are Constitutional*, p.7.
113. *Ibid.*, p.23.
114. Hall, *Commentaries. In Three Parts*, p.12.
115. Hall, *Diseases which are Constitutional*, p.2.
116. Hall, C., *Memoirs*, p.137.
117. Hall, *Diseases which are Constitutional*, pp.18–19.
118. Hall, *Descriptive, Diagnostic and Practical Essay*, p.6.
119. Hall, 'Postscript', *Commentaries. In Three Parts*, p.150.
120. Priestley, *Victoria's Heyday*, p.30.
121. Hall, *Diseases which are Constitutional*, pp.25–6.
122. Corry, J., *The Detector of Quackery*, 1802, pp.81–2. On the subject of sweat, Corry, in a 2nd edition of his work published in the same year noted, 'There is no such thing as sweat now-a-days; even the coal-porter and the butcher's boy perspire – elegant creatures!' *Ibid.*, 2nd edn, 1802, p.110.
123. What she actually said was, 'In all his habits he was particularly neat, being simple in his dress and rather indisposed to change with the fashion.' Hall, C., *Memoirs*, p.245.
124. Hall, *Commentaries. In Three Parts*, p.13.
125. Rattray Taylor, G., *The Angel-Makers*, 1958, pp.xvi–xviii.
126. Hall, *Commentaries. In Three Parts*, p.x.
127. Clarke, C. M., *Observations on Some Diseases of Females*, 2 vols, 1814, 1, p.1. (A second edition was published in 1821.)
128. *Ibid.*
129. In 1830 a 'member of the medical profession' published *An Important Address to Wives and Mothers on the Dangers and Immorality of Man-Midwifery*, 1830. The Address is accompanied by a splendid coloured cartoon, Plate IV, of a male midwife in drag, plus mobcap, with a dish

and spoon in one hand, forceps in the other, a pair of scissors dangling from the waist, and a syringe (clyster) protruding from one pocket.

130. Clarke, *Observations on Some Diseases of Females,* pp.4–5. This has resonances with the current debate on the allegation that unknown to themselves, women under anaesthetic have been observed/handled by students.
131. Hamilton, J., *Selected Cases in Midwifery,* p.vi.
132. *Ibid.*
133. Hall, *Commentaries. In Three Parts,* p.297.
134. *Ibid.,* p.170.
135. Hamilton, J., *Selected Cases in Midwifery,* p.135. *Idem, Observations Relating to Midwifery,* 2nd edn, 1840, p.113.
136. Hamilton, A., *Treatise on the Management of Female Complaints,* pp.413–14. He also specified the age range 20–35 years because '... when too young, they cannot pay the attentions which the infant may require, and when too old their milk is seldom sufficiently laxative'.
137. Acton, W., *Complete Practical Treatise on Venereal Diseases,* 1841. *Idem, Practical Treatise on Diseases of the Urinary and Generative Organs,* 2nd edn, 1851.
138. *Idem, Prostitution, Considered in its Moral, Social and Sanitary Aspects,* 1857.
139. Hall, *Diseases which are Constitutional,* p.5. It may not be coincidental that both men had been among Hall's proposers at his election to a Fellowship of the Royal Society. One good turn deserves another!
140. Hall, M., 'VIII. Sexual Abuse or Excess', *Diseases and Derangements of the Nervous System,* 1841, pp.339–40. The Section IX, which follows, is on Abortion but there is no reference at all to the fact that it was frequently illegally and dangerously induced, sometimes with horrendous consequences. He calmly considers it as often being, like birth itself, due to a reflex action.
141. *Ibid.,* p.340.
142. *Ibid.*
143. [Underwood], *Diseases of Children,* pp.468–9.
144. Percival, T., *Medical Ethics,* 1803, pp.231–3.
145. *Ibid.,* p.233 and [Underwood], *Diseases of Children,* p.469.
146. Hall, *Commentaries. In Three Parts,* p.354. Hall retained this observation in *Diseases which are Constitutional,* p.243.
147. Hall, *Diseases which are Constitutional,* pp.252–3. This new observation is contained in the enlarged chapter, 'Of the Period of the Final Cessation of the Catamenia' of this 1830 edition.
148. Hall, *Commentaries. In Three Parts,* p.356.
149. *Ibid.,* p.311.
150. [Cullen], *Works,* vol.1, p.194. The Marquis de Sade is said to have derived sexual pleasure from watching the blood issuing from the arms of pale young women being bled for a range of ailments.
151. Risse, *Hospital Life,* p.204.

152. Mollett, J. W., *Sir David Wilkie,* 1881, p.70.
153. *Ibid.*
154. Hall, M., *Cases of a Serious Morbid Affection,* 1820.
155. Hall, M., *Observations on Bloodletting,* 1836, p.116.
156. Hall, M., 'On the Effects of Loss of Blood', *Med.-Chir. Trans.,* 13, 1827, pp.121–51. The paper was read 23 November 1824. This paper in amended form appeared as 'Essay Second. On Some Effects of Loss of Blood', Hall, M., in *Medical Essays,* 1825, pp.34–69.
157. *Ibid.,* p.44.
158. *Ibid.,* p.43.
159. *Ibid.,* p.68.
160. Hall, M., *Morbid and Curative Effects of Loss of Blood,* 1830.
161. *Ibid.,* pp.182–3.
162. Hall, *Observations on Bloodletting,* p.273.
163. *Ibid.,* p.314.
164. *Ibid.,* p.249.
165. Hall, M., 'An Experimental Investigation of the Effects of Loss of Blood'; Read 10 April 1832, *Med.-Chir. Trans.,* 17, 1832, pp.250–99.
166. *Ibid.,* p.262.
167. *Ibid.,* p.169.
168. The papers are referred to again in Chapter 4 on the Royal Society.
169. Hall, C., *Memoirs,* p.79.
170. Hall, M., *Essay on the Circulation of the Blood,* 1831.
171. 'Introduction. Of the Principles of Investigation in Physiology', *ibid.,* pp.1–11.
172. Power, Sir D'Arcy, 'Dr Marshall Hall and the Decay of Bloodletting', *The Practitioner,* 82, 1909, p.320. This paper (pp.320–31), given by Power as President of the Harveian Society is a tribute to Marshall Hall who was the Society's first President.
173. Louis, P. C. A., *Recherches sur les Éffets de la Saignée,* 1835.
174. [Louis, P. C. A.], *Researches on the Effects of Bloodletting,* 1836.
175. *Ibid.,* p.64.
176. *Ibid.,* pp.7–8.
177. *Ibid.,* p.8.
178. *Ibid.,* p.63, in a footnote.
179. *Ibid.,* pp.xx–xxi.
180. *Ibid.,* pp.22–23.
181. Louis, *Recherches sur les Éffets de la Saignée,* pp.21–2.
182. [Louis], *Researches on the Effects of Bloodletting,* p.99.
183. *Ibid.,* p.105.
184. *Ibid.,* p.xiv.
185. *Ibid.,* p.140.
186. *Ibid.,* pp.170–1.

3

Medical Education in London

As an Edinburgh educated and trained physician moving to London in 1825, Hall had joined the top brass of the hierarchical medical profession. This group, the physicians, was generally marked out by its University background. Beneath the physicians were the more recently respectable surgeons, many of whom in London were very successful and rich, and beneath them the trade-tainted apothecaries. Members of these three groups could be specialists in a particular field as might those in the more fringe groups of the medical community around this élite trio.[1] These included the definitely commercial druggists and chemists whose success undermined the apothecaries.[2] More peripherally still were those from what the Porters might call the alternative 'irregulars' to the out and out quacks offering between them a range of therapies including mesmerism, galvanism, hypnosis, cupping, bone setting, corn cutting, stay making and more besides. Some of these healers and so called nostrum mongers were women.[3] The total amounted to a motley crew of mainly medical men – and dentists were included – with generally low status and income for the majority and absence of professional unity or sense of collegiality among the group as a whole.

Modes of Training

Introduction

The physicians, surgeons and apothecaries should have been qualified and licensed to practise according to the standards of their respective Colleges and Society and, if general practitioners, according to the provisions and demands of the Apothecaries Act of 1815. But many were not. The Apothecaries Act marked a real leap forward to regulate training and licensing for the medical profession.

Its roots lay, at least partly, in opposition from a specially formed group of apothecaries, to a glass tax imposed in 1812 increasing the price of phials in which so much medicine was dispensed. But more importantly the group extended its scope to deal with the simmering discontent about standards of training and the monitoring of license granting for general practitioners.[4]

Many provincial general practitioners had claimed that their practice was being undermined by unqualified druggists. So, 'like the sportsman who sets out to kill snipes and ends by bagging an elephant' the group was influential in getting the Apothecaries Act of 1815 on to the statute book. As one commentator of 1841 put it, before the passing of the Act 'A herd of uneducated men assumed our duties without having qualified themselves to the task.'[5] Although it certainly contributed to the increasing prestige and professionalization of the medical profession, the 1815 Act did not solve all its problems and there were still many unqualified practitioners operating as doctors. Whilst it was the more numerous, more modestly educated and trained and less expensive apothecaries who generally attended the population at large,[6] being paid for the medicines dispensed but not for the diagnoses made, the much smaller band of physicians (in 1600 London had only 50 physicians) attended the rich, and both physicians and surgeons staffed – unremunerated – the hospitals, where they were answerable to lay governors.

The range and standard of medical education offered, the institutions involved, the examinations and qualifications available and whether or not, and how, these were monitored varied greatly within the metropolis, between London and the provinces and between England and Scotland. Hall's whole London period was set against a background of institutional reform activities at all levels, and one aspect of medical reform for which he claimed enthusiasm, was the unification of the profession as a whole under a National College of Medicine. Such a College it was envisaged, as well as wresting power from the London-based élite institutions, that is the Colleges of Physicians and Surgeons and the Society of Apothecaries, would regulate standards of education and the number of licensing bodies as well as curbing quackery. There were 19 licensing groups in the UK by the early 1800s.

But Hall had in mind some '*central Council*, some *board of control* and revision, some *visitors*, specially appointed for the superintendence of medical affairs ....'[7] as the *Medical Gazette* expressed it in its series of editorials on medical reform during the session 1841–2. Hall certainly would have considered himself to be a gentleman physician and was no leveller in the sense of wishing to abolish the three corporations. He did, however, wish to widen access to the privileges bestowed on the few by the College of Physicians. He deeply resented for instance that as an Edinburgh graduate, although qualified to practise throughout the metropolis, 'barring the small circuit of seven miles from St Paul's',[8] he was only entitled to the non-voting Licentiate and not to the full Fellowship of the College of Physicians. Through his involvement with the British Medical Association once it became London based, Hall did make some contribution to medical reform. But he died the year before the next major landmark, the Medical Act of 1858, reached the statute book.

Hall's own medical education and training in Edinburgh had included University classes in the Practice of Physic, the Practice of Medico-Chirurgical Obstetrics, Materia Medica, Anatomy and Chemistry[9] as well as direct bedside clinical training from University Professors in the Edinburgh Royal Infirmary which was associated with the University.[10] During the final decades of the eighteenth century Edinburgh was held to be the best medical school in the English-speaking world.[11] At the bedside Edinburgh graduates had learned the discipline of careful, critical, diagnostic judgement based on informed observation.[12] Their training had been highly scientific and well organized. In a real sense they had already benefited from many of the activities which came to characterize the 'clinic' structure of medical education in France in the 1820s and 1830s. And as Hall's work on diagnosis indicates, the practice of careful case compilation was part of the established Edinburgh tradition and practice. But in this early publication Hall was, as we have seen, also trying to advance medical knowledge by questioning old nosologies and seeking to replace them with new frameworks.

In London, where the élite Colleges made little contribution, medical education in the 1820s for physicians, surgeons and apothecaries was provided in a variety of ways.[13] Only a small proportion of London's medical students destined to become physicians, had actually started their studies with a Classics biased undergraduate course at Oxbridge. The two ancient Universities produced fewer medical graduates in the eighteenth century than a century earlier. If students had been to Oxbridge, they came to London with more classics but less science than Hall and, on the whole, did not redress the balance in the metropolis.

Training could range from excellent to decidedly dubious and from methods already on the wane to that provided by the brand new London University which was not even quite ready when Hall first came to London. It could be pursued through apprenticeships which could range from association with someone as eminent as Charles Bell to working in an apothecary's shop, in dispensaries, in private medical schools or in hospitals. Naturally there would be a price differential reflecting quality. By the 1830s the University of London, which from its inception in 1828 had placed great emphasis on medicine in its curriculum, was available as a non residential, less expensive University alternative to Oxbridge. Thus practising medical men in London could hold degrees with or without one of a range of medical licences, or they could be practising with no qualifications at all. The *Lancet,* which Wakley launched in 1823, is an invaluable source of information on the complex pattern of London institutions providing medical education. The journal published annual staff lists, curricula, selected lectures and the fees charged for different courses.

Apprenticeships

Hall had started out on an apprenticeship with Mr. Moor, 'a chemist of some repute and wealth' at Newark.[14] He was, therefore, either originally intended to qualify as an apothecary or become a chemist, or he was embarking on his apothecary's apprenticeship before going on to study physic and surgery in preparation for his College of Surgeons' examinations. These would qualify him as a superior sort of general practitioner. Training through apprenticeships, which was still the method for apothecaries and many surgeons, could be in the household of an experienced medical man. It had for centuries been the method of instruction for all intending doctors and although it was disappearing by 1800 several eminent medical men of the first half of the nineteenth century had been trained in this way.

These included James Paget, who as late as 1830 was apprenticed at the age of 16 to Mr Charles Costerton a General Practitioner in Yarmouth, at a premium of 100 guineas for five years.[15] The last six months were excused to allow Paget to attend hospital and study in London. Clause 15 of the Apothecaries Act 1815 laid down that to qualify for admission to the examination for a license to practise as an apothecary, a candidate had to serve an apprenticeship of at least five years. And the 1825 Amendment to the Act allowed, for one year, apprentices to surgeons to enter for the examination and thus qualify as general practitioners.[16] Towards the end of his life in the 1880s, James Paget explained how his apprenticeship came about,

> When the intention to enter the Navy was abandoned, it was decided that I should be a 'Surgeon' – meaning a general practitioner, or

> something in the Medical Profession – and that to this end I should be apprenticed to Mr Charles Costerton, an active, energetic, and well educated practitioner in the town.[17]

Details of Mr Costerton's own education are not given. Paget himself took a house pupil, at a fee of £10 per month, when he first set up in practice around 1836,[18] and Charles Bell took in a small group who also attended the private school of anatomy where he was teaching.

Several of the first-appointed staff of St Mary's Hospital in Paddington, opened in 1850–1, had served an apprenticeship in their youth.[19] Senior hospital surgeons unlike physicians could command high fees from apprentices. But an 1841 editorial in the *Medical Gazette* questioned the whole practice of apothecary apprenticeships on the basis that most young people could acquire what they needed in 3–6 months. Most of them spend the rest of their apprenticeship period in,

> ... tying up phials, weighing out powders, beating pill masses, preparing day by day sundry infusions of rose buds, orange chips and senna leaves, to the manifest advantage of the master but without the smallest perceptible amount of profit to the apprentice.[20]

The same editorial called for a shorter apprenticeship preceded by a longer general education including studies in Latin and more hospital and dissection experience. The same editorial also called for physicians and surgeons to be included among the examiners since general practitioners were expected to practise every branch of the profession and in any case the larger number of students who came to London to study for the examinations at the College of Surgeons as well as at the Apothecaries' Hall spent their two and a half years studying physic and surgery.

Dispensaries

Dispensaries, like many hospitals, resulted from the eighteenth-century expansion in philanthropic charitable donations designed to increase public-health facilities. They provided advice and free medicine and were primarily to meet the needs of the poor, the indigent rather than the destitute, who 'would receive care without charge, thus, it was hoped, confirming ties of deference, gratitude and paternalism'.[21]

Although just starting to wane, dispensaries were still available for medical education in London in the first half of the nineteenth century. Students could do their clinical work in them and during the 1830s many students still did. The development of the dispensaries played a crucial part in the evolution of the role of apothecaries who were often trained in them and whose numbers had increased during the seventeenth century. They acquired the right to see, examine and prescribe for the large numbers of patients who turned to them.

The first three dispensaries in London had been set up by the College of Physicians in 1697 in Warwick Lane, Gracechurch Street and St Martin's Lane. Others followed, in an attempt to stop the increasing practice of apothecaries undercutting physicians and surgeons by giving free advice to patients who could not afford to pay the fees demanded by physicians, but who paid for the medicine prepared by the apothecary. Physicians at dispensaries offered free advice to anyone who consulted them and since the apothecaries refused to make up medicines at reduced prices, they engaged dispensers for dispensing medicines at low rates.[22]

In 1703 an apothecary named Rose was accused by the physicians of prescribing medicines when in fact he was not qualified to do so.[23] Rose had indeed been called in by a patient John Seale, butcher, had diagnosed his condition free of charge, but had sent him several parcels of medicine for which he did charge. Initially judgement was given for the College of Physicians; their regulation that only physicians are entitled to assess a patient's condition and prescribe medicines was upheld. However, the case went for appeal to the House of Lords who, in March 1704, reversed the verdict.[24] Thus, the apothecaries could now practise although still could not charge for consultations. Nevertheless their numbers and importance grew while that of physicians remained fairly static and the dispensaries opened by the Royal College of Physicians closed in 1725.[25]

The increase in population throughout the second half of the eighteenth century and its attendant rise in prosperity for a growing middle class, contributed to a rise in status of the medical profession; it led also to an increase in poverty, overcrowding and sickness for the poor. Their needs could not be met through existing resources for public-health provision, in spite of a number of new hospitals in London. By 1800 London's hospitals were handling 20,000 – 30,000 patients a year. But these hospitals did not have regular out-patients' departments and so further dispensaries were needed. In 1769 a Scottish doctor, George Armstrong, started a small dispensary for children in Red Lion Square[26] and in the following year another Edinburgh graduate and a Quaker, John Lettsom, opened the Aldersgate Street Dispensary.

Patients were seen daily at the dispensary or, and this was a new development, if too ill were visited in their own homes. What an eye-opening experience for a dispensary doctor going out on his first call among the poor. But from their ranks was derived some of the impetus for improved public housing, sanitation and health protection measures. Much of the drive for these reforms nevertheless sprang from outside the medical profession, from committed, concerned, philanthropic lay individuals. Similar dispensaries were established in other parts of London. By 1800 there were 16 dispensaries in London taking up to 50,000 cases a year and by 1830 the number had increased to 35.[27] The dispensary serving the clinical needs of the new University of London, prior to the establishment of the University College Hospital was set up in the late 1820s. From the middle of the nineteenth century the educational function of dispensaries rapidly diminished as that of the hospital medical schools increased.

Private Medical Schools – Origins, Development, Demise

The century 1740–1840 was the era of the London private medical – often called anatomy – schools and during the 1830s Hall did part-time teaching in three of them, the Aldersgate Street School, 1834–6,[28] the Webb Street School, 1837–9 and Sydenham College, 1838–9. In 1745 the Surgeons had by Act of Parliament separated from the Barbers and were freed from prohibition on dissection. The way had been opened for the founding of anatomical schools. They were conducted in the homes of individual entrepreneurs such as William Hunter or in hired premises such as the Schools in which Hall taught.

The influential personalities in the London hospitals which provided medical education for students, over the period of the private hospital schools, were the physicians and surgeons to the hospitals and not, as in Edinburgh, the teachers in the medical schools associated with hospitals. The staff lists[29] of the private schools reveal the significant role of Scottish-trained doctors, who like Hall found it difficult to obtain hospital posts. They reveal also the existence of professional dynasties between masters and pupils, a number of whom established prestigious reputations, in these private schools as well as in the hospitals.

In London the hospital staff had largely confined their teaching to clinical subjects associated directly with their work with patients and there developed a need for tuition in other aspects of the curriculum. The private schools, many of which were established close to the teaching hospitals, met this demand and initially they concentrated on anatomy. They often provided a higher standard of anatomical teaching than that found in certain of the outmoded, cramped, insanitary, understaffed hospitals.[30] In addition they taught anatomy all the year round in spite of the difficulty of preserving bodies during the summer. But their increasing student numbers did give rise to the problem of securing enough cadavers.

William Hunter, a Scot who had been an apprentice of Cullen and who had moved to London where he probably earned about £10,000 a year, was the first to take full advantage of the new opportunity provided by the separation of the Surgeons from the Barbers. In 1766 he rebuilt a large house in Great Windmill Street and established a School of Anatomy with accommodation for dissecting and ample space for his extensive collection of preserved specimens. This was the first building ever to be constructed in this country for the primary purpose of teaching anatomy.[31]

Over the years William Hunter had a number of partners including his nephew Matthew Baillie who took over the school on his death in 1783. Other well known figures also taught in private schools. In 1812, by which time chemistry, surgery and medicine had been added to the curriculum of Hunter's school and Brodie and Roget were already giving lectures there, Charles Bell from the Middlesex Hospital joined the school. Bell's association with the Windmill Street School lasted until his appointment in 1828 to a professorship at the newly formed University of London. The Windmill Street School was taken over by Herbert Mayo[32] and Caesar Hawkins[33] who themselves left it in 1830 and 1831 respectively to take up hospital appointments. Thereafter the Windmill virtually ceased, but in its 70 years of existence several thousand students had been trained there. Hall had hoped that his son might have received some of his London medical training, following Cambridge, in this prestigious school.

In 1787 another private school of anatomy was established in Blenheim Street, near Oxford Circus by Joshua Brookes, a former pupil of William Hunter. James Paget spoke admiringly of the quality of help in anatomy which he received during his apprenticeship. He attributed this to the fact that his master had dissected under Mr Joshua Brookes, referred to by James Paget as a 'grinder' who 'taught anatomy far better than did the teachers in the Hospital-schools'.[34] Brookes, who had paid up to 16 guineas to resurrectionists for bodies for dissection,[35] taught for 40 years and put together a valuable museum of human and comparative anatomy. He had depended mainly upon getting large classes during the summer. When in 1822 the Royal College of Surgeons decided not to recognize students' certificates for anatomical classes held in the summer months, because it was believed that they would not have received adequate hands-on experience due to the inadequate supply of cadavers, Brookes was seriously affected. His classes dwindled and he was obliged to dispose of his collection at considerable financial loss.[36] Poetic justice perhaps for one who had dealings with the London body snatchers and who is said to have admitted that he once kicked a sack containing a newly delivered corpse down a flight of stairs at his anatomy school.[37]

Following a period of apprenticeship to his father, a surgeon in Birmingham, Edward Grainger (1797–1824) moved to London in 1816 to study at the combined St Thomas's and Guy's Hospitals. He hoped for a demonstratorship in anatomy at Guy's where he had been dresser to Astley Cooper. He was not successful probably because he was not yet part of the London hospital network. At the early age of 22 he set up his own school at the home of a tailor in Southwark. Soon afterwards he moved to a building in Webb Street, close to Guy's Hospital, where he was joined by his younger brother Richard Dugard Grainger who took over the school following Edward's early death from consumption in 1824.[38]

Edward had been a good anatomist and his school flourished, widened its curriculum and expanded from 30 pupils to almost 300. He had moved twice to find a larger lecture theatre. It is said that he too enjoyed the favour of the resurrection men and thus had an assured supply of subjects for dissection. He attracted students away from the hospital schools in spite of the opposition from hospital surgeons in London, especially those on the Council of the College of Surgeons, whose restrictions had so affected Brookes.

Richard Grainger, interested in physiology as well as anatomy, maintained the good reputation of the Webb Street School, securing the cooperation of several able teachers including Marshall Hall. He lectured on the Practice of Medicine giving a series on the nervous system during the session 1837/8. The association between Hall and Grainger could have stemmed from their mutual Christian interests for in 1854 Grainger took a prominent part in the founding of the Christian Medical Association.[39] Hall's lectures, published in the *Lancet* were held on Monday, Wednesday and Friday evenings at a quarter to seven. Charlotte Hall implied that the Grainger lecture theatre was large, when she claimed that Hall found lecturing at the school a great exertion to his voice which was not powerful enough for its theatre. He did indeed have voice trouble which became so marked that he could not finish his course at the Webb Street School in 1839.

Sydenham College, Hall's final school, was the last of the better-known private medical schools to be established. It survived for only a dozen years, to be closed when its founder Erasmus Wilson accepted a post as lecturer in one of the hospital medical schools. The College was established in 1837 as a rival to the medical school associated with University College founded in 1828. Here Wilson had acted as demonstrator of anatomy under Jones Quain until Quain's retirement from the Chair of Anatomy and Physiology in 1836. Wilson opened his College in Gower Street and attracted the services of Dr Sigmund, Thomas Alcock, Richard Hoblyn, Marshall Hall and others. Again Hall was unable to complete the 1839 session due to the troublesome throat symptoms which affected his voice.

According to Pettigrew, Hall had been involved in the formation of Sydenham College[40] and a comment from James Paget following a meeting with Marshall Hall might corroborate this. Paget, who knew German, had been introduced to Hall to discuss Müller's ideas on reflex action and in a letter dated 29 November 1836 Paget said of him,

> He is restlessly anxious to be lecturing again and tried to establish a new school near the University (University College). He told me there would certainly be one in less than two years and asked me if I should like to join anything of the kind.[41]

Paget had not been anxious to join Hall even though he was at the time finding it difficult to secure a London post for himself. He told his correspondent,

> I did not say no, but I should certainly refuse. We might, I think, make a good assortment of Lecturers to publish 'Rejected Addresses' – M. Hall, Budd, Johnstone, and myself.[42]

It seems nevertheless that Paget knew Hall well enough to borrow his books. In correspondence from Manchester Square, Hall said he would be happy to meet Paget and his friend the next day and added, 'Will you also bring Van Deen (which I sent to you and Dr Baly some time ago) with you.'[43]

Reports in the *Lancet* show that the private schools declined during the first half of the nineteenth century as the London hospital medical schools grew in size and importance. There was considerable mobility of staff between the private and the hospital schools and when the private schools closed, their resources and goodwill were sometimes transferred to the hospital schools.[44]

The demise of the private medical schools by 1840 was also associated with the tightening up, especially for those going into 'general practice', of standards of medical instruction, and examinations as laid down in the Apothecaries Act of 1815. Furthermore, the increasingly sophisticated science courses which came to be included in the curriculum for the training of doctors became ever more threatening for the private medical schools. Finally the Royal College of Surgeons threw in its demands for a component of surgical study which, of course, could only be properly provided in a hospital setting.

Hospitals and Hospital Medical Schools

Hospitals as Teaching Institutions

Medical training became ever more expensive and the private medical schools could not bear the costs. Training, therefore, moved increasingly to the teaching hospitals and the hospital medical schools. The starting of lectures by the hospital staff made it possible for medical students to take out tickets for lectures without necessarily incurring the great expense of becoming the private pupils or apprentices of individual physicians or surgeons.

At the beginning of the nineteenth century the London hospitals which provided medical education, mainly surgery, were St Bartholomew's, the United Hospitals of St Thomas's and Guy's which separated again in 1826 and the London Hospital. To these in the first quarter of the century were added St George's and the Middlesex Hospital, and in 1831 the Charing Cross Hospital which had begun as an accident station in 1818. A hospital appointment, preferably combined with teaching, had been on Hall's personal career agenda when he first came to London. Being on the staff of a London hospital, where physicians claimed precedence over surgeons, had by the eighteenth century, already become a hallmark of success. But a permanent post for Hall was not to be and worse still, during the period 1830–50, he had to bear the success of so many fellow London medical colleagues who did obtain appointments. For instance Richard Grainger, who was very interested in research, joined St Thomas's Hospital. And Paget returned to his alma mater St Bartholomew's as a lecturer in physiology, a post Hall could well have coveted.

However, the London teaching hospitals had not all enjoyed a steadily rising reputation in the field of medical education. In 1834 Paget wrote critically of his student days at St Bartholomew's and he mentioned the rising reputation of London University, in spite of having given as one of the reasons for entering St Bartholomew's, that 'The University of London was then young and seemed contemptible.'[45]

Paget, a surgeon, criticized the work of the staff at St Bartholomew's, among whom there was constant discussion and mischievous rivalry especially following the retirement of Abernethy. He had combined good teaching and administration with being a surgeon of high repute. According to Paget, William Lawrence was disliked and though he could be strong in controversy he avoided it whenever possible. This might explain Lawrence's somewhat defensive manner when later he had dealings with Hall at the Royal Society over a memoir on the nervous system.

Christison was a medical pupil[46] at St Bartholemew's during the time when its rival, the private Aldersgate Medical School was flourishing. His reminiscences of those days confirmed the depressed state of medical teaching in one of the leading London hospitals and the greater emphasis placed on surgical teaching. Christison observed,

> It was a frequent subject of wonder to me, that so little use was made of the medical wards of St Bartholomew's for the purpose of instruction, and generally that education in medicine proper was almost entirely neglected. While the medical students were only three in number, and indeed all being graduates were no longer 'pupils', though called so, the surgical students, amounting to several hundreds never entered a medical ward, and, though pupils in reality, got no more information in medical practice than the few crumbs they might pick up now and then during the medical treatment of surgical cases.[47]

Christison deplored the fact that in spite of this inadequate training, hundreds of students supposedly capable of tackling the whole range of patients' medical problems, were passed annually into the ranks of general practitioners by the London College of Surgeons. He could not resist commenting on the superior nature of general practitioners with an Edinburgh background,

> I could easily understand subsequently the superiority of the general practitioners educated at Edinburgh where medicine proper held a prominent place in the system of hospital instruction, the preference in which they were held in England, and their success and reputation, especially in most of the large English towns.[48]

It is possible that had Hall remained in Nottingham or settled in another provincial city especially one with facilities for medical education his career and reputation would have benefited greatly.

According to McMenemey the profession recognized the advantage of hospitals for teaching and also the value of those intangible assets which went with a hospital appointment.[49] But during the eighteenth century and beyond, the teaching in hospitals was pursued primarily for the personal gain of individual doctors, who were allowed to take a certain number of paying pupils for training. Their hospital work was generally not orientated towards research into more effective diagnosis or therapies. The physicians had, until the 1830s, very few pupils compared to the surgeons. Thus although in the first half of the nineteenth century consultants in teaching as in non-teaching hospitals were still not paid, the post was a very worthwhile sideline alongside a private practice, since pupils did pay fees and, when qualified, recommended to their chiefs the sort of patients who could afford a consultation. Students represented a good source of future profit. In addition a staff member of a hospital could give his pupils the opportunity of gaining practical experience in the surgical and medical wards, which the private medical schools could not provide. And a hospital appointment conferred social status and an entrée to useful social and professional circles.

Whereas the number of private pupils was originally very restricted, it became much larger in the second quarter of the nineteenth century when hospital education showed great development. The very high pupil fees became for a short time a considerable source of income to the younger members of hospital staffs. But in devoting time to lucrative private consultations the consultants failed to promote hospitals as places for clinical research or to promote a more scientific basis to medical education. Hall, well aware of the research carried out by Scottish medical figures, felt that he himself would have appreciated and exploited the research facilities offered by a London hospital.

Hospital Medical Schools – St Thomas's

According to Clark-Kennedy, hospital governors had begun to object to hospital practices in which medical students served only one chief; arrangements were made by which they served all members of staff equally.[50] This led to the pooling of fees and then to the setting up of a staff committee to administer the teaching. Thus began the development of the medical schools proper which became associated with the London hospitals. The development of organized medical education within the hospitals was at first opposed by the governors who felt that the hospitals had been founded to treat the sick and that there were in any case other sources of medical education outside the hospitals. But since hospital staff were not paid, the teaching could not be stopped. Furthermore, the students carried out tasks such as dressing for which the governors did not pay.

In the event, most hospital governors came to cooperate with the medical staff in maintaining, furthering and financing medical education. The degree to which they granted financial aid varied from hospital to hospital. By 1851, Mr Curling, surgeon to the London Hospital, wrote to the House Committee of the hospital pointing out how much less generous it was than the governors of other London hospitals.[51]

In some cases the private medical schools were taken over by the hospitals. For instance Grainger's Webb Street School was taken over by St Thomas's in 1842 when Grainger was offered an appointment by the hospital as lecturer in anatomy and physiology. The take-over led to a major reorganization in St Thomas's medical education and Grainger was rapidly elevated to the Chair of Anatomy. Soon, outside lecturers, not on a hospital staff, were coopted generally in areas such as botany, chemistry, dentistry and obstetrics. In this way Hall joined the part-time staff of St Thomas's. He was recruited by his friend, the Quaker Thomas Hodgkin, himself recently recruited to re-invigorate medical education in the hospital. Thereafter Grainger, who did not have a University background and who had been turned down for a Chair at University College London in 1836 on the grounds that there was 'no evidence that Mr Grainger is a gentleman of cultivated mind and extensive general knowledge' and for good measure it was also noted that his intonation in lecturing was particularly unpleasant, was fully accepted into the medical fraternity in London.[52] In 1845 he was elected to the Council of the Royal College of Surgeons, which had put such pressure on his brother.

Throughout his own period at St Thomas's, 1842 to 1846, Hall lectured on the Theory and Practice of Medicine including lectures on the nervous system, alongside Dr Barker.[53] At the beginning, Thomas Hodgkin also lectured with them. During the summer term of the first two years Hall additionally gave lectures with Barker on pathological anatomy. The thrice weekly lectures on Medicine took place at 3 o'clock and cost four guineas a session. Hall's published lecture material reveals, probably uniquely, that he illustrated his lectures both by reference to animal specimens and to patients who were specially taken to the lecture theatre. Pointing out on one occasion the advantage to be gained from this practice, he wrote,

> I shall endeavour to make these Lectures in a certain degree *Clinical;* that is I shall bring such patients before you, as may admit of being conveyed here, without injury to themselves .... And remember that, as one book well read is worth infinitely more than whole libraries run over superficially, so one patient well examined instructs more than 'walking' whole hospitals.[54]

Hall was clear that one book well worth the attention of his audience was a work by himself. Referring to his *On Diagnosis* Hall described it as,

> ... a work I hope you will allow me to recommend to you as a *guide* to the study of the Practice of Physic generally, and especially in your attendance on these lectures. It is placed in the Library.[55]

Hall had known Hodgkin since at least 1830 for in January that year William Stroud, the Edinburgh-trained apothecary at Guy's, wrote to tell Hodgkin that Hall was accompanying him on a visit to the Guy's Hospital Museum which Stroud described as 'your Museum'. Stroud also referred to a work he was preparing on the four Evangelists. Hodgkin's role in both the reorganization of St Thomas's medical school around 1842 and Hall's recruitment to it had been significant. Writing to Dr Foville 26 December 1842 he generously said of Hall,

> The Doctor has just given ten lectures at St Thomas's as a part of the cooperative plan which I projected when the School was reformed, and I am glad to say that it appears to have answered well – Thou wilt find two of the Doctor's lectures in the Medical Times.[56]

Hodgkin was uncertain if Hall's lectures would continue. They did, at the expense of Hodgkin's own but not through any connivance on Hall's part. The *Lancet* noted the reorganization at the hospital,

> The medical school here, which was thoroughly changed last year has this year undergone some further alterations. Dr Marshall Hall replaces Dr Hodgkin and now has a more prominent position in the department of medicine .... Mr Grainger continues to fill the chair of anatomy and physiology, in which branches of medicine he has for so many years held a high reputation in this metropolis.[57]

On 29 July 1843, R. J. Whitfield wrote curtly on behalf of the hospital to inform Hodgkin that 'The Medical School Committee have appointed Dr Marshall Hall and Dr Barker to give the lectures on the Theory and Practice of Medicine and also the lectures on Pathological Anatomy.'[58] This news was completely unexpected. A couple of days later Hodgkin drafted two letters[59] to Hall explaining that although he had not been officially informed, he had heard about Hall's forthcoming classes in the lecture programme. Considering his part in developing the school in the previous year, the present step seemed to him 'uncourteous'. The second letter revealed how hurt but also how civilized Hodgkin was. He told Hall 'there are different ways of making one's exit and I should wish mine to be both honourable and amicable'. He enquired about Hall's opinion and intentions. Between 22 and 28 July 1843, four letters passed from Hall to Hodgkin.[60]

At the time Hall was suffering the frustration of not being able to enlist the support of any Royal Society officers whose cooperation he felt he needed if he were to get papers published in *Philosophical Transactions.* He had also been involved in a controversy within the Royal Society and it is understandable that he was anxious to avoid any confrontation with someone he regarded as a friend. He claimed he too had heard only rumours about the new arrangements. In fact Barker had recently visited Hall because he also was anxious about his own future at St Thomas's. Hall had even delayed the preparation of his display of visual aids on the nervous system, 'not knowing whether I shall be called upon to give my lectures on that subject next winter'. Hall then leaked what he thought was the real cause of the changes, which has about it a contemporary ring. It was falling student numbers. His next letter was more explicit. As a result of some enquiries he had just made, Hall confessed,

> I am told that, as the object of all the arrangements of last session was to raise the School, that, as your class did not keep up and that as mine did, there is in order to carry out the original plan, a wish that you should resign your office as Lecturer and that I should undertake it.[61]

Hall hoped Hodgkin would not be offended at his candid manner and advised him, in strict confidence, to accept the arrangement provided the Governors allowed Hodgkin to retain his work on the Clinical Ward and his superintendence of the Museum as Director, on which he had spent much time and money. Hall, who had coveted a permanent hospital appointment for himself consoled Hodgkin, telling him that his talents were 'of an infinitely higher order than those required in a Lecturer'. Hodgkin was invited to discuss the matter but he declined and on 28 July, Hall wrote definitively of his recent appointment as part-time Lecturer on the Practice of Medicine. Throughout all the negotiations Hall, who clearly was a success at St Thomas's, had made it clear that he did not want personal preferment at Hodgkin's expense. But he added, 'I have long been aware that the failure in the *medical* class, *as a whole would* lead to changes ....'[62] Hall was clearly embarrassed about the whole situation and in a further letter a year later he wrote to Hodgkin telling of his retirement from St Thomas's. He told Hodgkin, 'I am fast receding from the public eye and long for quiet repose.' In time, Hodgkin withdrew from medical practice never having built up a really successful one. He devoted himself increasingly to philanthropic activities including work on behalf of Negro emancipation.

Le Gros Clark, another supporter of Hall on the staff of St Thomas's, publicized Hall's views to his students and through his own publications, to the wider audience of young practising doctors. He held daily courses in Anatomical Demonstrations and Dissections[63] and, in 1836 published *The Practical Anatomy and Elementary Physiology of the Nervous System*,[64] especially for use in the dissecting room. In a generally favourable review of the work, including praise for Le Gros Clark's exposition 'in a plain and perspicuous manner' of Marshall Hall's doctrine of reflex function and excito-motory actions, the *Edinburgh Medical and Surgical Journal* did, however, question the wisdom of emphasizing, in a student text, a doctrine which was not firmly established. It justifiably pointed out that students 'must not be taught exclusively any single doctrine but should be instructed in all and be taught to choose the best ....'[65]

While Hall must have derived some satisfaction from his association with St Thomas's, the experience cannot have been without its pain since he had to tolerate the greater preferment not only of Richard Grainger. In the year he joined the school, Le Gros Clark was appointed Assistant Physician to the Hospital and William Carpenter, whose reviews of his work in the medical journals sometimes offended him, was promoted from part-time Lecturer in Comparative Anatomy and Natural History, to the Chair in the same subjects. With Grainger already holding the Chair of Anatomy and Physiology,[66] Hall was occupying only a temporary post alongside these younger men who had achieved the kind of success and recognition he had hoped for. It is not clear why Hall ceased lecturing at St Thomas's in 1846 although he was working on another memoir on the nervous system. He may have needed more time for research but it is also likely that at 56 he saw that he would not gain a permanent appointment in the hospital.

The University of London – A Case Study in Medical Education

Background

In the early part of the nineteenth century, a group of Nonconformists associated with Jeremy Bentham, had been active in establishing the present-day University College London, originally the University of London. The first statement of the aims of the University Council referred to the effect on 'a very large proportion of the youth of England' of the religious tests and the heavy fees, together with the undesirable feature of young people living separated from their parents.[67] Religious and financial limitations generally excluded them from Oxford and Cambridge.

The University of London was established in 1828 to serve the sons of the middle class whose admission would not depend upon religious tests such as those imposed by Oxford and Cambridge. As John Conolly, one of its early medical professors expressed in his introductory lecture, the new University was,

> ... an institution in which it is professed, as I solemnly believe, without reserve or equivocation, that no sect, no party, no persuasion, no difference of rank, or fortune, or opinion, will be a bar to all the academical honours which a pupil may merit, or which can here be bestowed.[68]

A further feature of the new University was that it was non-residential for both staff and students. Undergraduates would thus be able to reside at home and receive their education at a much lower cost than at Oxford and Cambridge.

From its inception, the University of London had placed great emphasis on the study of medicine and indeed its opening on 1 October 1828 was marked by an Inaugural Lecture from Charles Bell one of the first medical faculty professors. On the following day John Conolly delivered his own Inaugural Lecture. The intention to ensure medicine a significant role in the new institution was high on its first list of aims, which also included an unequivocal reference to the need for a hospital.

> An Hospital capable of containing a sufficient number of patients to afford opportunities of Clinical Practice, both medical and surgical, and of illustrated Lectures, will be provided, as being an essential requisite of a Medical School; and a connection will be formed with some Dispensary at a convenient distance.[69]

Emphasis was laid on the fact that at Oxford and Cambridge a professional course of study in Law or Medicine was not one of the primary objectives, whereas London provided special opportunities for linking theory and practice in these branches of knowledge. In other words, the University of London would provide direct professional training.

One of the first and most prominent individuals to be involved in launching the new College was Isaac Lyon Goldsmid, a wealthy financier of German origin who was closely associated with the influential, élite Sephardim Jews in England at that time.[70] In addition to his philanthropic interest in educational reform Goldsmid, like the Benthamites, had taken an active part in criminal law reform and Negro emancipation. He came to be Treasurer to the new Hospital and later, as Sir Isaac Goldsmid, its Chairman.

Goldsmid provided the really essential support for such a radical and initially uncertain reform as the University of London – financial backing. In May 1826, following the February signing of the Deed of Settlement and formation of the Council of London University, a general outline of the plan of the University was laid before the public. At the same time, it was announced that a piece of freehold ground, about eight acres, at the northern end of Gower Street had been purchased for £20,000 from Mr Bevan, a Lombard Street banker.[71] This purchase was financed by Goldsmid and two colleagues John Smith and Benjamin Shaw, also on the Council and who had been persuaded to share the financial responsibility. The three men held the land until the shareholders of the University had raised enough additional money to take it over.[72]

One group who played a significant role in establishing the University of London, had a Scottish background. Thomas Campbell, poet and writer, to whom is attributed the original idea of the University of London, had studied at Edinburgh as had Henry Brougham whom Isaac Goldsmid had introduced to Campbell. Others from Edinburgh included Charles Bell and his brother George, Sydney Smith, George Birkbeck, Anthony Todd Thomson, Robert Grant, John Conolly and the Marquis of Lansdowne. Five of these became members of the first Council of London University and four its first medical professors – Bell, Thomson, Grant and Conolly. Edward Turner, the first Professor of Chemistry, had also obtained an MD in Edinburgh in 1819 before studying chemistry.

Early Medical Appointments to the University of London

Appointments to the University were much sought after; up to 1 October 1828 there were about 180 applicants for fewer than 30 chairs. Hall might have felt less keenly his own failure a decade later, if he had known that among the unsuccessful earlier candidates had been another leading physician Herbert Mayo. He later refereed and contributed to the rejection of one of Hall's papers submitted to the Royal Society. Relations between the two men were competitive rather than cordial. Peter Mark Roget, a candidate for the Chair of Anatomy and Physiology, had withdrawn his application.[73]

Robert Grant, three years younger than Marshall Hall, had graduated from Edinburgh in 1814. Having met as students, Hall and Grant became friends in London. Since the first public meeting about the proposed new College took place in 1825 and Grant's appointment as Professor of Comparative Anatomy and Zoology in the University of London dated from June 1827, it is likely that Hall's hurried departure from Nottingham in 1826 had been motivated by a desire to join the Edinburgh group in the new institution. During his 46 years at University College, Grant is said never to have missed a lecture. But his lectures,[74] which seem to have changed little over the years, were not compulsory and were not well attended thus bringing him poor remuneration. And when, on Jones Quain's retirement from the Chair of Anatomy and Physiology, Wakley championed Grant's suitability as a successor, the College blocked his transfer.[75] Grant lectured at a number of other institutions including the Royal Institution and at private medical schools such as those in Aldersgate Street and Windmill Street, but he was not greatly admired. It is not known to what extent, if any, his alleged homosexuality disadvantaged him.[76]

A university post such as that held by Grant would, Hall imagined, accommodate his aspirations especially as he would have anticipated large audiences. With the security of regular payments from the College he could give his lectures and pursue his research. And although he would be involved with patients in the hospital which eventually came to be associated with the College, he could decrease the burden of maintaining a practice.[77] Hall's early ambitions to join University College came to nothing. His association in 1837 with the rival but short-lived Sydenham College, can hardly have enhanced his chances of success.[78]

In addition to Grant, Bell[79] and Conolly, the first Professors appointed by July 1827, there were others including John Meckel from Halle who was originally to have brought his museum with him, and Granville Sharp Pattison who with Bell and Meckel was appointed Professor of Surgery.[80] The range of Professors and their subjects indicate the breadth of the medical curriculum envisaged. They also emphasize that Physiology was not yet perceived as a separate scientific, experimental subject in its own right, but as a branch of Anatomy. Most of the early appointments were made by invitation; later vacancies were advertised. It was agreed that the Professors of Clinical Medicine and Surgery were to be paid £250 a year and most of the other professors £300 each, for three years. After this, salaries were to cease and Professors would receive two-thirds of the fees of such students as they could attract;[81] the remaining sum would go to the University. Leonard Horner, F.R.S., was appointed Warden of the College at an annual salary of £1,200.

From the start, relations between the Professors themselves and between some of them and members of the Council were not harmonious. Bell and Conolly were active in these disputes involving some rearrangements of teaching duties which the

Professors were left to arrange among themselves. Charles Bell soon felt the disadvantage of the continuing close association between anatomy and surgery. He called upon Council members to delineate the duties of the Professor of Anatomy from those of the Professor of Surgery, reminding them in a letter on 12 December 1829 that,

> ... I pointed out the necessity of defining the duties of the Professor of Surgery and the Professor of Anatomy (Pattison). The Professor of Anatomy has by degrees assumed a place which it was not intended originally he should occupy, and which I conceive is injurious to the School. Steps should be taken to prevent the necessity – I should say absurdity – of your Professor of Anatomy bringing a body into the theatre to perform an operation at 2 o'clock and the Professor of Surgery (Bell) doing precisely the same thing at 5 o'clock.
>
> I have the honour to be Gentlemen with great respect your very obedient servant,
>
> *CHARLES BELL*[82]

Perhaps it was this generally disputatious atmosphere which led Dr Smith, a member of the Faculty of Medicine to write tetchily in the margin of the minutes of 26 October 1827,

> I protest against a female, or any other person, not of or not authorised by the Medical Professors being made privy to their proceedings. This is a lady's handwriting.
>
> *J. G. SMITH*[83]

Hall, who had felt himself to be the victim of arrogant criticism in the published reviews of certain young men in the profession, might have attributed this lack of harmony in the early days of the University of London to the youthfulness of many of these early professors. As Hale Bellot noted, 'Perhaps the most conspicuous characteristic of the staff thus appointed was its youth. Setting aside Merlet, of the remaining twenty-seven professors, twenty-one were, when the University opened in October, 1828, under forty years of age.'[84]

When the Council was looking for possible sites for a hospital, Bell favoured a permanent union of the University with the Middlesex Hospital where he had held an appointment. This was opposed by Birkbeck a member of Council who claimed that only Charles Bell would benefit from the merger. Bell was also impatient in his insistence that the College should be granted powers to award degrees. From 1836 it did award degrees but Bell had already resigned in 1830 to rejoin the staff of the Middlesex.[85] His desire for a merger of the two hospitals was fulfilled, but not for another century and a half.

John Conolly, who like Hall had been President of the Edinburgh Medical Society, also soon resigned. He spent most of his professional life working with the insane and mentally disturbed. This was, even then, an unpopular specialism for which there was little public or professional sympathy. He left his Chair at the University of London less than four years after taking it up and in his valedictory lecture on 29 April 1831 gave reasons for this early resignation. Having forsaken a secure practice in Stratford to take up the London post, Conolly had been honoured to join the University and occupy a position which older men of higher professional reputation would have been proud to fill. But, the ambition of a medical professor should be either,

> ... to prosecute some original branches of practical investigation, or to be so habitually engaged in surveying the relations of all the parts of theoretical and practical medicine, as to leave behind him some memorial of his thought and his research.[86]

Clearly, Conolly too had envisaged his Chair giving him time to pursue research. But he had been severely disappointed by the realities of the post. Already in 1830 he had co-signed a letter, from a number of the professors to the shareholders and Council of the University emphasizing the difficulties of pursuing research. The letter highlighted poor organization and inadequate funding as the main sources of the problem. The professors also complained of high costs, unnecessarily long courses and uncertainty about salaries. Of this uncertainty and the low level of remuneration, the view of the particularly damaging effects on any research endeavour was that,

> No one can properly discharge his duties as a professor, or expect to recommend himself to public attention, if he do not especially devote himself to the study of the subject he is appointed to teach. This he cannot do unless he derive, from his situation as professor, an income at least sufficient to secure him against want.[87]

By 1831, Conolly felt that this plea had fallen on deaf ears as had his enlightened request that students be allowed to gain what he regarded as essential experience on the wards of lunatic asylums.[88] Furthermore in a series of charges against Warden Leonard Horner the previous year by nine of the professors, Conolly had complained forcefully of his intervention in the running of the Dispensary without reference to the Professors and in particular to the appointment of John Hogg as Resident Apothecary.

Conolly was also dissatisfied with the meagre time allocation for the study of medicine and the low esteem in which, he claimed, it was held compared to other subjects, including physiology,

> Anatomy is held in honour; chemistry and physiology promise discoveries and fame; surgery holds out a certain and defined rank; even pharmacy has its evident utility; but medicine, as a distinct study – medicine, for those who, without being called physicians, are to occupy almost all medical practice – seems to the student to lead to nothing. It offers no immediate advantage; it confers no rank .... It is not, in fact, recognized as a study in London, ...[89]

Conolly's sense of extreme frustration led him to claim that none of the advantages initially envisaged had been available,

> ... or were very likely to be enjoyed by myself as professor here; and I had, consequently, no prospect of succeeding in a noble design to which I would gladly have devoted my mind, and my time, and even my life.[90]

Having resigned his chair at the University of London, Conolly returned to the country to work in the asylums of Warwickshire. In 1839 he was appointed Resident Physician to the Middlesex County Asylum at Hanwell where his most significant innovation was the abolition of mechanical constraints commonly used in the management of lunatics.[91] Conolly was well read and particularly fond of Shakespeare. He published extensively on the nature and treatment of insanity and in *A Study of Hamlet* addressed the question of whether Shakespeare intended simply to portray the feigning of madness by Hamlet or if he deliberately drew a representation of a mind really disordered.[92]

In 1849 Conolly issued a remonstrance against the Right Honourable Sir Frederick Pollock, Lord Chief Baron of the Court of the Exchequer for his view that no lunatic should be confined in an asylum unless dangerous to himself and others. Pollock advocated that the Commissioners in Lunacy ought to liberate into the community all other persons so confined. According to Conolly this was both mistaken and mischievous and he supported his stand by reference to the legal case, Nottidge versus Ripley, which involved one of Hall's patients, Miss Nottidge, at the privately run Moorcroft Asylum formerly owned by Mr Stilwell.[93]

Through his reputation as an investigator of the nervous system and its diseases, including mental disorders, Marshall Hall had become consulting physician at Stilwell's institution where he worked for 14 years from 1839. According to Stilwell, Marshall Hall was involved in the management of the patient Miss Nottidge. She was the 42-year old religious maniac daughter of an elderly, wealthy widow whose four daughters all seem to have been mentally disordered. In 1846 a young curate named Prince had come to their parish and with the help of an accomplice had lured away the daughters, each of whom had a considerable dowry, to a colony in Somerset called the Agapemone or Abode of Love.

The colony was ruled by Prince. Three of the daughters 'dazzled by its luxury, charmed with its games and pastimes and sustained by glorious assurances of judgement being passed, and heaven come,' had married three young male members of the colony, whom they hardly knew and had handed over their dowries of eighteen thousand pounds. The fourth daughter had been rescued by members of her family, but was lured back to the colony and so successfully brainwashed, that she regarded Mr Prince as God. According to Conolly, while her sisters had been only fanatical and did not see Prince as the Almighty, Miss Nottidge was actually insane. She was placed in Moorcroft Asylum where her condition improved and where Marshall Hall regularly visited her on his fortnightly visits. He, along with Conolly and other specialists, declared her of unsound mind.

Following the new law on lunacy and against which Conolly argued so forcefully, Miss Nottidge had later been released on the ground that her bodily health was now declining and might be seriously impaired by longer detention. Within a few days she had surrendered her personal fortune of six thousand pounds to Prince. Conolly's insistence that even if insane people were not actually dangerous to themselves or others they might still, in the interests of themselves and their families, need to be retained in institutions, was vindicated.

Details of this particular case are not given by Hall but Charlotte Hall does note how much her husband enjoyed his visits to Moorcroft. She said 'I doubt whether anything in his whole practice gave him equal pleasure.' Although it was sometimes difficult for him to be free, he always aimed to take the four o'clock train to Uxbridge, which he found refreshing after the turmoil of his London practice.[94] One inmate composed a birthday ode to him,

> Oh lend your help, my friends, I pray,
> To welcome in this happy day!
> We hail it as the birth of one
> Whose kindness every heart has won.
> A day of joy, a day of glee,

> This day to us will ever be.
> May blessings every hour be shed Upon his path and on his head;
> True joy and peace his portion be,
> And days unnumbered may he see![95]

The resignations from the University of London of Bell and Conolly, who had both made invaluable contributions to the development of the Medical School, were accompanied by that of Horner, blamed by the majority of Professors for the lack of harmony. Pattison was actually expelled on grounds of incompetence.[96] These early setbacks of the University were openly discussed in newspapers and pamphlets and student protesters had made their contribution in opposing Pattison. Nevertheless the University survived and its much needed hospital came into being. Despite the disenchantment and departure of Bell and Conolly in the early 1830s, before the decade closed Hall had applied for a Chair.

Hall's Application for a Chair at University College

The voice trouble which had adversely affected Hall's 1838/9 session of lecturing in the private medical schools subsided and on 1 May 1839 he felt able to apply for the Chair of the Theory and Practice of Medicine at University College. Hall must have known of Conolly's dissatisfaction a decade earlier. Perhaps he hoped there had been improvements or believed that despite the difficulties he would be able to fulfil his own aspirations. His letter of application to the Council was a model in terms of its content and presentation.[97] In marked contrast to other three-line, successful applications for chairs at the College, it contained a lengthy *curriculum vitae* giving details of career and publications. It also provided brief synopses of the numerous papers which had been read before the Royal Medical and Chirurgical Society and the Royal Society. Of his lectures Hall said,

> For four years I have been engaged in giving Lectures on the Theory and Practice of Medicine in the several Schools in Aldersgate Street, Webb Street, and Sydenham College with a degree of success which has always amply repaid me for my exertions.[98]

Copies of his works were sent to Council along with certificates in support of the application, from Benjamin Brodie, Dr Holland, Mr Louis, Mr Flourens and Professor Müller. The latter three individuals were described as the first physiologists in France and Germany. There was some anxiety in Hall's mind about the prevailing prejudice against physicians engaging in physiological research, jeopardising his chances of obtaining the Chair. So he added,

> Two points I beg to impress upon your minds, with the utmost earnestness: the *first* that I had been nearly twenty years a *practical* physician, before I touched upon the subject of *physiology* at all; the *second,* that I have constantly, since I engaged in that study, made it subservient to pathology and practice.[99]

The letter ended with details of how Hall intended to fulfil his role should he be appointed to the Chair.

The Council Minutes of University College for 1 June 1839 record the names of all the applicants for the Chair of the Principles and Practice of Medicine[100] which also conferred appointment as physician to the hospital. Hall thus hoped to achieve two of his ambitions, teaching and clinical work. Ten days after submission of his application Hall wrote to Atkinson, the College Secretary, asking for the address of certain gentlemen whose names were, he said, given on a separate paper.[101] In the meantime all the applications were passing through Senate and on the 5 June a committee of eight Professors was drawn up to consider them.[102] By 13 July Professor Jones Quain acting for the Dean of the Faculty of Medicine presented the report of the Committee to Senate which resolved that it should be passed to Council, as the report of Senate.[103] However, on 10 July displaying what seems precipitate behaviour, Hall wrote briefly to the members of the Council withdrawing his application,

> My Lords and Gentlemen,
>
> Various circumstances induce me respectfully to request that you will allow me to withdraw my application for the Chair of the Theory and Practice of Medicine in University College. I have the honour to be My Lords and Gentlemen Your obedient Servant,
>
> *MARSHALL HALL.*[104]

This letter was read at the session of Council held three days later on 13 July.[105] At the same meeting the Senate reports on the other candidates were read. However, it is clear that the structure and duties of the post had been under discussion within the Select Committee, for a paper signed by Professors Quain and Sharpey was also read to the Council. They proposed that the appointment to both the Professorship and Physician to the hospital should proceed provided a revised apportionment of patients within the hospital could be made. In addition new arrangements were needed for Clinical teaching within the hospital. In the circumstances it was decided to postpone further consideration of the applications until the next Council Meeting a week later, 20 July.[106]

A reference in the Senate minutes of 3 August 1839 to the Council Meeting of 20 July 1839 shows that a Dr Williams had been appointed Professor of the Theory and Practice of Medicine and Physician to the Hospital, subject to the conditions laid down by Quain and Sharpey.[107] But Williams' name had not even been among the original list of candidates. Thus, an allegation by Hall, in a more detailed letter dated 23 July 1839 to the Council of University College, claiming that a candidate was admitted after the limit tacitly implied by the advertisements for the reception of applications and testimonials, seems to have been justified.[108]

Hall asserted that this late applicant claimed that an actual offer of the Chair had been made to him. Hall gave this as the immediate reason for his withdrawal. He also accused the Council of University College of persecuting him personally for his religious views.[109] Hinting that he regretted his action in withdrawing his application and that he would have appreciated the opportunity to rescind it, Hall added that he had been advised that it would have been wiser to discuss his situation with a member of Council before withdrawing.

Hall's letter was mentioned in the Council Minutes of 7 August as was a letter from Dr C. J. B. Williams dated 20 July 1839, the day on which the earlier Council Meeting had already announced Williams' appointment, announcing that he accepted the appointment of Professorship of the Principles and Practice of Medicine and Physician to the Hospital. In spite of this reference to Williams's acceptance, the Council Minutes explain that at the meeting on 7 August applications were considered and the Members proceeded to name the candidate they preferred. It was recorded that Dr Williams was named by everyone.[110] Thus, there was as Hall claimed, undoubtedly some irregularity over this appointment.

In his autobiography Williams later complained bitterly of his workload and of the impossibility of fitting in any research. Hall himself might well have been equally disappointed had his application succeeded. Four years after his appointment Williams, like Conolly before him, made clear the deleterious effects of such excessive clinical workloads in hindering those who were also supposed to be researchers,

> ... science and practice have been rarely pursued by the same parties. Scientific men are not and cannot be practical, because they have had no experience and practitioners know little of science and therefore derive little good from it.[111]

Williams retired from University College in 1849 after only ten years in post. According to Hall's wife 'a strong wish was expressed by many of the students of that institution that Dr Marshall Hall should again become a candidate.'[112] She said that Hall declined. Perhaps Hall had a more realistic perception of the post or maybe it was his age – now 59 – and the fact that his throat problem had recurred. In the event Williams was succeeded by Dr Walshe whose appointment was noted in a *Lancet* editorial which offered its view that 'no other physicians applied for the appointment' and implied that (as had happened a decade earlier) the appointment had once more not been open to competition. The journal claimed, 'We are assured the Chair was already settled even before the actual departure of Dr Williams.'[113] Perhaps once more the University was determined to block Marshall Hall and this time acted well in advance.

But Walshe had already held a Chair in the College since 1841. In January of that year he had applied for the Chair of Pathological Anatomy and Clinical Medicine.[114] Three months later he submitted a revised application to Atkinson, clearly in response to a letter informing him that the Council of University College had decided to make separate appointments to the Professorships of Clinical Medicine and Pathological Anatomy.[115] Walshe now applied for the Chair of Pathological Anatomy and his letter of acceptance of this post, which was subsequently offered to him, was dated 8 June 1841.[116] However, the two posts were later again combined. The *Lancet* explained that on his subsequent retirement 'Dr Walshe vacated two Chairs, those of Clinical Medicine and Pathological Anatomy.'[117] The journal speculated that two of Walshe's junior colleagues, Dr Parkes and Dr Garrod, would succeed him acquiring one Chair each.

Provision for Clinical Training at University College – The Dispensary

Since University College Hospital was not open until 1834 some years after medical training had begun in the newly opened University of London, temporary arrangements had to be made for the clinical training of students. These arrangements included the use of Middlesex Hospital, but not the merger Bell had wanted, where students could witness clinical practice. They also involved the setting up of a Dispensary which was specially opened to provide students with out-patient instruction. The University Dispensary opened in Euston Square, conveniently close to the University, in September 1828, three days before the opening of the University and six years before the opening of the hospital. John Hogg's appointment as Resident Apothecary to the Dispensary carried an annual salary of £50 per half year and he remained there until 1834 when the Dispensary closed. During his tenure Hogg, who was described as a surgeon and graduate in Medicine of Edinburgh, acted under orders of the Dispensary Committee and certain of the Medical Professors. He was not allowed to engage in private practice. However, according to Merrington, he omitted to display details of the Professors' lecture programmes thus depriving them of income.[118] Hogg himself held classes for the students but he certainly did not use Hall's nosology,[119] for in 1830 he published from the University Dispensary, a set of chemical, pharmacological and nosological tables which included Cullen's nosology. It was Hogg's appointment to which John Conolly had objected when he resigned as Physician to the Dispensary.

The objects and rules for the Dispensary were laid down by the Council of the University on the recommendation of the Dispensary Committee. It was to provide medical and surgical assistance to the sick poor who applied in person and were recommended by Proprietors of the University, subscribers to the Dispensary and its Medical Officers. If necessary, patients including women in childbirth were to be visited in their own homes since, like most other dispensaries at the time, it had no beds for patients, being more like a contemporary Casualty Department. The funds for running the Dispensary came from annual subscriptions and donations and from the fees paid by each pupil attending. The Council of the University of London made up any deficiency.

Hogg has indicated the nature and number of cases which were treated at the University Dispensary during 1829 and he compared these cases with those treated in Bateman's Public Dispensary in Gray's Inn Lane in 1816.[120] He noted the lack of consensus among medical writers of the time in designating illnesses but the most common diseases seem to have been the same in both places: catarrh, indigestion, fever, rheumatism and asthma. According to Hogg, between 1828 and the time the Dispensary closed, 1,200 patients were treated annually.[121]

Student Medical Societies

As already existed in Edinburgh there arose in London a number of medical societies, sometimes associated with teaching hospitals. Their later appearance in London was probably due to the private medical schools having delayed the development of hospitals as centres for medical education. Since the private medical schools were institutions in which pupils came and went as they wished, there existed little opportunity for the development of a corporate spirit among the student population. Neither did they foster further professional activities such as the presentation of papers or cultural activities, the general absence of which marked the non-residential medical education of London at that time.

However, the eventual development of hospital medical schools and of medical education within the University of London with its markedly more scientific curriculum, as demanded by the examining bodies, led to the formation of groups of students studying together for several years. Under these conditions it was possible for societies to flourish. The Pupils' Physical Society was founded at Guy's Hospital only in 1850.[122] At University College, on the other hand, the student Medical Society was an early development and already in March 1838 John Durance George, one of its Presidents, read his paper 'Contribution to the History of the Nervous System' with its attack on Hall and an implied charge of plagiarism against him.[123] Joseph Lister was also an active member of the Society during his student days at the College, being in turn its Librarian, Secretary, Treasurer and President. In January 1853 he read a paper to the Society on 'The Use of the Microscope.'[124]

These Medical Societies provided opportunities for discussions between practising medical men, the staff of medical schools and students. They also provided a channel for discussions on the political issues in medicine during this period, which was marked by the passing of a number of significant medical Acts profoundly affecting the structure and practices within the profession.

Student Behaviour

Writing at the end of the nineteenth century, Sprigge was highly critical of the lax system which had allowed the private medical schools to grow up and of their deleterious effect in promoting the unflattering image of medical students, still abroad in the twentieth century. According to Sprigge the Royal College of Surgeons and the Society of Apothecaries accepted, to a certain extent, work done in these private schools and attendances registered at them as evidence of medical study. But no formal signatures were required to testify that an individual had duly heard a certain number of lectures or had been in the wards for a specific length of time. Furthermore, since there were no precautions against personations at the examinations, this was thought to occur. Sprigge compared the life of an early nineteenth-century medical student to that of a Parisian art student,

> ... the looseness of the requirements of the examiners reacted upon the schools, so that they became hot-beds of a Bohemianism that was as undoubted in cold fact as it is, and ever will be, celebrated in comic fiction.[125]

Student behaviour certainly came under scrutiny by the Faculty in the University of London. The Minute Book kept by the Dean of the Faculty of Medicine records that on 23 March 1853 a letter from Atkinson, College Secretary, to Sharpey then Dean of the Faculty of Medicine reported the student Mr C. Martin. According to College Beadle Morrison, he had infringed College regulations by smoking in the Refreshment Room.[126] Smoking seems to have been widespread among medical students, probably on account of the unpleasant odour in the dissecting room. In 1866 a regulation at St Mary's Hospital Medical School prohibited smoking in the Medical School and Hospital and even in the adjoining yard. One student of the period indicated the difficulty, even for staff, of enforcing this regulation, when he observed, 'The office of Assistant Demonstrator was not an easy one unless he smoked....'[127]

Another student misdemeanour at University College in April 1857 involved a student named Tilbury Fox who was reported to the Dean, William Jenner, by the Matron of the hospital. She complained that Mr Fox had grossly insulted her at table and had 'applied to her the expression "mongrel bred sneak"'. Fox was admonished.[128] In the following year another student, Frederick Pearson, brought a charge of assault against a fellow student named Leonard Manuel. In the event three students defended Manuel and made a charge against Pearson for annoying Manuel 'in the exercise of his duties by making insulting statements respecting him in presence of patients in the casualty room of the Hospital'. Manuel was further supported by one of the patients and Pearson was severely reprimanded and excluded for one month.[129]

Hall's Public Comments on Medical Reform

Appeals for the reform of attitudes and practices within the medical profession occur in many of Hall's publications and lectures. He was never an overt political campaigner like the vigorous Wakley, who naturally made frequent use of the *Lancet* to publicize his criticisms, polemics and demands for reform. Loudon talks of Wakley's use of ridicule and satire as being 'that of a crude bruiser rather than a polished fencer' and claims that Wakley later regretted this style.[130] Although Hall may occasionally have submitted letters to journals under pseudonyms, he was generally not afraid to stand up and be identified among those appealing for a unified profession and openly criticizing practices based on the vested interest of institutions and their small élite cadres, such as the Royal College of Physicians.

Hall had been a member of the group which formed the British Medical Association (BMA) in 1836. George Webster of Dulwich, a lifelong friend was its first president and Hall was elected to its Council. However, although not initially working in unison the two principal founding activists of the BMA had been Charles Hastings (1794–1866)[131] and Thomas Wakley. Before embarking on his medical studies at Edinburgh, Hastings had for a period assisted Wilson Philip, another Edinburgh graduate, with his work on the nervous control of the heart carried out at the Worcester Infirmary[132] where Wilson Philip was a physician. Wilson Philip's work on the circulatory and nervous systems brought him into conflict with Marshall Hall in the Royal Society, where Wilson Philip was more successful than Hall.[133]

Hastings who had begun his medical studies at Edinburgh in 1815 the year following Hall's departure, was a member of the student Medical Society and held one of the Clinical Clerkships at the Royal Infirmary. He later replaced Dr James Home who died of a fatal fever contracted in the Infirmary. Some years after his return to Worcester, Hastings started the *Midland Medical and Surgical Reporter* [134] a quarterly journal catering for a section of the provinces which acutely resented the domination from London. The journal was published between 1828 and 1832.

Some years earlier in 1826, a protest meeting at the Royal College of Surgeons, possibly organized by Wakley, was presided over by William Lawrence one of Hall's referees at the Royal Society. It was attended by 1,200 members who voiced their grievances against the College which was accused of neglecting their interests following the passing of the Apothecaries Act of 1815. Wakley succeeded in proposing and carrying a resolution to petition Parliament for a new Charter for the College. In 1830 the general practitioners in London inaugurated the Metropolitan Society of General Practitioners and a year later the concerted efforts of many including Wakley and Hall, to bring about unification of the different branches of medicine led to the formation of the London College of Medicine.

Neither of these two new bodies prospered long and by 1832 Hastings' *Midland Reporter* recorded its own forthcoming demise. The formation of the Provincial Medical Association for Medicine and Surgery which Hastings, with the support of Edward Barlow and John Conolly, had set up as an alternative avenue for medical men in the Midlands[135] to share their professional views, was announced in the *Reporter*'s final number. Wakley had been offended by criticism from Hastings of medical reform in the metropolis. In the columns of the *Lancet* he criticized the separate reference to medicine and surgery in the title of the new association and also the use of the term *Transactions* to its publications, which contained papers that had not been read to the association. Nevertheless, reflecting the need for a voice in the provinces, the association's membership rose in its first year from 150 to 316. But since Wakley, Hall and others wanted a national rather than a regional college of medicine based on the two principles of equality of title and equality of rights, Wakley extended his criticism to the term 'Provincial' in the title,

> Let us hope that this great Society will henceforth be known by some such designation as the 'British Medical Association' thus contradistinguishing the institution from any other that may be formed essentially belonging to the metropolis, or to any one of the large cities or towns.[136]

The Provincial Association did become the British Medical Association and eventually moved its headquarters to London.

At the first anniversary meeting and dinner of the British Medical Association held in London in September 1837 Robert Grant had replied to the toast to 'The professors and teachers of Medical Science.'[137] Marshall Hall, newly elected as Vice-President, acknowledged the toast to the Vice-President and Council. The occasion was used for making yet another appeal for reform of the profession. Focusing this time on the nepotism in its ranks, he declared,

> I do not happen to be the nephew of, or connected with, any of the great men, nor have I sought their favours, I believe I never made a bow to one of them in my life. I am content with my station, which has been the work of my own industry, and I trust that every man who expends great pains, labour and unwearied industry in the profession, will henceforth take that station in society, which ought to result from them.[138]

In November of the same year, Marshall Hall was asking for membership on behalf of two or three surgeons from great provincial hospitals. These individuals had been in practice many years before the Apothecaries Act of 1815 and were not members of either the College of Surgeons or any other College. Despite this they were no less respected nor less able as practitioners and in Hall's view would undoubtedly be eligible to become members of the British Medical Association.[139] One biographical sketch in referring to Hall's involvement in the early BMA, nevertheless admitted that,

> The strife and contention almost inseparable from medical politics were not suited to his taste; but a strong sense of right and justice urged

> him to lend his powerful aid to the operations of this Association, as often as his professional engagements and his scientific researches would allow.[140]

The time then available to Hall for reform activities must have been limited because during the late 1830s he was struggling with his confrontations within the Royal Society in connection with his rejected *Memoir* (1837). He was also conducting in print a dispute with George who had accused him of plagiarism from Prochaska. However, this activity calmed down and in October 1840 Marshall Hall gave the Association's Annual Oration, celebrating the fourth anniversary of the British Medical Association. It was entitled 'Medicine, its Divisions, its Rewards, and its Reforms'.[141]

The Royal College of Physicians was openly criticized for its two-tiered qualification of membership, awarding Fellowships to graduates of Oxford, Cambridge and Dublin but offering only the lower grade of Licentiate to Edinburgh graduates. Although both qualifications licensed their holders to practise, the Licentiate did not bestow the facilities and privileges of full membership of the College.[142] Hall outlined the case of Dr Wells, author of the philosophical 'Essay on Dew' who had in his own right unsuccessfully sought a Fellowship at the College. When much later it was offered to him he chose to reject it. Hall did not approve and claimed that in the unlikely event of being offered a Fellowship himself, he would accept in order to attempt some reform within the institution. In the following year Hall was offered and accepted a Fellowship.

Turning to the main section of his Oration Hall, in a highly injudicious manner, made frequent reference to his own sufferings. He bitterly criticized the unjust manner in which, through unfair modes of appointment, youthful inexperienced individuals – and he had in mind William Carpenter – were enabled to pass dismissive judgements on the work of highly experienced writers[143] or, having J. Durance George in mind, to accuse them of plagiarism.[144] He criticized the Royal Society and its method of awarding the Royal Medals for Physiology. And, referring to his own experience at University College, he was especially critical of the method of appointing professors and medical staff to the hospitals. 'The affair of "testimonials" is an absolute mockery. Intrigue is predominant.' A public Concours or examination was the only way of making such appointments, since this would be 'at once honourable to the individual, and beneficial to the public, assuring to the former a just reward of his talents and industry, and to the latter the ablest medical officer'.[145] And yet he knew that in France, Louis had in

1831 failed, under the same system, to gain a Clinical Chair for which he had competed.

Finally, Hall called for the extinction of grades such as physician, surgeon and general practitioner, for an improvement in the standard of education and, attacking one of his chief *bête noire,* for a complete segregation of the sale of drugs from the practice of medicine and surgery. Nevertheless success in any one of the possible specialisms in medicine should be acknowledged and recognized. There may have been more than a touch of hypocrisy and ambivalence in Hall's stand. He was always aware of his own elevated status as a graduate physician, an educated gentleman physician at that, rather than one of the rank and file. And it is likely that he really sympathized with George Guthrie's successful efforts to prevent the establishment of a College of General Practitioners in the 1840s. Guthrie had called the group 'General Practitioners of nothing'.[146] The Oration was concluded by the presentation of 12 'Principles of Medical Legislation' for obtaining what was described as 'the best medical advice and care for the sick public'.[147]

This Oration was one of many on medical reform. Professor Williams at University College, who had described himself as a '*conservative* reformer' and who admitted his reluctance to 'meddle with medical politics', was in 1841 urgently calling for the profession to rid itself of its real evils, 'the oppressions, the collisions, the invidious and unjust divisions' which vex and divide its members and to unite on the basis of 'true principles of incorporation and representation'. Williams was concerned that the profession had no real representation in Parliament and acknowledged the valuable contributions of the Colleges of Physicians and Surgeons and the Society of Apothecaries. Nevertheless, he had a proposal for reform of the College of Physicians. This would see the end of the demeaning Licentiate grade and 'the surrender of certain empty privileges by a few, to the end that justice and protection shall be administered to the many'.[148]

In February 1841 the *Lancet* reported the reading at another meeting of the British Medical Association of more pamphlets on medical reform presented by Professor Kidd, the Regius Professor of Medicine at Oxford and by Dr Sinclair.[149] Wakley for his part sustained in the columns of the *Lancet* a constant campaign on the issue, tackling now one aspect of the profession now another. In December 1839 he had already devoted a leading article to the question of rank among members of the medical profession. He was unequivocally opposed to it or rather he disliked the term 'General Practitioner' which included himself, regarding it as less prestigious than Surgeon–Apothecary.[150]

One wonders if, had Hall gained a more elevated status in the medical public arena in London, he would have continued to be quite so opposed to the hierarchical structure within the profession. What most of the reformers really wanted was a more highly educated basic grade of doctor and that all in the profession should have attained at least a common level of education, training and licensing and have been subject to compulsory registration. These measures would, it was hoped, increase the professionalism and hence the status of the group as a whole and reduce or eliminate the growing competition from the wide range of irregulars, including apothecaries, some of whom were in fact successful healers.

As part of his call for reform Wakley concentrated on the consumers. He highlighted the inadequacies of the social and medical-care provision for the poor, pointing out that in the 1840s there were more people in workhouses than in prisons and that the mortality rate was higher in the former than in the latter. The existing medical districts were too large and the remuneration for providing the poor with medical services and medicines was inadequate. In April 1841 the British Medical Association drew up a petition embodying the reforms sought to present to the Poor Law Commissioners.[151]

Although at the end of his Annual Oration to the British Medical Association, Hall had told his audience 'I think I may retire from all active participation in your affairs, with the consciousness of having done my duty', he did not in fact withdraw his services and on 20 October 1840 was elected Vice-President for the ensuing year.[152] During the first half of 1841 there were regular meetings between a delegation of the BMA and delegates from other associations including some from the provinces. Hall was one of the London delegates and on at least two occasions he chaired the meetings.[153] At the Conference of the British Medical Association on 15 June 1841, Webster and Hall were among those thanked for 'the efficient and able manner in which, as a deputation to Parliament, they carried ... the views of the British Medical Association'.[154]

Hall's continuing interest in the reform debates, which for him included concern for medicine in the dominions, was expressed in a letter dated 16 February 1841 to the Right Honourable Sir Robert Peel.[155] Peel was provided with a detailed analysis of the effect upon existing institutions of incorporating the profession into one faculty which was referred to as the Faculty of Medicine. Having re-rehearsed some of his well known criticisms of the Royal College of Physicians Hall turned his attention to the other two institutions involved and said of them,

> The Royal College of Surgeons does not even grant a licence to practise, and its Council are self-elected and irresponsible. The Society of Apothecaries, with power to grant a licence to practise, does not confer a diploma, which, generally speaking, entitled its possessor to be eligible to the office of surgeon to an hospital or dispensary.[156]

The Faculty of Medicine was envisaged as an institution for examination and licensing only and one which would unite the whole profession into a single community with uniformity of qualification, with equality of privilege, rights and immunities and without, as he put it, invidious and injurious distinctions, and with a government at once representative, responsible and protective. But Hall was no leveller; he did not see the Faculty as a threat to the existence of the individual colleges. On the contrary both would continue to prosper. And predicting to some extent the contemporary structure, he acknowledged that the members of the Faculty would be divided into those who would on the one hand be general practitioners and those on the other who would be consultants, either as physicians or surgeons. It would be interesting to compare the hierarchy between these groups in Hall's day and in contemporary medicine.

The real target, the Society of Apothecaries and its association with trade and commerce was then singled out for attack. A soothing, flattering preamble, with reference to the Society's significant 1815 Act was offered,

> This *new* Society – new, at least, in its powers – I boldly assert, had done more for the real and substantial improvement of the profession than all our other institutions put together .... It has raised the scale of medical education and consequent acquirement, and in proportion it has raised the utility, the respectability, and trustworthiness of the profession and its members.[157]

But then came the real thrust of Hall's withering attack. The Society was dismissed because,

> ... the Society of Apothecaries is a *trading* society. It has mingled the trade in drugs with the profession of medicine and in this it has done as much to degrade the profession, as, by raising the standard of education it has done to raise it.[158]

The separation of demeaning trade from the profession of medicine was demanded. Until this was effected the profession could never be respectable or respected.

Naturally, these changes had implications for the Society of Apothecaries which could not continue in what was described as 'its present amphibious conditions'. Some hope was expressed that its most distinguished members might become active members of the new Faculty but the general members would have to choose between the trade in drugs and the profession of medicine. In a postscript, it was noted that at a recent meeting of the Chemists and Druggists a resolution was passed that Parliament should be petitioned to uphold their 'right to prescribe'. With some acerbity Hall questioned whether these gentlemen should be allowed to combine with this right, that of being ignorant of anatomy and diseases, whether they should be immune from the '*education* and *examination*' legally appointed for the practitioner of medicine.

In August 1841 Hall delivered yet another address on medical reform to the Provincial Medical Association then assembled at York.[159] While reiterating his main points, Hall's chief target was again the Society of Apothecaries. The Royal College of Physicians was now treated briefly alongside the College of Surgeons with both institutions simply being described as in need of reform. This newly muted reference to the Royal College of Physicians was clearly due to the fact that, as Hall put it 'I have not hesitated to accept the Fellowship of the Royal College of Physicians, recently offered to me.'[160] Having joined them, Hall was now much less keen to beat them.

He had been arraigned for this action and although he did not feel obliged to give the motives for his behaviour he proceeded to do so. The Fellowship had been accepted because it was his right, because it gave him the opportunity of pleading in the right place for medical reform, because he had pledged to do so in his Annual Oration of 1840 and because he felt that unity within the ranks of the profession would be more effective. Finally, and somewhat provokingly, Hall made it quite clear that in any case he had no intention of allowing others to dictate his actions to him.

Misgivings expressed by physicians and surgeons concerning the possible effects on the survival of their Colleges of the proposed inauguration of the Faculty of Medicine were well known to Hall. They were reassured that the Colleges would prosper more than ever and indeed might well be the institutions to which Members of the Faculty, however high their diploma and title, would turn for higher honours. In other words, the Faculty was envisaged as issuing initial awards while the Colleges might award postgraduate certification.

The Society of Apothecaries was attacked yet again. As before the criticism began with some words of flattery with a call for even higher standards of education for the profession, especially for the most numerous section, the general practitioners. The single qualification as an apothecary was really held in contempt,

> Is there any one here who thinks that the diploma, or title of apothecary is a worthy one? Is there any one here who does not think a diploma and a title conferred, or rather imposed, by a half-professional half-trading society, an indignity?[161]

The station of the respectable tradesman was not to be despised but it was deemed unworthy for a medical practitioner to be engaged in a pursuit below his exalted and liberal profession. The selling of drugs must be made illegal on the part of the Members of the Faculty.

On the other hand the druggist must be protected in his right to sell drugs and while prevented altogether from prescribing, 'the trade in drugs must be secured to him'. In what the Society must have found an insulting gesture, it was proposed that the Society of Apothecaries should be renamed the Society of Druggists. Hall did wonder where Pharmacy should be ranked since it needed to be considered in both its scientific and commercial aspects. In its scientific aspect it involved the important subjects of the Pharmacopoeia, or the code of directions for the preparation of medicines, deduced as Hall put it, from science and careful experiment. In 1849 Hall himself put forward a suggestion for reform in this area, namely a National Decimal Pharmacopoeia.[162] In its commercial aspect Pharmacy involved the mass preparation of drugs for sale.

In a final comment there was introduced an innovatory plan for a professional pension scheme. As in the East India Service and in the College of Surgeons of Edinburgh it was suggested that each member of the profession should be called upon,

> ... to make an original, or annual deposit, which should secure his widow or his orphan, by *right,* against the possibility of want or poverty – an object far more exalted than that of all the *benevolent* schemes which have been devised.[163]

In spite of all this activity, the Medical Reform Bill was shelved. According to one editorial of 1854 this was because the reformers themselves were disunited.[164] It was not until the year after Hall's death that, by Act of Parliament, the Medical Act of 1858 was passed. This was intended to regulate the qualifications of practitioners in Medicine and Surgery. A General Medical Council was set up to prepare a Medical Register of all who were qualified to practise and the Council was empowered to remove from the Register the names of individuals found guilty of a crime or offence which was held to be infamous conduct in any professional respect.

Notes

1. 'Introduction', in Bynum, W. F. and Porter, R. (eds), *Medical Fringe and Medical Orthodoxy,* 1987, pp.1–4.
2. Loudon, I., *Medical Care,* 1986, p.133.
3. Porter, D. and Porter, R., *Patient's Progress,* 1989, p.99, pp.177–183.
4. 'Medical Reform – Apothecaries Act', editorial, 24 December, 1841, *Lond. Med. Gaz.,* new series, 1, 1841–2, pp.515–20. See also pp. 117–23 for editorial referring to the views on reform of Professor C. J. B.Williams, University College, and the response of a retired 'General Practitioner' (Surgeon/Apothecary) supporting his call for improved standards of medical education for general practitioners. He also called for improved remuneration for them.
5. *Ibid.,* p.516.
6. George Eliot (1819–80) and Gustave Flaubert (1821–80) present in their novels of middle-class life, including Eliot, G., *Middlemarch,* 4 vols, 1871–2, and Flaubert, G., *Madame Bovary,* 1857, detailed portraits of medical men of their period. Flaubert's father was a surgeon and clinical professor in Rouen.
7. *Lond. Med. Gaz.,* new series, 1, 1841–2, p.120.
8. *Ibid.*
9. MS *List of Students Attending Lectures in the University of Edinburgh from 1790,* University of Edinburgh, p.246. MS *Lists of Students, General and Medical,* 1811, 1812, 7, University of Edinburgh. MS *Lists of Students, General and Medical,* 1812, 1813, 8, University of Edinburgh.
10. Risse, G. B., *Hospital Life,* 1986, p.240.
11. Emblen, D. L., *Roget,* 1970, p.18.
12. The following criticism should therefore rarely have applied to an Edinburgh undergraduate, 'The pupil who on entering a sickroom is in such an exceeding hurry that he has no time to sit down or take his hat off, had better wait, as the vulgar say, till his hurry is over.' *Lond. Med. Gaz.,* new series, 1, 1837–8, p.383.
13. Newman, C. E., *Medical Education,* 1957. Peterson, M. J., *Medical Profession,* 1978.
14. Hall, C., *Memoirs,* p.8.
15. Paget, S. (ed.), *Memoirs of Sir James Paget,* 1901, p.19.
16. The 1825 Act was a temporary Amendment Act to that of 1815 and it expired after 12 months due, according to the *London Medical Gazette,* to 'pressure of more important business' rather than to wilful neglect. *Lond. Med. Gaz.,* new series, 1, 1841–2, p.516.
17. Paget, *Memoirs of Sir James Paget,* p.19.
18. *Ibid.,* p.70.
19. Cope, Z., *History of St Mary's Hospital Medical School,* 1954, p.2.
20. *Lond. Med. Gaz.,* new series, 1, 1841–2, p.517.
21. Porter, R., *Disease, Medicine and Society,* 1987, p.35.

22. Cope, Z., 'The History of the Dispensary Movement', in Poynter, F. N. L. (ed.), *Evolution of Hospitals,* 1964, p.73.
23. Loudon, *Medical Care,* pp.22–3.
24. Cope, Z., *General Practitioners,* 1961, pp.35–6.
25. Cope 'Dispensary Movement', p.73.
26. Franklin, A. W., 'Children's Hospitals', in Poynter, *Evolution of Hospitals,* p.103. This was the first specialist institution for sick children in London and it survived 16 years. It had no successor until 1816 when John Bunnell Davis founded the Universal Dispensary for Children.
27. Cope, 'Dispensary Movement', p.74.
28. An MS student notebook of lectures corresponds with the published syllabus of Hall's Aldersgate lectures indicating that the writer was a member of his class. MS The *Denis Notebook,* n.d., *Student Lecture Notebook,* St. Thomas's Medical School, London. The notebook appears to have been written by Peter Denis from Yorkshire. Hall, M., *Introductory Lecture on the Practice of Physic,* [1830], p.51. [Note: the British Library Catalogue gives 1830 but internal evidence within the lecture indicates a publication date post 1833.]
29. See for example the annual presentation of 'List of the Schools of Medicine in London, Open to Students in the Session 1835–36'. *Lancet,* vol.i, 1836, pp.1–13. Details for each 'school' include subjects offered, lecturers involved, times of lectures and the fees for each course. In a spirited and critical editorial following this article, Wakley included the disclaimer, 'We are not answerable ... for what we have inserted .... Heaven forbid that we should be answerable for the promises and pretensions of the lecturers!' *Ibid.,* p.13.
30. Paget, *Memoirs of Sir James Paget,* pp.21–2.
31. Cope, *General Practitioners,* pp.62–73.
32. Mayo later joined the staff of King's College Hospital and School which had recently been established as an Anglican foundation within the University of London.
33. Hawkins was elected to the staff of St George's Hospital.
34. Paget, *Memoirs of Sir James Paget,* pp.21–2.
35. Richardson, R., *Death, Dissection and the Destitute,* 1989, p.57.
36. Cope, Z., 'The Private Medical Schools of London' in Poynter, F. N. L. (ed.), *Evolution of Medical Education,* 1966, p.96. By 1835, following his death, Brookes's school was renamed Blenheim Street School of Medicine, a reflection of its broader curriculum which included Anatomy, Physiology and Morbid Anatomy, Medicine, Surgery and Surgical Operations, Chemistry and Pharmacy, Materia Medica and Botany, Midwifery, the Diseases of Women and Medical Jurisprudence.
37. Richardson, *Death, Dissection and the Destitute,* p.97.
38. 'Grainger, Richard Dugard', *Dict. Nat. Biog.,* 22, 1890, pp.367, 370.
39. Grainger delivered 'An Address ... to the Christian Medical Association on the Occasion of its First Meeting, November 10,

1854' London, 1855. *Ibid.* Since C. J. B. Williams, appointed to the Professorship in Medicine, University College in 1839 (and for which Hall was initially also a candidate) was in the Chair on the occasion of this Address it would suggest that, contrary to Hall's claim, his christianity *per se* had played no significant part in what he perceived as discrimination against him.

40. Pettigrew, T. J., *Medical Portrait Gallery,* 4 vols, [1840], 4, p.14.
41. Paget, *Memoirs of Sir James Paget,* p.93.
42. *Ibid.*
43. MS Letter, Hall to Paget, n.d., *WMS/ALS,* Wellcome Institute for the History of Medicine, London.
44. St Bartholomew's Hospital Medical School flourished at the time the Aldersgate School declined and some of its best staff moved to St Bartholomew's. Among them was Paget who had been a student at St Bartholomew's and to whom the governors offered the Lectureship in Physiology, a post Hall must have coveted.
45. Paget, *Memoirs of Sir James Paget,* p.39. It was usual for graduates of Oxford and Cambridge to pursue their clinical studies in one of the London hospitals. Paget referred to two friends of his brother, who had himself recently been elected to a medical fellowship at Caius College, Cambridge. Paget spoke disparagingly of these two friends as being 'idle men, expecting, as many then did, that their University education would give them sufficient claim to as much success as they cared for'. Nevertheless, Paget indicated that belief in the privilege of Oxford and Cambridge was decaying in the 1830s. *Ibid.*
46. Students in hospitals at this time were commonly called 'pupils', even if they were graduates, because they were still largely the private pupils of members of the staff, being attached to them as short-term consultant apprentices. See Newman, C., 'The Hospital as a Teaching Centre' in Poynter, *Evolution of Medical Education,* p.197.
47. [Christison, R.], *Life of Sir Robert Christison.,* 2 vols, 1885, 1, pp.93–4.
48. *Ibid.*
49. McMenemey, W. H., 'The Hospital Movement of the Eighteenth Century and its Development', in Poynter, *Evolution of Hospitals,* p.64.
50. Clark-Kennedy, A. E., 'The London Hospital and the Rise of the University' in Poynter, *Evolution of Medical Education,* p.111.
51. *Ibid.*
52. Taylor, D. W., 'The Life and Teaching of William Sharpey (1802–1880) "Father of Modern Physiology" in Britain: Part I' *Med. Hist.,* 15, 1971, p.134.
53. According to the *Dictionary of National Biography,* Hall lectured on nervous diseases at St Thomas's 1842–6, *Dict. Nat. Biog.,* 8, p.965. Le Gros Clark, a keen supporter of Marshall Hall pointed out that Hall's name appeared as a lecturer in the Hospital prospectuses over this period. Documents covering these years were lost during the

1939–45 war. However the *Lancet* annual entries confirm Hall's role in the Hospital. *Lancet,* vol.i, 1842–3, p.18; *ibid.,* vol.ii, p.925; *ibid.,* vol.ii, 1844, p.8; *ibid.,* vol.ii, 1845, p.338.

54. Hall was, therefore, in some degree, trying to introduce the Parisian method of 'The Clinic' to his teaching. Hall, *Introductory Lecture on the Practice of Physic,* p.51.
55. *Ibid.,* p.25.
56. MS Letter Thomas Hodgkin to [Dr Foville], 12 December 1842, *Hodgkin Papers,* Ilmington, Warwickshire.
57. *Lancet,* vol.ii, 1842–3, p.934.
58. MS Letter R. S. Whitfield to Thomas Hodgkin, 20 July 1843, *Hodgkin Papers.*
59. MS Letters Thomas Hodgkin to My Dear Friend (Marshall Hall), 22 July 1843, *Hodgkin Papers.*
60. MS Letters, Marshall Hall to Thomas Hodgkin, 22 July 1843, 23 July 1843, 24 July 1843, 28 July 1843, *Hodgkin Papers.*
61. MS Letter Marshall Hall to Thomas Hodgkin, 23 July 1843, *Hodgkin Papers.*
62. MS Letter Marshall Hall to Thomas Hodgkin, 28 July 1843, *Hodgkin Papers.*
63. *Lancet,* vol.i, 1842–3, p.18. The fee for Le Gros Clark's lectures was 4 guineas (£4.20) per session.
64. Clark, F. le Gros, *Practical Anatomy and Elementary Physiology,* 1836.
65. *Edin. Med. and Surg. Jour.,* 48, 1837, pp.477–85. Also *Brit. and For. Med. Rev.,* 3, 1837, pp.203–4.
66. *Lancet,* vol.ii, 1842–3, p.934.
67. *Statement by the Council of the University of London,* 1827, p.7.
68. Conolly, J., *Introductory Lecture,* 1828, p.27. Conolly, who had worked on insanity, was appointed to the Professorship of the Nature and Treatment of Diseases. His Inaugural Lecture was on mental disorders. He tried unsuccessfully to gain admission for medical students to the wards of a lunatic asylum. Following disputes at the University, Conolly left in 1837 to take charge of the Asylum at Hanwell, Middlesex where he abolished mechanical restraint on patients. During his term of office at the University, Conolly published *Inquiry Concerning the Indications of Insanity,* 1830.
69. *Statement by the Council of the University of London,* p.15.
70. Goldsmid, whose main cause in life was that of Jewish emancipation, was the first Jew to be created a baronet, in 1841. *Dict. Nat. Biog.,* 22, 1890.
71. Merrington, W. R., *University College Hospital,* 1976, p.3.
72. *University College Report for 1859,* 1860.
73. Bellot, H. Hale, *University College London,* 1929, pp.46–7.
74. Grant, R. E., MS *Lectures on Comparative Anatomy. Notes for a Course of Lectures Delivered Oct.–Dec. 1830* (plus printed copy of examination paper on comparative anatomy, dated 8 Jan. 1831)

MS.ADD.1, University College London. MS *Lectures on Comparative Anatomy Delivered in the University of London, session 1832–33,* University College London. MS *Notebook Containing a Student's Abstracts of Grant's Lectures, 1832–33. Strong Rm. MS.ADD.39,* University College London. MS *Lectures on Comparative Anatomy delivered in the University of London, session 1833–34,* University College London. MS *Notebook Containing a Student's Abstracts of Grant's Lectures, 1833–34. MS.ADD.38,* University College London.

75. Taylor considers Grant's application in his article on the successful applicant for the Chair, William Sharpey. Taylor, 'William Sharpey', pp.131–7.
76. Desmond describes Grant as having flouted convention and observes that, 'He never married and is thought to have been a homosexual ....' Desmond, A., *Archetypes and Ancestors,* 1982, p.117.
77. But see comments of C. J. B. Williams who was appointed at University College to a Professorship for which Hall applied, pp.136–7 *Infra.*
78. See note 41 *Supra.*
79. Bell, who was friendly with several of the founders of the University including Brougham, had been elected to the staff of the Middlesex Hospital in April 1814, a connection he maintained after joining the University of London. By the time of his University appointment Bell was an anatomist, surgeon and physiologist of established fame. He built up a good collection of museum specimens and is said to have done business with the 'body snatchers' or 'resurrectionists' until the Anatomy Act, 1832. Several of his specimens are held by University College Hospital Medical School.
80. For list of Professors appointed by 26 October 1827, see MS *Minute Book of the Faculty of Medicine of the University of London, 1827–1830,* University College London. The minute reads 'The Faculty of Medicine of the University of London consists of the following Professors, who formed themselves into a deliberate body at the formal request of the Council of the University, communicated through the Warden by a note dated the 26 of October, 1827.'
81. *Statement by the Council of the University of London,* p.18. Arrangements for a retirement fund were also made and Council 'will feel it their duty to make such allowances as liberal as the means at their disposal will admit'. *Ibid.,* p.17. Student fees were initially envisaged to be around £25–£30 for non-medical students who were nominated by a Proprietor of the University, with a higher fee for medical students and a considerably higher fee still for all students not nominated by a Proprietor. *Ibid.,* pp.23–4. In the event, even the guaranteed initial salary dwindled as funds grew short. Merrington, *University College Hospital,* p.7.
82. *Ibid.,* p.8.
83. MS *Minutes 26 October 1827. Minute Book of the Faculty of Medicine of the University of London, 1827–1830.* University College London.
84. Bellot, *University College London,* pp.46–7.

85. Merrington, *University College Hospital,* pp.8–9.
86. Conolly, J., 'Lecture IV. Valedictory Lecture, on Retiring from the University', in *Four Lectures on the Study and Practice of Medicine,* 1832, p.156.
87. Conolly, J., Morgan, Augustus de, Lardner, Dionysius, Long, G., M'Culloch, J. R., *Letter to the Shareholders,* 1830, p.15.
88. Clark, Sir James, *Memoir of John Conolly,* 1869, p.8. Merrington, *University College Hospital,* pp.10–11.
89. Conolly, 'Valedictory Lecture', p.163.
90. *Ibid.,* p.157.
91. Clark, *Memoir of John Conolly,* p.16. According to Clark 'Dr Conolly, by his judicious and energetic conduct while Resident Physician to the Hanwell Asylum, succeeded in establishing *non-restraint* as the invariable practice ... and in satisfying the ... medical profession who visited ... not only of its perfect safety but of its happy influence ....' See also, Scull, A., 'A Victorian Alienist: John Conolly, FRCP, DCL (1794–1866)' in Bynum, W. F., Porter, R. and Shepherd, M. (eds), *The Anatomy of Madness,* 2 vols, 1985, 1, pp.102–50.
92. Conolly, J., *A Study of Hamlet,* 1863.
93. Conolly, J., *Remonstrance Touching the Case Nottidge v. Ripley,* 3rd edn, 1849. In this remonstrance Conolly and Stilwell provide details of the case especially as they affect the insane victim, Miss Nottidge. Hall's fortnightly visits to her are also recorded. Ibid., p.25.
94. Hall, C., *Memoirs,* pp.142–4.
95. *Ibid.,* p.144.
96. Merrington, *University College Hospital,* pp.10–12. According to Merrington, Bell had plotted with Horner to replace Pattison. Horner himself wrote to the Council 1 June 1830 concerning its action in giving private audience, without Horner's presence, to nine of the Professors who made and subsequently published charges against him. Horner answered these charges. Horner, L., *Letter to the Council of the University of London,* 1830.
97. MS Letter Marshall Hall to the Council of University College, 1 May 1839, *Archive 4564a,* University College London.
98. Hall sent these separately with a letter to Mr Atkinson, Secretary to the Council of University College. The letter read: 'Dr Marshall Hall presents his compliments to Mr Atkinson and will be greatly obliged to him to lay the accompanying packet before the Council of University College.' MS Letter Marshall Hall to Mr Atkinson, 1 May 1839, *Archive 4538a,* University College London. The list of contents of the packet appeared on a separate sheet. It reads, 'Received from Dr Marshall Hall, One packet containing letters, Testimonials, Dr Baillie's letters. Eight copies of Printed books. Four ditto Pamphlets. Printed Table.' MS Testimonials, *Archive 4564b,* University College London.
99. *Ibid.*

100. The candidates in addition to Hall were Beamish, Bradley, Copland, Clendinning, Craigie, Jefferson, Ryan, Sanders and Venables. The Council Minutes record that the applications and testimonials were referred to the Senate. MS *Minutes of Council,* 1 June 1839, vol. 3, 1835–43, University of London.
101. MS Letter Marshall Hall to C. C. Atkinson, 11 May 1839, *Archive 4538b,* University College London. The list of names has not been traced.
102. The Professors were Carswell, Cooper, Malden, Quain, Sharpey, Thomson, De Morgan and Graham. MS *Minutes of Senate,* 5 June 1839, vol. 2, 1835–43, University College London.
103. MS *Minutes of Senate,* 13 July 1839, vol. 2.
104. MS Testimonials, *Archive 4564b.*
105. MS *Minutes of Council,* 13 July 1839, vol. 3, 1835–43, University of London.
106. *Ibid.*
107. MS *Minutes of Senate,* 3 August 1839, vol. 2.
108. MS Letter, Marshall Hall to the Council of University College, 23 July 1839, *Archive 4604,* University College London. On the previous day Hall had sent a messenger to collect from Atkinson the packet of books and testimonials sent earlier. MS Letter Marshall Hall to C. C. Atkinson, Esq., 22 July 1839, *Archive 4602,* University College London.
109. MS Letter, 23 July 1839, *Archive 4604,* University College London.
110. MS *Minutes of Council,* 13 July 1839, vol. 3, 1835–43, University of London.
111. Williams, C.J.B., *Memoirs of Life and Work,* 1884, pp.141–7; pp.239–41. Williams describes Hall as one 'of my good friends' and explains that he had treated Hall's family. Williams believed that Hall, whom he considered more eminently qualified for the post, withdrew on health grounds. *Ibid.,* p.138. *Idem, Principles of Medicine,* 1843, p.VI.
112. Hall, C., *Memoirs,* p.146.
113. Editorial, *Lancet,* vol.i, 1849, p.620.
114. MS Letter, W. H. Walshe to the Council of University College, 6 January 1841. *College Correspondence Applications,* University College London.
115. MS Letter, W. H. Walshe to Charles C. Atkinson, 8 April, 1841. *College Correspondence Applications,* University College London.
116. MS Letter, W. H. Walshe to Charles C. Atkinson, 8 June 1841. *College Correspondence Applications,* University College London.
117. *Lancet,* vol.i, 1849, p.620.
118. Merrington, *University College Hospital,* p.11.
119. Hogg, J., *Chemical and Medical Tables,* 1830, final spreadsheet.
120. Hogg, J., *London As It Is,* 1837.
121. *Ibid.,* pp.128–34.

122. Newman, *Medical Education,* p.37.
123. See Chapter 4 *Infra,* where the Hall–George controversy is presented.
124. Merrington, *University College Hospital,* p.44.
125. Sprigge, S. Squire, *Life and Times of Thomas Wakley,* 1897, p.18.
126. MS Minutes 23 March 1853. *Minute Book to be Kept by Dean in Conformity with the Bye Laws relating to his Jurisdiction Over Students,* Faculty of Medicine, Section 14, vol. 22, 1853–8, University College London.
127. Cope, *History of St Mary's Hospital Medical School,* p.38.
128. MS Minutes 30 April 1857, *Minute Book Kept by Dean.*
129. MS *Minutes* 27–9 October 1858, *ibid.*
130. Loudon, *Medical Care,* p.321.
131. McMenemey, W .H., *Life and Times of Sir Charles Hastings,* 1959.
132. McMenemey, W. H., *History of the Worcester Royal Infirmary,* 1947.
133. See Chapter 4 *Infra.*
134. McMenemey, *History of the Worcester Royal Infirmary,* pp.189f.
135. *Ibid.,* pp.121f.
136. *Lancet,* vol.i, 1836–7, p.173.
137. *Ibid.,* vol.i, 1837–8, pp.56–68.
138. *Ibid.,* p.67.
139. *Ibid.,* p.266.
140. Hall, C., *Memoirs,* p.157.
141. Hall, M., *Medicine, its Divisions, its Rewards, and its Reforms,* 2nd edn, 1840. The reading of Hall's Oration was noted in the *Lancet* which referred to the 'marked attention' with which it was listened to and the applause it elicited. *Lancet,* vol.ii, 1840–1, pp.755–6.
142. Hall reminded his audience that the origin and basis of this binary qualification was religious. The University of Dublin was Episcopalian, that of Edinburgh Presbyterian. Thus graduates of Dublin were admitted at Oxford and Cambridge and hence also to the Fellowship of the College of Physicians.
143. Hall, *Medicine, its Divisions, its Rewards, and its Reforms,* p.9. In referring to the fact that 'pert boys, during their pupillage, and totally incompetent to the task, are encouraged to sit in the chair of the critic; ....' Hall was referring to Carpenter, to whose reviews in the *British and Foreign Medical Review* he sometimes took great exception.
144. Hall was referring to his extremely unpleasant 'battle' with John Durance George in the pages of the medical press, over George's implied charge of plagiarism by Hall from Prochaska. See Chapter 5 *Infra.*
145. Hall, *Medicine, its Divisions, its Rewards, and its Reforms,* p.13.
146. See Loudon, *Medical Care,* p.2. in a note. *Report of the Select Committee on Medical Registration,* 1847, p.ix.
147. Hall, *Medicine, its Divisions, its Rewards, and its Reforms,* p.15.
148. Williams wrote forcefully and eloquently on the issue of reform, in letters to the *Medical Gazette,* which he later reproduced in his memoirs. Williams 'On the State of the Medical Profession, and on

the Kind of Reform that it Requires', *Memoirs of Life and Work,* p.165–77.
149. *Lancet,* vol.i, 1840–1, p.621.
150. *Ibid.*, vol.i, 1839–40, p.414.
151. *Ibid.*, vol.i, 1840–1, pp.481–4.
152. *Ibid.*, p.158.
153. *Ibid.*, vol.ii, pp.134–9; p.167.
154. *Ibid.*, p.453.
155. Hall, M., *On Medical Reform, February 1841,* 1841.
156. *Ibid.*, p.4.
157. *Ibid.*, p.8.
158. *Ibid.*
159. Hall M., *Observations on Medical Reform, August 1841,* 1841.
160. *Ibid.*, p.14.
161. *Ibid.*, p.12.
162. Hall, M., 'Suggestions of a National Decimal Pharmacopoeia', *Lond. and Edin. Mon. Journ. Med. Sci.*, 9, 1849, pp.441–3. Hall began this article by relating a personal case in Newark when opium instead of jalap was added to medicine for a child who subsequently nearly perished. More recently there had been a fatal case where 3 grains of strychnine had been administered instead of 3 grains of salicine. Hall criticized the complexity and diversity of existing Pharmacopoeias and alleged that an English physician would scarcely know how to prescribe in Edinburgh or Dublin.
163. Hall, *Observations on Medical Reform, August 1841,* p.16.
164. See *Coffin's Bot. Journ.*, 5, no.157, 1854, p.192.

4

Marshall Hall and the Royal Society: 1830-50

A fellowship of the Royal Society was one of Hall's early ambitions when he came to London. An FRS, Hall hoped, would give him status and admit him unreservedly to the capital's leading scientific institutions. He did gain what he desired – his Fellowship in 1832.[1] But relations with the Royal Society which extended over the following 20 years were throughout most of this period marked by bitterness and extreme disappointment. Hall was not alone in his criticism of the institution which was not exactly enjoying its reputation at the hands of a number of other men of science within its own ranks. They felt that the Royal Society was not promoting science to the same degree as were corresponding Continental institutions.

To some extent Hall was active in and ambitious about science at the wrong end of the century. While he complained bitterly about his treatment within the Royal Society, and although he is sometimes described as a reformer in the nineteenth-century movement to professionalize science and to give medicine a more scientific foundation, Hall actually played a fairly marginal role as an activist. He may have lacked the stomach for overt politicking; more likely he begrudged, as did his contemporary John Herschel, the sheer time needed for the endless committee and other administrative work involved. Hall was shrewd and wanted fame even if it had to be posthumous and even if it involved – as it came to – bitter confrontations and controversies about his research and his contributions to knowledge. For this, Hall needed research time; he preferred to do science and to publish rather than sacrifice this time to campaigning on behalf of science.

But the status and prestige of science was not high at the time. According to one reviewer in the *Westminster Review* in 1825, of a work by George Jardine, Professor of Logic and Rhetoric in the University of Glasgow, 'Science, on which the wealth and power of Britain depend, is not honoured in Britain.'[2] Nor was it for some time to come. Nearly half a century later W. H. Flower, a protégé of Huxley, complained that he 'could make more thousands as a surgeon than he ever would hundreds as a man of science'.[3]

His real enemy within the Royal Society, Hall believed, was Peter Mark Roget (1779–1869) who had succeeded Sir John Herschel to the influential office of Secretary in 1827. Roget held the post until 1848, when he departed under something of a cloud. According to Hall it was Roget who played a major role in the rejection of most of his papers, submitted for publication in *Philosophical Transactions*. It is significant that it was only following Roget's departure that Hall was appointed to a seat on the Council of the Royal Society and was nominated a referee of papers.

Grounds for personal animosity between Hall and Roget certainly existed. Roget, although descended from a French protestant emigré family, was a member of the Established Church of England. Like Hall he was educated in the Edinburgh Medical Faculty. But he was connected to the influential Romilly family in London and hence, unlike Hall, had immediate access to London's social and professional circles. And in spite of his Edinburgh background Roget was elected *speciali gratia* a Fellow of the Royal College of Physicians in 1831 at an age ten years younger than Hall was when he achieved this status. Roget held high office in many of the scientific institutions in which Hall also gained success and recognition. But these were at a more modest level for Hall, even though his scientific work was more extensive and more distinguished.

Roget had spent his early years in practice in Manchester where he was vice-president of the Manchester Literary and Philosophical Society. This was one of numerous similar societies founded in the early nineteenth century and from their ranks were drawn many of the members of the future British Association for the Advancement of Science. In London, Roget lectured at a number of these 'improving' institutions including the Russell Literary and Scientific Institution in Bloomsbury. And following the humane and reform activities of his uncle Romilly, who was associated with movements for the abolition of capital punishment, for the relief of the poor and for the abolition of slavery, Roget was a founder member of the Society for the Diffusion of Useful Knowledge. He wrote for it a series of treatises on electricity, galvanism, magnetism and electromagnetism.[4] This

Society was concerned to improve the scientific literacy of the lower middle and working classes.

Roget was the first Fullerian Professor of Physiology at the Royal Institution, which has been described as the laboratory of the Royal Society,[5] and had held high office in the Medico-Chirurgical Society before transferring his time and his enthusiasm to the Royal Society. For more than 30 years he was a frequent participant at the meetings of the British Association for the Advancement of Science even after his major débâcle at the Royal Society, and at one of its early meetings had been chairman of the physiological section.

Regardless of his views about Roget, Hall was trying to make a name for himself within the Royal Society at a period of intense social and institutional activity, marked by the passage of the Reform Bill in 1832. This was directed towards major reform of English society in general and of its institutions including the Royal Society in particular – even though the Royal Society had acted as adviser on a wide range of scientific issues during the 1820s. A number of other national institutions such as the British Museum, the Public Record Office, the Rolls Chapel and the Record Commission were also singled out as in need of major reform. Roget himself was an early target of attack[6] from within the Royal Society by Charles Babbage, sinecurist Professor of Mathematics at Cambridge (1829–39). Babbage, referred to as 'the militant man of science',[7] was a leading figure among those Fellows seeking reform of the constitution and the practices of the Royal Society.

Reform and Conflict: Cliques and Counterciques

Like any other institution, the Society contained numerous cliques or networks, some of which overlapped, within its community of Fellows. Also, and unfortunately for Hall, the number of medical men within this group was significantly high at the time. This was a source of criticism as indicated in a pointed letter to the *Lancet* by one 'FRS'. This Fellow claimed that 'the glory of our Society is fast fading' and declared that the Society needed rescuing,

> ... from the degrading condition to which it is now reduced – that of a medical advertising office, a very puff shop for the chaff of medical scribblers, where you may see prospectuses of patent medical inventions; ....[8]

Other leading personalities among the reformers included individuals from a range of the hard sciences. For example, such figures as Charles Bell, David Brewster, Benjamin Brodie, Charles Lyell, Herbert Mayo,

Adam Sedgwick, A. B. Granville and William Whewell, as well as Sir John Herschel, were all friends and associates of Roget in other places including the Royal Institution and the Medico-Chirurgical Society.

While Marshall Hall allied himself in spirit with this group he did not receive help when he turned to them with his own problems within the Society. At the same time he was careful not to associate himself directly with Babbage and his closest cronies who were described by Granville, of the less extreme wing of reformers, as 'these noisy ones'.[9] Many of these same individuals were also members of the exclusive Royal Society Club with its limited membership of 40, not including the officers, conferred by ballot.[10] Neither Hall nor Robert Grant, Robert Lee and Richard Grainger, the group with whom Hall had cordial personal relations, was admitted to this club even though its object was the promotion of inter-community and good fellowship.

The efforts of Babbage and others were by no means the first episodes of attempts to reform the Royal Society. Sir Joseph Banks, botanist and traveller and President for 41 years from 1778, had provoked unsuccessful moves to oust him, when he sought to remove the indiscriminate admission of candidates to Fellowship of the Society.[11] In a sense this was an early move towards professionalization. He wished to ensure that only fit persons should be admitted, that is, only those with an appropriate scientific background. And during his reign a number of men of real scientific merit – including Davy, Wollaston, Brodie, Rumford and Brewster – gained the Royal Society medals. Granville took up this issue of admission in the 1830s, but its limitation to those with genuine links with science was not achieved until 1847. On another occasion Banks had insisted that the foreign secretary, then residing at Woolwich in connection with his duties as Professor of Mathematics at the Royal Academy there, should live in London.[12]

Banks, who had found favour with both Court and government, survived in spite of his critics. Towards the end of his term these included younger men who believed he was too entrenched in the eighteenth century. His Sunday evening soirées had become prestigious and renowned as a meeting place for learned people in Britain and for visiting foreign nationals. But the President was accused of favouring botany and neglecting mathematics and the sciences generally. In Hall's day and to his distress and frustration, it was experimental physiology, developing apace on the Continent, which was neglected.

In 1830 Babbage, who wanted a smaller and more focused Royal Society and definitely not a perpetual President, published his

Reflections on the Decline of Science in England in which he particularly criticized the education system[13] and the Royal Society for their failure to promote science, to press its claims for financial support from government and to encourage the development of a professional career structure for scientists. In his criticism of schools Babbage claimed that, 'A young man passes from our public schools to the universities, ignorant almost of the elements of every branch of useful knowledge; ....'[14] The lack of a distinct profession devoted to science, such as existed in France and Germany with their research laboratories, Babbage saw as a great handicap since it meant that posts requiring competent scientists had to be occupied by individuals who could only be described as amateurs.

The government and some of its institutions did have a modest system for supporting the practical application of discoveries, such as the prize Hall's father had gained for his crane design. And, despite his 'declinist' views, Babbage had over a number of years received thousands of pounds for work on the construction of his first calculating machine – a project he himself abandoned.[15] This type of support was forthcoming because of the tangible rewards expected to follow. But, according to Babbage, with regard to pure scientific research, 'all abstract truth is entirely excluded from reward under this system' even though, as Babbage rightly pointed out, there is a necessary interdependence and 'for one person who is blessed with the power of invention, many will always be found who have the capacity of applying principles; ....'[16]

Following the publication of Babbage's *Decline of Science,* a group of Roget's friends discussed the possibility of ejecting Babbage and this led to more criticism and counter criticism. Sir James South, an accomplished astronomer and member of the Royal Astronomical Society, supporter of Babbage and at times an irascible individual, produced his own set of 36 charges against the officers of the Royal Society. These charges included the withholding of access to the Society's records, the deletion from the Minutes of criticism from Fellows, irregularities in the award of medals – Hall was a victim of this – and in the appointment of officers, and the squandering of funds on entertainment. (South later claimed that £236 of the Society's money had been, 'devoted to the *scientific* purposes of *eating* and *drinking* ... [over] six years!!!'). Money had also been wasted on useless publications, the complete disregard of a report on reform produced by a committee on which Babbage had served was also criticized.[17] Commenting on the possibility that publication of his *Charges Against the President and Council of the Royal Society* might precipitate his own expulsion, South wrote 'where ADMISSION IS NO HONOUR, EXPULSION CAN BE NO DISGRACE'.[18]

There were now moves to expel South and to exclude his pamphlet from the library of the Royal Society.[19] The ensuing further aggressive politicking, no doubt exacerbated by the fact that astronomy did attract some government support because of its importance to the Royal and Merchant Navies, provoked considerable debate. This demonstrated the somewhat fluid nature of the various cliques within the Royal Society. Thus, one admirer of Herschel and Babbage did not admire South who, although married to an heiress, was the son of a dispensing chemist. In a spitefully entitled pamphlet, *The Knighthood of Sir J. South, A Death Blow to Science*[20] the writer belittled South's scientific achievements and revealed his own class and professional snobbery with the comment that, 'It would perhaps be unfair to expect from Sir James South that liberality of spirit which is the natural inheritance of gentle birth, and the inherent quality of a refined mind.'[21] South was accused of using his money – actually his wife's – and his influence to secure his knighthood, rather than to promote science. His title was awarded in 1830 when he threatened to emigrate to France.

According to A. B. Granville, in his unacknowledged *Science Without a Head,* published in 1830 in support of the recent appointment of the Duke of Sussex, rather than Sir John Herschel, as President of the Royal Society,[22] Babbage had exaggerated the inferior status of science in England.[23] And Cuvier's *Éloge* on the death of Sir Joseph Banks in 1820, in presenting an impressive catalogue of scientific achievements, gave a foreigner's perception of British science which was certainly at odds with that of Babbage.[24] Nevertheless, Granville did want reforms and in his own plan for the reform of the Royal Society – through changes in its management and election of officers – Granville produced data on the background and publication record of Fellows. These indicate that Berman's description of the Royal Society as being, around 1700, the institutional embodiment of the 'gentlemen amateur tradition', still had relevance over a century later.[25] Further, Granville's own support for the Duke of Sussex served only to perpetuate this perception.

The British Association: Origins and Aims

The British Association for the Advancement of Science, formed in 1831 through the initiative of a number of declinists including the Edinburgh physicist David Brewster as well as Charles Babbage, and with the ready support of the Reverend William Vernon Harcourt, son of the Archbishop of York, was part of the positive response of a number of the reformers to give science a new impetus but one outside the Royal Society. Although Babbage withdrew from the Association, following a dispute over the Presidency and what he described as a potential threat to his long friendship with Herschel, Babbage still remained a commentator on the scientific scene. In 1851, two decades after the formation of the British Association, the year of the Great Exhibition and the year in which Hall retired from the medical scene in London, Babbage published *The Exposition of 1851; or, Views of the Industry, the Science and the Government of England*.[26]

In a chapter entitled 'Intrigues of Science', Babbage was still critical of the state of science in England. Despite changes in the leadership of the Royal Society, he again attacked the lack of professionalism in British science compared to that on the Continent. As part of his tirade Babbage wrote,

> Science in England is not a profession: its cultivators are scarcely recognised even as a class. Our language itself contains no *single* term by which their occupation can be expressed. We borrow a foreign word (*Savant*) from another country whose high ambition is to advance science, and whose deeper policy, in accord with more generous feelings, gives to the intellectual labourer reward and honour, in return for services which crown the nation with imperishable renown, and ultimately enrich the human race.[27]

Hall appears not to have attended either the first meeting of the British Association in York, at which all who attended were automatically entitled to membership, or the second 1832 meeting in Oxford. But he was an active participant at the Cambridge meeting in 1833,[28] the year in which his first nervous system memoir on 'The Reflex Function of the Medulla Oblongata and the Medulla Spinalis' had been read at the Royal Society and published in *Philosophical Transactions*.[29] He was, at the Cambridge meeting, already seen as something of a specialist on the nervous system.

As is the case for all institutions, there are differences of opinion about what motives, which men and, in this case, which region really initiated the establishment of the British Association, which was founded at a time when a number of specialist scientific societies such as the Royal Astronomical Society had already been spawned from the parent Royal Society.[30] According to some it was provoked by dissatisfaction among the thwarted would-be reformers of the Royal Society then run, they claimed, by a bunch of amateurs interested in preserving a forum for the publication of the work of individual real men of science, the 'professionals', but not concerned actively to promote new lines of scientific research nor to press the government for funding either for individuals or for group research programmes. For others the Association represented a response to Babbage's doom-laden message about the state of science in Britain. But for all commentators on its origin, there is some agreement that the founders of the British Association were seeking a means of gaining recognition, status and patronage for science and its serious practitioners. Although the term professionalization has been used as an over-arching one to describe what it was these founders were after, the term itself is open to a variety of interpretations some of which Cannon presents in *Science in Culture*.[31] Macleod draws attention to the fact that recent scholarship has shown the professionalization model to be 'at best oversimple, and at worst untenable'.[32]

For most commentators, the Association's origin was centred around the metropolis – including Oxbridge. But according to Morrell, himself a Northerner, its roots are to be found among northern men and certainly the Edinburgh-based Brewster seems to have played no small part. Morrell believes that Babbage's pronouncements hindered rather than promoted the founding of the British Association.[33] A spot check of the 1838 List of Life Members of the British Association reveals a strong provincial element in the early membership of the Association and shows that few of the London scientific-medical luminaries were not paid-up members.[34] Nevertheless Roget, despite his difficulties within the Royal Society was on the list together with numerous Reverends and the Archbishop of Canterbury and a strong representation from Dublin, Ireland. Someone seems to have accomplished a good recruiting job on the Principal, Fellows and others at both Magdalen College and Christ Church, Oxford – probably at the 1832 Meeting in the City.[35] The founders themselves, 'gentlemen of science', were not all practising or even involved in science. Individuals like the Reverend William Vernon Harcourt who was a man of the church, were what Cannon calls '"cultivators" of science'.[36] But many were well established professionally and others, for instance the sons of bankers, were wealthy. Furthermore, almost all were associated with the Royal Society and other London institutions. Many of them were high office holders in these institutions as well as in the British Association. Membership of the Royal Society was, in the view of some, a disqualification for the British Association being able to act as an agent for the professionalization of science.

William Whewell who did not agree with Babbage's declinist views saw the Association as a means of, on the one hand, bringing together British scientists and on the other of facilitating communication between British and foreign scientists. 'The researches of our men of science have been too much insulated from each other and from what is doing in other countries; ....'[37] Whewell was greatly in favour of setting up specialist groups or subject committees to deal with individual branches of science and he favoured the production, by experts in their field, of progress reports on that science and the highlighting of areas where research was needed. According to Harcourt, President of the York Philosophical Society and chairman of the management committee of the British Association founding meeting at York, in 1831, the subject committees would be 'the instruments by which ... the impulse of the Association must be principally directed'.[38]

The highly individualistic Sir John Herschel, whom Cannon sees as a much more central figure than Babbage in the founding of the British Association, had recently suffered during the Presidential elections within the Royal Society when he the 'scientist' had lost to his Royal Highness the Duke of Sussex the 'gentleman' and an amateur although someone genuinely interested in Astronomy. Herschel was not entirely in favour of trying to set up a large, centralized power base for the whole of science.[39] In any case he felt that while the idea of progress reports – which might originally have been his idea and which was the practice already in many foreign societies – was excellent, the new body should not attempt to control the direction of research or to influence the government. He certainly did not want anyone representing him to the government. Neither did he, who had helped Babbage gain government grants for his calculating machine, think the government could be attacked for its lack of financial support to science.[40] What he criticized was the system of choosing good candidates and the self-centredness of scientists fighting only for their own specialist subject. He also feared the mediocrity of a large group and believed that since the membership perforce would comprise a wide range of levels of expertise it could not be professional enough to promote science. Nevertheless he wished the Association well! – and sat on the Sectional Committee for Mathematics and General Physics in Cambridge in 1833.

Committees and Membership: Promenades and Punch

Autonomous subject committees were set up, after the general committee had been nominated. By the second meeting in 1832 a Council had been formed. Membership of the committees was drawn up at each meeting according to the rules of the Association, although members must have had advance warning that they were going to be nominated. The committees consisted of scientists, cultivators of science and friends of science. And whereas the professional scientists dominated the Council and some of the committees in some periods, they were out-numbered by the non-professionals on others. Thus Committee Two – chemistry, mineralogy, electricity and magnetism – had eleven professionals and six non-professionals.[41] The committee in which Hall had greatest interest was Committee Four – Zoology, Botany, Physiology and Anatomy; it had only five professionals and eleven non-professionals.[42] This state of imbalance between the groups is reflected in Whewell's ignorance about physiology as revealed in correspondence with Harcourt on the membership of the committees. Having assumed that Harcourt could find suitable people to represent Botany and Comparative Anatomy he went on, 'I should suppose that one of the most difficult of the sciences to speak distinctly and impartially about must be *Physiology*, but it is very important and apparently a very progressive one.'[43] Having made these brief but telling comments about the state of physiology for which he personally was not at this stage able to name a single specialist, he passed on to more familiar ground in recommending Mr Wheatstone as one of two people who 'appeared to be the persons who principally attend to the subject of *Sound* ....'[44]

Membership of the British Association was a fluid affair and as the local Secretaries of the 1838 Newcastle meeting pointed out, while the list of 540 members did include 'local Members [who are enrolled for one year only] there are many in each neighbourhood who are drawn permanently ....'[45] and who were willing to travel around with the Association on its annual roadshow to different leading cities. The Newcastle meeting had turned out to be more expensive for the Association and for visiting members than originally calculated. Entrepreneurial northern landladies cashed in with exaggerated rent demands for accommodation and the prices for a range of social events organized by the Association did not cover the costs. For instance the receipt for centrally provided meals amounted to £879 whereas the cost to the Association was £1,151. In addition the evening Promenades where tea, coffee, ices, fruits, punch and other refreshments were provided and which were highly popular were to be discontinued and replaced by something much simpler on account of the exorbitant additional cost to the Association and it seems, because of the 'behaviour of parties whom it is impossible to exclude from so large and promiscuous an assemblage'.[46] Perhaps it was the punch but Sir John Herschel had issued his warning about the riff-raff who might seek to join or attend.

Physiology: a State of Flux

Whewell was indeed not at all comfortable about physiology and was élitist about the pure sciences versus the practical experimental sciences. In his own Presidential Address at Cambridge in 1833,[47] having devoted much time and enthusiasm to the theoretical sciences, extolling mathematics and astronomy, he became – almost at the end of his address – somewhat apologetic, acknowledging that 'it would little become us here to be unjust to practical science. Practice has always been the origin and stimulus of theory: Art has ever been the mother of Science; ....'[48] Having made the briefest reference to knowledge of the laws of motion in fluids he finally turned to Natural History and Physiology and blamed the sections for his own ignorance declaring that these subjects had received less attention 'because the Reports on other subjects appeared more easy to obtain in the first instance'. He hoped the Cambridge meeting would redress the balance.[49] Little did the audience know that during the previous year there had been considerable difficulty in obtaining any Report on Physiology, the original nominee not having his Report ready until the meeting of 1834.

In later correspondence between Whewell and Harcourt it was the little known William Clark, Professor of Anatomy at Cambridge and Dr William Charles Henry, physician and son of Hall's father's 'adviser' in Manchester, who are referred to as the specialists in physiology. Clark, an anatomist, was obviously making heavy weather of his report on physiology and, not surprisingly in view of developments on the Continent as well as in Britain, he was so overwhelmed by the task[50] that he was only too willing to share the work with another and to let his future help-mate choose whichever aspects of the subject he preferred.[51] In the event 'young Henry' a provincial, 14 years older than Hall and clearly well entrenched in the science network, opted to report on Hall's field of interest, the physiology of the nervous system, leaving the general topic of animal physiology to Clark. Henry's Report was completed first and so he was invited to submit it for the 1833[52] meeting at Cambridge while Clark's Report did not appear until the following year at Edinburgh.[53]

Even if Hall is ignored, there were other Fellows of the Royal Society such as Charles Bell, a member of the 1833 subject committee, Mayo, whose name had been on the provisional list of the sectional committee for Anatomy, Medicine and Physiology for 1833,[54] and Wilson Philip, who were all more widely recognized as physiologists than either Henry or Clark; Clark was the deputy chairman of the sectional committee under an equally little known chairman, Dr Haviland. But these others were no better liked as individuals and perhaps this is why Mayo's name, along with those of a few others was then dropped from the specialist committee list. At the heading to the provisional list had been the note 'with power to add to their number' but there was no mention of eliminating names.

In the event physiology had a very chequered career within the committee structure of the British Association, appearing and disappearing, in name at least, over the years and being one year subsumed under another subject only to reappear in its own right the next. This state of flux reflects the fragile position of physiology in the parent Royal Society and hence generally in Britain at this time and the corresponding lack of powerful proponents among the founders of the British Association who might have striven to establish and maintain it as an experimental science in its own right.[55] In the 1831 committees – only sub-committees in this first year – the name physiology did not appear as part of the nearest familial grouping of zoology and botany. Zoology was represented by an exhibition of birds mounted by a member of the Zoological Society of London,[56] although by 1832, the year in which the British Medical Association was formed, Physiology was part of Committee IV alongside Zoology, Botany and Anatomy.[57] In 1833 there was a separate Section V for Anatomy and Medicine; and Zoology and Botany became 'Natural History'. By 1835 the names Zoology and Botany had reappeared to form Section D.[58] From 1836 until 1844 Section F was named Medical Science and then retitled as Physiology until 1847 when it was once more combined with Zoology and Botany in Section D.[59] It remained thus combined until 1865, several years after Hall's death. By this time there was a physiological group within both the Biological Sciences and Anatomy and Medicine. The committees handling Medicine and Physiology were rather sluggish for whereas at a number of the annual meetings the titles of papers to be presented within the subject groupings were printed in advance, at the Plymouth meeting in 1841 there are complete blanks on some days for the Medical Science section meetings indicating that this section definitely had not got its act together.[60]

Hall and Problems of Finance: Insider Dealing

And apart from the Reports by Henry and Clark in 1833 and 1834 on the nervous system, few other reports[61] on the nervous system in general appeared during the first 30 years of the Association. Little wonder that Hall did not join and, since he was having difficulties with the Royal Society during the 1830s and 1840s and the British Association was essentially, at least initially, being controlled by the Royal Society mafia, it is perhaps not surprising that the Association did not actively seek out Hall's expertise. Also Hall, a provincial by origin, had chosen London as his territory of professional activity and the British Association remained always through its annual meetings something of a provincial institution. Similarly although the Provincial and Medical Surgical Association was founded in Worcester in July 1832[62] and although Hall gave time to this association, his support was wholehearted only when the Association became the more metropolitan London-based British Medical Association.[63]

In spite of Herschel's reservations the British Association did commission research taken up from the suggestions or 'desiderata' put forward at the Annual Meetings. For example at the 1833 meeting, the Committee for Anatomy and Physiology (the specialist groups were allowed by the rules of the Association to sub-divide if they wished and this group had dropped Medicine), made a recommendation, 'That an experimental investigation should be made of the sensibilities of the Nerves of the Brain; and that a sum not exceeding £25 should be appropriated to this object.'[64] Marshall Hall and Mr S. D. Broughton, also a member of the subject committee at the meeting, were requested to undertake this investigation.[65]

But acute problems of obtaining financial support for scientific research persisted across the middle of the century, despite the setting up of the British Association in 1831. During the period of Hall's research years, 1831–51, the grants made by the British Association to encourage scientific research were modest. Sums of £200 were exceptional and £20 more typical. Hardly a fortune. The first research grant was made in 1834 and over the years 1834–1931 only £3,367 out of £92,000 were devoted to physiology.[66] For all its declared intention of promoting scientific research the British Association seems hardly to have over-exerted itself searching for finance.

Sussex, the 'Gentleman Scientist' and the Scientific Community

Within the Royal Society Hall might have fared better from the start if William Hyde Wollaston, who had been Secretary to the Royal Society 1804–16, had properly succeeded Banks on his death in 1820, rather than seeing himself only as a caretaker President for a few months until the Anniversary Meeting in 1820. As a practising physician Wollaston had made original contributions in physiology as well as in other branches of science.[67] His appointment as President had been supported by Babbage and Herschel, he had received a Copley Medal and was – a prerequisite for the job – very rich. But unlike the coterie of real contenders, he did not covet the position. In the event, and following a certain amount of insider dealing, the post went five months later to Sir Humphry Davy who possessed a number of excellent qualities and his wife's money but who, in the opinion of some, lacked 'class' and showed his origins as a Cornish lad made good. According to Leonard Horner who had favoured Wollaston, Davy looked ridiculous in his semi-Court dress and resembled 'the porter at a shabby nobleman's gate'.[68]

Despite his detractors Davy positively enjoyed his position and did at least make some contribution to making the Royal Society less eclectic in its membership and more 'scientific', but not in the direction of Hall's specialism. It was Davy who recruited as Secretaries for the Royal Society both Roget and Brande, a favourite of Banks, from the Royal Institution.[69] This was itself undergoing institutional changes and along with, for example, the medical corporations was caught up in the network of public debate.[70] Davy served for seven years and in July 1827 following a series of strokes he resigned and a fresh episode of wrangling over a successor ensued.

This time Davies Gilbert, a strong contender either for the Presidency itself or for the role of 'éminence grise' in whom would be placed the real power behind the throne, took on the mantle of caretaker President. He served for more than three years before being replaced in the contested election in 1830 by the Duke of Sussex. He in his turn was succeeded by two further men of the aristocracy. A major change associated with the presidency of Sussex was his introduction of specialist committees.

A specialist committee structure was not new to the Royal Society and indeed its earliest committees might have been the model for the Continental developments so much admired by Whewell who supported their formation at the British Association. In March 1664, during the early days of the Society, a number of committees were formed to take charge of the several branches of science. These included the Mechanical, Astronomical and Optical, Anatomical, Chymical and Georgical (Agricultural) committees. But they did not last long.[71]

Early on in his presidency, Sussex had to adjust to the existence of the British Association for the Advancement of Science with its newly formed Committees, and to presiding over a Society whose senior officers were really trying to maintain the *status quo* amid considerable social and political upheaval all around. In his 1831 Annual Address[72] the Duke of Sussex made slight reference to 'some circumstances' which attended his election in the previous year. He also expressed sympathy for Herschel of whom he spoke with great admiration. But he did not feel himself inadequate to the position of President of the Royal Society. On the contrary, while acknowledging the superior scientific skills and knowledge of the Fellows, he felt that his own upbringing and especially his membership of the Royal Family, fitted him well to act as their voice and on their behalf in dealing with the government and other major state institutions. And while on the one hand he elegantly distanced himself from the idea that his membership of the Royal Family gave him additional clout he seemed to see no contradiction in acknowledging that nevertheless when he expressed the opinions and wishes of the Royal Society, the institution 'should experience both the respect and the prompt attention to which it is so justly entitled'.[73]

Thus the Duke of Sussex, sixth son of George III, was the first of a line of noblemen who presided over the Royal Society for almost the next 30 years and who steered it through its passage from being a scientific club to becoming a scientific Society.[74] He was genuinely interested in promoting science and was concerned about the lack of a proper full-time, adequately remunerated career pattern, which would place science alongside other professions for serious consideration by ambitious young men. In his anniversary address of November 1832 in which he expressed admiration for the recently formed British Association and its promise for British science, Sussex touched on this concern,

> In this country, where wealth is the general measure of the social rank of families at least, if not of individuals, men of science must either possess an independent fortune, or they must pursue it, as is most generally the case, in connexion with a laborious profession; for we have few establishments which afford them support, independently of other employments; and even in those very rare cases, the provision which is made is so small, that no man of superior education can look forward to the attainment of the advantages which science and learning offer, in forming his scheme of life, unless he be prepared to make the most serious sacrifices.[75]

Sussex compared what he perceived as a cause for concern in Britain, with what he claimed were much more favourable conditions, including state support for research, on the Continent. There, he warned, rapid advances were threatening to leave Britain behind. He firmly believed that the scientific character of Britain was 'most intimately associated with the scientific character and estimation of the Royal Society'.[76] Like Davy, Sussex saw the procedure for the handling of scientific papers as a direct way in which the Royal

Society could affect the quality of science in Britain. More serious though was the mode of election of Fellows which was still a pretty casual affair. It required only three Fellows to sign a certificate that the candidate was likely to be a useful and valuable member of the Society, whatever that might mean, and for the certificate to remain unchallenged after hanging in the meeting room for ten weeks. There was no annual limit on the numbers who could be elected and there were usually more than 30 elected, sometimes over 40 and on one occasion it touched 50.[77] This state of affairs persisted until the election procedures were revised in 1847.

The Duke recommended the retention of the Committee of Papers which judged papers for publication in *Philosophical Transactions.* But he tightened up procedures and formalized the optional extra of calling on specialists to referee papers outside the competence of the committee. Sussex felt that the Society should adopt the practice of many foreign societies, for example the Academies of Science and Medicine in Paris, which required written Reports produced by veterans in their respective areas on every paper submitted to them. In the annual address of 1832, the year in which Hall had successfully submitted his first memoir on the nervous system, the Duke emphasized a resolution adopted by Council 'to allow no Paper to be printed in the Transactions of the Royal Society, unless a written Report on its fitness shall have been previously made by one or more Members of the Council, to whom it shall have been especially referred for examination'.[78] This procedure was in addition to the practice of preparing straightforward abstracts, which were published in the Society's *Proceedings.* This formulation by the President was technically more limiting than the current Statutes which had been printed the previous year. They stated that a majority of the Committee of Papers 'shall be at liberty to call in to their assistance, ... any other Fellows of the Society, who are knowing and well skilled in the particular branch of Science to which the Paper under deliberation relates; ....'[79] The Statutes also stated that such additional referees could vote on the fate of the paper as if they were members of the Committee of Papers. Perhaps the Duke had just let the cat out of the bag in disclosing that the Committee of Papers intended to trawl no further afield than Council – possibly to avoid the awkward squad. Or it could have been a deliberate move by the Duke to induce Council to open up its ranks since, as he pointed out, 'the Council should, in all cases, include men eminent for their proficiency in all those branches of science which usually come or are likely to come, under the notice of the Society'.[80]

Concerning the specialist Reports which would be confidential, Sussex valued the fact that they would be produced by referees who, as well as being members of the Committee of Papers, a sub-committee of Council, would also be members of one of the special subject committees which Sussex envisaged but which were only set up following his departure in 1838.[81] Many advantages were claimed for the new procedure with its expert referees, for which the Duke realized people would have to be willing to give up their time. It would lead to the production of reports which could be even more valuable as representative of contemporary science than the original communications. It should also result in decisions on publication by prestigious referees who although unidentified would, it was naively supposed, be above the influence of personal feelings of rivalry or petty jealousy. Wishful thinking indeed by Sussex. Furthermore, it was expected that the authority of the scientific committees which Sussex wanted would, when considering work submitted to them, be sufficient to establish the importance of a discovery, to indicate its relations to existing knowledge and to define its probable effects upon the future progress of science in England. Interestingly though, the committees were not envisaged as initiating and supporting new lines of scientific development.[82]

Nevertheless, in the 1836 Address following his return after a two-year absence due to threatened but averted blindness, the Duke demonstrated the conservative side of his nature through his preference for certain aspects of the *status quo*. Thus, he refused to accede to a request that there should be discussions on the papers read before the Royal Society. As justification for his attitude, he pointed to 'the abstract and abstruse nature of many of the papers which come before us, and which no single reading can make perfectly intelligible, even to the best-instructed hearer ....'[83] But Sussex revealed his and no doubt Council's real fear – that of provoking more confrontation, uproar and strife in the Royal Society, concerning such a new departure – with,

> When we likewise take into further consideration the irregularities and personalities to which such debates would on some occasions give rise ... such an experiment would be at least dangerous to the peace, as it very possibly might prove ultimately injurious to the scientific character, of the Royal Society.[84]

It is possible that Hall would have been among the 'personalities' the Duke had in mind, provoking and participating in lively debate, no doubt defending his own papers. For instance, in the following year,

1837, Hall had submitted his second Memoir on the nervous system and the anonymous referees who happened to be Herbert Mayo and William Lawrence had rejected it. Hall protested vehemently and at length here, there and everywhere within the Royal Society; he even tried to enlist the support of Herschel. All to no avail. Despite Sussex's confidence and pious hopes, Hall's major and significant criticism, both on this occasion and even more so a decade later when he submitted a paper on the newly emerging subject of electro-physiology, was that the Royal Society simply lacked members with the relevant expertise and competence to judge his work. In fact he was right.

The Committees Re-Visited: The Physiological Committee

New scientific committees were set up within the Royal Society in the autumn of 1838. For Hall it was the Committee of Zoology and Animal Physiology, commonly referred to as the Physiological Committee, which was established 'to promote ... subjects of experimental research',[85] where his papers were discussed and their fate decided, and which played a major role in Hall's fortune and misfortunes within the Royal Society. The first printed list of the Committee of Zoology and Animal Physiology contained 15 members, of whom ten were surgeons – Sir Charles Bell, Sir Benjamin Brodie and Messrs Thomas Bell (no relation of Charles Bell), William Clift, Joseph Henry Green, Thomas Kiernan, William Lawrence, William S. MacLeay, Herbert Mayo and Richard Owen. Only five were physicians. These were Drs John Bostock, Alexander Wilson Philip, James Cowles Prichard, Robert Bentley Todd and Peter Mark Roget.[86]

Of the surgeons Richard Owen, 14 years younger than Marshall Hall, was the representative par excellence of Zoology.[87] He was a Northerner who before completing his medical studies at Edinburgh had moved to London. Having been at school with Whewell he enjoyed a life-long friendship with him. He had also recently acquired Clift as a father-in-law whom he succeeded as Conservator of the Hunterian Museum. Owen was closely involved in the launching of the Natural History Museum in Kensington, of which he became the first professor. He enjoyed prestige and standing in London and among his many activities he met Cuvier, was President of the British Association for Science in 1858 and advised Peel on museum management. Throughout Hall's years in the Royal Society the imbalance against physicians prevailed although by the late 1840s the Physiological Committee comprised two-thirds surgeons and one-third physicians.

On the printed list of members of the Physiological Committee the names of the zoologist W. S. MacLeay and the surgeon Sir Astley Cooper, whose name had in any case been added in MS, have been deleted.[88] The effective committee size, therefore, was 14 the majority of whom, including some physicians and especially Roget, Hall would have classified as 'non friends'. Certainly the group, including Owen with his interest in comparative anatomy, contained little in the way of natural support for experimental physiology in general or for Hall and his personal work involving vivisection in particular. On the contrary they saw it as in their interest to block the development of physiology and its advance within, for example, the medical education curriculum or as a discipline in its own right within universities. It might be thought that Charles Bell, one time vivisector but who now renounced the practice, might have contributed the not unfamiliar convert's zeal to oppose experimental physiology from within this particular committee. Not so, for Charles Bell, who had in 1836 decided to return to Scotland and take up the Chair of Surgery in Edinburgh, seems not to have attended any of the Physiological Committee meetings even though his name was retained for some years. So, in this context at least, he cannot be included among Hall's opponents.

Sir Benjamin Brodie was elected the first chairman and in spite of an unspectacular attendance record – he was not even present on the day of his election on 23 November 1838[89] – and in spite of his own acknowledgement of his frequent absence and his request to be released from the chairmanship he was re-elected to the chair and again, thereafter, over a number of years. A range of committee members stood in over the years for Brodie including Owen and Lawrence and more rarely a physician such as Wilson Philip. In general, the attendance of other members of the group was patchy, and on more than one occasion, for example 29 May 1841 when only four members attended,[90] it was sometimes extremely low. William Lawrence was an erratic attender in the early years and while he did initially send letters of apology, even this courtesy petered out. Lawrence's erratic attendance played a crucial part in what Hall perceived as a deep, dark plot against him within the committee when one of his papers was being dealt with. Lawrence, although chairman in the late 1840s, was probably thankful that he could point to his generally irregular attendance as a way of avoiding responsibility for a committee decision against publishing Hall's papers in 1847.

These low turnouts at meetings reduced cohesion within the group, inhibited continuity from meeting to meeting and certainly contributed to the committee's dilatory efforts over recommendations for awards and financial and other support. On 28 October 1840 only three members attended, Sir Benjamin Brodie, Dr Roget and Mr Bell.[91] Not surprisingly their only piece of business seems to have been to pass the resolution, 'That it does not appear to the Committee that any discovery in Physiological Science has taken place during the past year which deserves the award of the Copley Medal.'[92] Certainly this committee as revealed in its Minutes seems to have lacked the enthusiasm, drive and above all commitment which would have been needed to firmly establish Experimental Physiology as the important subject which people like Whewell had already acknowledged it to be.

The business of the meetings was highly circumscribed and concerned itself mainly with referring papers submitted to the Society and directed to the committee, and with hearing the reports and then deciding on publication or rejection. Voting details were not recorded. In line with the Society's tradition, endorsed by the Duke of Sussex two years earlier, there was no discussion of the papers submitted – at least no such reference appears in the Minutes. Nor were any discussions initiated on the state of Physiology nor for that matter on the state of Zoology despite Owen's presence, and he was a better attender than most, over the repercussions of Cuvier's work. The committee was simply not sufficiently committed or motivated to promote the subjects for which it was responsible.

One of the very few occasions on which the committee had been sprung into action and forced to discuss its deeds, or more appropriately in the particular circumstances its misdeeds, was on the 27 November 1845. In the Minutes of this meeting there appears, atypically, a lengthy appraisal of a paper by Snow Beck on the nervous system of the uterus, for which a Royal Medal had already and quite improperly been awarded.[93] The award had been challenged by Robert Lee, one of Marshall Hall's associates who had also presented a research paper on the same subject for publication in *Philosophical Transactions.*[94] The whole issue became a *cause célèbre* within the Royal Society and there is clear evidence that Roget played no small part in what was clearly shady business, with Lee the immediate victim. But Roget paid for his less than fair behaviour and was obliged to resign his Secretaryship of the Royal Society.[95]

This event obliged the committee to bestir itself again to take an initiative, rather than pursuing merely its usual reactive role to papers submitted, to requests for award nominations and the like. On 4 July 1848 the Physiological Committee sent a note to Council, which would appoint Roget's successor, drawing attention to,

> ... the large proportion of Papers communicated to the Royal Society on subjects of Anatomy, Physiology and other branches of Natural History and the loss which the Society is about to sustain in the retirement of a Secretary eminent for his acquaintance with these sciences.[96]

The committee then respectfully begged Council to pay attention to the advantages and desirability of having as one of the Secretaries someone conversant with those branches of knowledge. The committee favoured Thomas Bell, its faithful and long-time Secretary, as Roget's successor and in spite of an intrigue involving the Council nominee, the chemist W. R. Grove who was rumoured to be keen on removing Physiology and physiologists from the Royal Society, Thomas Bell won the day.[97]

Another serious indication of the lack of energy and commitment of the Physiological Committee was seen in its discharge, along with the other committees, of its duty to submit nominations for the medals and monetary awards which, as part of the Society's merit award system, had originally been introduced as mechanisms for encouraging improvement in the quality of British science. The medals were highly coveted but came to be rewards for work already done. Of the early nineteenth-century medals, the senior honour was the Copley Medal first awarded in 1736 and founded on the donation of Sir Godfrey Copley. He ordained that a gold medal of the value of as much money as the interest on £100 sterling would produce (about £5) should be given annually to the author, from any country, of the most important scientific discovery or contribution to any of the sciences by experiment or otherwise and either published or communicated to the Society.[98]

Later, in a letter to the Earl of Rosse (who succeeded Sussex as President of the Royal Society) Hall claimed[99] that Brodie had told him the Physiological Committee had recommended him to Council (in November 1842) for the Copley Medal 'for his various researches in the physiology of the Nervous System'. This was the case and those present at the meeting, the majority of whom must have supported the nomination, were Drs Todd, Roget and Barry and Messrs Owen, Lawrence, Kiernan and Bell with Sir Benjamin Brodie in the chair. Hence Hall must have received the support of some he perceived as his 'non-friends'.[100] But he lacked supporters on Council itself and he never received the Copley or any other award. Not surprisingly Roget was blamed for this lack of success. Hall averred, 'Dr Roget feels towards me like a man who has injured another; and whilst he remains on the Council, no justice, far less any honour, will be awarded me at the Royal Society.'[101]

The Rumford Medal, or rather two medals one of gold the other of silver, was a more costly award to be paid for out of the interest (about £60) on a donation of 'One thousand pounds in three per cents'. This award, open to any European, was inaugurated by Count Rumford and, in contrast to the Copley Medal, awarded only biennially and exclusively for research in the field of Light and Heat.[102] However, Rumford had hoped to promote a humanitarian approach to science by advocating that the recipients of his award should, if possible, be working to promote the good of mankind. The two later annual Royal Medals (about £52) were the donations of George IV and were first established in 1825.[103] They were actually the brainchild of Robert Peel, Home Secretary, who was seeking strategies for improving English science especially Zoology, which at the time was thriving in France under the energetic drive of Cuvier.

Indeed Peel had originally intended that the Medals should be awarded under competitive conditions as in certain French academies. Peel fondly hoped that there would appear a Cuvierian counterpart who as well as putting British work in the biological sciences on the map – which in the event had to await Darwin's publications later in the century – would also provide impetus for educational reforms and for the establishment, as in France for instance, of a network of provincial universities. This was another development which occurred much later in England than on the Continent. Like the Copley Medal, the Royal Medals were associated with science in general. This of course only enlarged the constituency for the dissatisfaction and vexation which reigned among the annually disappointed *vis-à-vis* the annually delighted winners.

Hall definitely felt that he had been deprived of the second Royal Medal to be awarded. The first had gone to Charles Bell and, speaking in his letter to the Earl of Rosse of his rejected 1832 memoir on the nervous system, Hall asserted, 'I do not hesitate to proclaim here, that, had the value of my first paper been duly and justly appreciated at the Royal Society, the Royal Medal for 1833 would have been adjudged to me.'[104] The 1836 Medal was awarded to George Newport for his 'Series of Investigations on the Anatomy and Physiology of Insects'.[105] It is ironical that Hall had been involved in a wrangle with Newport in the forum of the Royal Society and, through the pages of *Philosophical Transactions,* concerning his role in Newport's work. So not only did Hall not win the Medal but he was drawn into a confrontation which did his reputation no good at all.

Thus, except over the case of Richard Owen, the Zoology and Physiology Committee never developed a forceful policy of putting forward nominations for medals. Towards the end of its period of existence, by which time Marshall Hall had resolved to submit no more papers to the Royal Society, its more usual stance was represented by the Minutes of 14 November 1849 which read, 'No recommendation was made for the award of the Copley Medal.'[106] A week later on 21 November at its last gathering, a similar entry reads 'No recommendation was made for the Royal Medal.'[107] The attendance on these occasions had been only slightly better than nine years earlier when Sir Benjamin Brodie, Dr Roget and Mr Bell had passed their similar, negative resolution about the Copley Medal.

In spite of his influential connections in London and in the Royal Society, Richard Owen had an uphill task towards his own prize medals. He was repeatedly and almost annually recommended for the Copley Medal. These efforts began in October 1841 when the committee resolved that in its opinion,

> ... Mr Owen's researches on the structure of the teeth, both recent and fossil, lately published, and his researches on the fossil reptiles of Great Britain, the latter forming the subject of a paper read at the last meeting of the British Association, entitle that gentleman to the honour of receiving the Copley Medal and that the Committee do accordingly recommend to the Council to award the Copley Medal to Mr Owen.[108]

Subsequently, in the autumn of 1842, 1843 and 1844 the resolution in support of Owen was repeated. Only in 1845 did there appear an explanation, or excuse, for these rejections of an individual who was, in some quarters, probably no more popular within the Royal Society than Marshall Hall. 'It appears that Professor Owen, being still a member of the Council, is ineligible for the Award of the Copley Medal ....'[109] So the committee had failed to do its homework on eligibility for the Medals and had wasted its time and its recommendations backing a non-starter.

But the committee did not give up and in October 1846 returned to the fray in support of Richard Owen, who was no longer on Council.[110] It underlined its support by also recommending him three weeks later for a Royal Medal in recognition of a paper on Belemmites published in 1844.[111] This medal he did win but Council rejected the recommendation for the Copley Medal. In 1847 the committee still rallied to Owen's support although enthusiasm for detailing his latest achievements had flagged. The recommendation simply read '... upon the same grounds as the committee made ... in previous years'.[112] Interestingly Richard Owen, who was another erratic attender of the

Zoology and Physiological Committee, had himself written up and signed the Minutes containing his proposed awards. He was certainly keeping an eye on his own interests. In 1848 the committee made yet another effort to recommend Owen and with more than a touch of acerbity added '... the committee takes the liberty of observing that this is the fourth time that they have recommended Mr Owen for this honour'.[113] If one includes the ineligible years it was actually the seventh. Owen finally gained the Copley Medal in 1851.

The other area of patronage vested in the committees of the Royal Society and which worked somewhat more successfully, was the promotion of new research rather than the rewarding of completed work. This was through recommendations for the allocation of Royal Society research money. Even before the inauguration of the British Association, amounts were small and it was part of the flavour of Babbage's tirade that the Royal Society was utterly complacent about this and did nothing to spur the government to establish scientific research institutes like those springing up in France.

During the late 1830s Hall offered to give up the emoluments of his practice in order to pursue a research programme on the nervous system, if the Royal Society would provide him with the kind of support P. C. A. Louis had enjoyed in Paris.[114] Hall's application was in vain, even though the Society had specific research money available. For instance, it had at its disposal the Donation Fund created in 1828 by the gift of shares from the fortune of William Hyde Wollaston. According to his will, Wollaston wished the fund to be used 'in promoting experimental research ....' although he did allow for 'rewarding those by whom such researches have been made'. He also urged Council not to hoard the dividends parsimoniously but to spend the money liberally and, to guard against insider dealings, he stipulated that no Council member should receive any award.[115]

Since like Hall, Wollaston had originally trained as a physician who then carried out physiological research, and since it was to be almost 20 years before Hall would serve on the Council of the Royal Society, Hall would seem to have been an ideal potential beneficiary. But he was obviously not part of the right network. Notwithstanding the fact that during the two decades following its formation, the Donation Fund was little used, Hall was not supported by the Zoology and Physiology Committee. According to Boas Hall the scant use of the fund was because Fellows found it difficult to think in terms of 'research grants for personal use'.[116] At least Babbage and Hall had been exceptions.

In 1842 a committee was set up to look at the low rate of applications to the Donation Fund. George Newport was a direct beneficiary of this initiative for on 1 February 1843 the committee of Brodie in the chair and Roget, Bostock and Lawrence recommended to Council that it 'grant the sum of Fifty Pounds out of the Donation Fund, to Mr George Newport in aid of the investigations on the Anatomy and Physiology of Insects, which he is now engaged in prosecuting'.[117]

It was not until 1849 and far too late to benefit Marshall Hall, that the government itself really took positive action to promote a research ethos among scientists. This was through its Government Grant, the proposal of Lord John Russell. This annual award of £1,000 was distributed by the Royal Society which was looked on merely as a vehicle rather than as the donor of the money.[118] Nevertheless, it was the case that, through the predilections and prejudices of members of the special committee formed to allocate the grant, the Royal Society could influence which branches of science were supported and which rejected. And this in spite of the expressed and obvious need for impartiality.

As an individual researcher throughout his professional life, Hall would have been eligible for consideration under the first and chief recommendation that 'the grant be awarded in aid of private individual scientific investigations'. The tradition by which individual investigators were isolated from one another was a marked feature of British science and was in sharp contrast to the centralization, through the formation of research institutes, which characterized French scientific research at that time. The second and third recommendations for the grant did suggest an emphasis on mathematics and the physical sciences and on astronomy rather than the biological sciences, despite the presence of Wilfred Owen on the committee. These further recommendations were,

> Secondly in aid of the calculation and scientific reduction of masses of accumulated observations.
>
> Thirdly, in aid of astronomical, meteorological, and other observations, which may be assisted by the purchase and employment of new instruments.[119]

Whewell was also a member of the committee and he already had experience of problems with research money at the British Association. He wrote to Murchison of these difficulties which were less concerned, he claimed, with procuring the money than with 'avoiding the appearance and the reality of waste, caprice, partiality and jobbing'.

Whewell admitted that while he could offer some suggestions on organizing the grant he was not quite sure that he would 'like the responsibility of handling, or directing the handling of parliamentary money'.[120]

As with the British Association, allocations from the Government Grant varied, ranging from £15 to, rarely, over £200, being more usually in the range £50 to £100.[121] Equipment purchased from the grant was technically to be the property of the Royal Society. If, as Boas Hall claimed, the grants were made in a judicious manner and 'there was never any sign of blatant favouritism' – but it need not be blatant to exist – it is possible that if Hall had been a decade younger and had missed those thankless thirties in the Royal Society he might have been a recipient of a Government Grant.

Following its formation, the Committee of Zoology and Physiology had taken some time to organize its refereeing procedures and over its first 12 months or so frequently named in its Minutes the referees to whom Papers were referred. In most cases the committee secretary had subsequently scratched out the names.[122] Perhaps it was only an oversight but on 25 April 1839 while the names of the referees for Dr Barry's Paper on embryology and Dr John Davy's on cartilaginous fish were deleted it is possible to make out that Clift was one of Davy's referees. Clift and Bostock were clearly named as the referees to decide on the suitability of a third paper on animal colour by Dr Alderin being read at the Society.

The committees had certain autonomy over the frequency of their meetings and at the Physiological Committee meeting of 25 April 1839, when those present were the physicians Wilson Philip in the chair in Brodie's absence and Roget, Todd and Bostock together with the surgeons Mayo, Kiernan, Clifton and Bell, it was decided to meet on the second Thursday in every month at eight o'clock in the evening. Either they failed to keep to this resolution or they failed to keep Minutes of every meeting, for there are gaps in the expected sequence of the Minutes.

That the Scientific Committees were not functioning well is manifest in a set of resolutions passed by Council at the end of their first decade, in January 1849. The first of these was 'That it is desirable to retain the system of the Scientific Committees, and to render them more efficient.'[123] The static nature of the committee membership was to cease. It was resolved that, on an annual basis, 'a certain proportion of each Committee be removed and be replaced by new members; ....' Furthermore, the committee size was to be limited to 12 members except for two committees including Zoology and

Animal Physiology which could appoint 20 members. This larger size, while reflecting the two discrete disciplines covered, also allowed for the low attendance record of many members over the years. Without a sizeable committee some of the meetings would not have been quorate.

The Committee of Zoology and Animal Physiology, which had undergone some change of membership was still heavily biased towards surgeons. It now consisted of Sir Benjamin Brodie plus eleven other surgeons and only seven physicians. These included young Dr Carpenter who reviewed Hall's work for the *British and Foreign Medical Review* and Dr Sharpey an experimental physiologist at University College[124] and who might have been expected to give physiology the boost which Hall wanted. Dr Grant's name had been on the original printed list but it was deleted and replaced by that of the surgeon George Newport who pushed the balance of the committee, therefore, even further away from the physicians.

Despite efforts at the beginning of the year, the Scientific Committees were still not working satisfactorily. On 20 December 1849, three years after some bitter editorial attacks by Wakley in the *Lancet* and two years after the rejection of Hall's final memoir submitted to the Royal Society, the committees were abolished.[125] This happened despite efforts to tighten up procedures for handling papers such as insisting that committees judging papers should return every paper to the Secretary within two months.

Hall's Quest for a Fellowship

With his desire to gain a Fellowship of the Royal Society Hall had, early on, changed his stance from that of observer and recorder of clinical conditions studied through work at the bedside, as for example his work on diagnosis, to that of true investigative researcher. He now addressed, initially at least, general physiological problems which required laboratory research using live animals. Hall was interested in probing certain physiological phenomena to arrive at some explanation of their underlying causes. At the same time, he believed he would be unravelling the secrets of Nature and displaying admiration of the Creator. The fruits of empirical physiological research – the new scientific culture in which Hall had such faith – could if properly applied in the arena of medicine provide genuine payoff for diagnosis and therapy. Successful medical practice depended on a sound scientific foundation to which his own contributions were the papers submitted for publication in a range of journals but especially the *Philosophical Transactions.*

The question of the proper path by which science might be assimilated into medicine was a crucial one. According to Maulitz[126] the early decades of medicine in America were characterized by just such a debate over the appropriate locus of scientific medicine, reflected in the tension between the ideology which saw the bedside as the proper place to study disease and that which was increasingly favouring the laboratory. Hall's Edinburgh training had placed high value on bedside and clinical work, where there were face-to-face encounters with patients; his increasing interest in experimental work on animals demanded a laboratory approach. Hall tried to play it both ways as did a number of German physiologists including the Berliners F. von Frerichs and E. von Leyden later in the century. In 1880 they published the first volume of *Zeitschrift für klinische Medizin*,[127] this represented medicine as a discipline which in its own right could participate in the ethos of science.

But Hall had set himself an even more uphill task. He came to specialize in the nervous system which was and remains an exceedingly difficult area of physiology and which extends into the complex territory of psychology with its concern to study the relationship between brain, mind and behaviour. Equally significant at the time, was the contentious scientific-religious debate about the nature and seat of the soul. As if this were not challenge enough, Hall's determination to put experimental physiology on the map in Britain, and to gain wide acceptance for it, led him into confrontation with the newly burgeoning anti-vivisection movement.[128]

Hall's Papers to the Royal Society

Between 1831 and 1847 Hall submitted seven papers to the Royal Society of which the first four dealt with his experimental work on a range of fundamental biological and physiological topics. These included the capillary system, the effect of temperature on metabolism in amphibia, the principle behind the relationship between respiration and irritability and a study of the application of this principle as demonstrated by hibernation. These papers were submitted during 1831 and 1832. They were followed over the next 15 years by three substantial memoirs on the physiology of the nervous system representing what Hall considered to be his real life's work. Of the first set of contributions only those on respiration and hibernation were published. Even more crushingly for Hall, of his nervous system papers detailing some unique and painstaking experimental work, only the first paper in 1833 was published. Hall did not accept quietly any of his rejections and his assertive counter-attacks involved him in a lengthy and stressful struggle during which his appeals for help and support from within the Royal Society more or less fell on deaf ears.

In view of Hall's earlier work on the effects of loss of blood and on the practice of blood-letting, it is not surprising that Hall's first paper to the Royal Society should have concerned the blood circulatory system. In the spring of 1832 this paper, 'On the Anatomy and Physiology of the Minute and Capillary Vessels' was submitted to J. G. Children the Junior Secretary and read on 28 April.[129]

The original MS is in Charlotte Hall's hand with corrections by Marshall Hall and further additions by the referee to whom it was sent. Joseph Lister, amateur opticist and father of the celebrated surgeon had recently, in 1830, established the theoretical principles that made possible the construction of lenses free of chromatic and spherical aberration. A Dolland achromatic microscope had been used by Hall to examine a range of tissues including lung, mesentery and skin from animals such as stickleback, salamander, frog and toad which Hall kept in his menagerie at home.[130] Two contemporary and highly controversial issues concerning the circulatory system were being probed by Hall. The first concerned the true nature of the capillary system and the fine ends of arteries and veins. Were the capillaries true vessels or merely canals in the tissues and did anastomoses or branching networks exist in all three types of vessels as had been claimed? The second issue concerned the contractile power of the blood vessels, especially the capillaries. Hence Hall was joining the contemporary debate about the source of propulsion of the blood.

Hall claimed that capillaries were distinct vessels and that despite 'the most diligent search', the minute arteries and veins in, for instance, the web of a frog, did not reveal anastomoses. In any case, it was reasoned, in view of the forcible and rapid circulation within them, anastomoses are not actually needed in small arteries. Contempt was expressed for contrary claims and the use of imperfect microscopes was blamed for the delusive appearances,

> Every account of the minute and capillary vessels and circulation which I had seen contains so many inaccurate statements, that they appear to have been written from the imagination rather than from actual observation.[131]

Access to a good microscope had certainly encouraged in Hall a feeling of certainty about his views.

The notion of contractile capillaries was dismissed, on the basis that observation of the blood in them revealed flow in both directions.

This would not be the case if the capillaries could contract. Capillary blood flow, Hall maintained, depended on the pulsation of the heart,

> From every observation which I had made, it appears that the capillaries, properly so called, have no power whatever in promoting the motion of the blood. There seems to be no reason, except the impulse of the heart and of the arteries ... why the globules should move in one direction in them, rather than in the opposite.[132]

This first paper was rejected by the Society. A week later, on 1 May 1831, a second equally unsuccessful paper by Hall was read at the Royal Society, 'On the Effect of Water Raised to Temperatures Moderately Higher than that of the Atmosphere, upon Batrachian Reptiles.'[133] Already in the previous paper, the issue of what influences might modify the capillary circulation had been raised and Hall had speculated that the nervous system was at least one possibility.

This second paper was probably prompted by the work seven years earlier in 1824 of William Frederick Edwards, 'De l'Influence des Agens Physiques sur la Vie.'[134] This was devoted to the wider issue of the effect of a range of external agents on the general physiological condition of animals. Edwards had investigated the effect of air, water, temperature, light and electricity on the grounds that 'Les agens dont je me suis proposé d'examiner les éffets nous environnent de toutes parts et exercent sur nous une influence continuelle.'[135] Thomas Hodgkin helped by a family friend Thomas Fisher, published in 1832 a translation of this work for which additional specialist comments had been sought on aspects of the research. Hodgkin expressed his indebtedness for 'the kindness of my friends'.[136] Marshall Hall was listed and hence he must have been familiar with both the original and the translation before 1832.

Since the French physiologist was seeking a grand generalisation applicable to vertebrates in general, he appreciated the need to investigate the effect of each agent on,

> ... diverses espèces dans chacune des quatre classes d'animaux vertébrés, pour donner plus de certitude aux résultats particuliers, dans les cas où un même agent exercerait une influence uniforme sur des constitutions si diverses.[137]

He also emphasized the need for controls, for the careful regulation of external conditions in order to vary only the agent under investigation. Describing his methodology as 'la méthode expérimentale' Edwards explained that,

> Il est une méthode qui règle les conditions extérieures, qui fait varier celle dont on veut apprécier l'action, et qui fait juger, par la correspondence entre ce changement et celui qui survient dans l'économie, du rapport de cause et d'éffet: ....[138]

Edwards' work was a splendid model for Hall. Following it, and using a range of amphibia in which he was investigating the effect of heat on nerve and muscle tissue, Hall found that the animals died at temperatures above 90°F. This situation was analogous to that in warm-blooded vertebrates where muscle tissue is irreversibly contracted by temperatures much above the normal body temperature.

Edwards' inclusion of electricity as an agent worth investigating reflected the current Continental interest in electro-physiology, including the work of Ørsted and Ampère. The reservation that the present state of knowledge on electricity was not 'assez avancées pour me fournir les moyens d'étudier son influence', as was possible with the other agents, was added.[139] It was another 15 years, in 1847, before Hall turned his attention to the then more widely researched field of electricity and its relevance to the physiology of the nervous system. The occasion was yet another paper rejected by the Royal Society.[140]

The rejection of his 1831 paper had been particularly painful since there occurred a month later, in June 1831, the reading followed by publication in *Philosophical Transactions* of a paper by Dr A. P. Wilson Philip, a Council member of the Royal Society.[141] Wilson Philip erroneously and contrary to Hall's view, claimed that capillaries maintain, through their own independent muscular power, the motion of their blood, long after the heart has ceased to beat. According to Wilson Philip,

> ... the following inferences appear to be unavoidable; – that the circulation is maintained by the combined power of the heart and blood-vessels; and that the power of both is a muscular power.[142]

Since one of the Royal Society rules concerning the selection of papers for publication was that the Society should not take sides on controversial, unresolved issues, either the papers of both men should have been rejected or both published. The incident indicates the capricious nature of the refereeing system and, since Hall's first paper contained innovative and accurate findings on the capillaries it illustrates the inhibiting effect the Royal Society had on developments in physiology.

The rejection of this paper was bitterly recalled a decade later. In the first of his Gulstonian Lectures, a coveted honour bestowed annually by the Royal College of Physicians on leading individuals, Hall quoted in support of his own views, those of Johannes Müller whose writings reveal that the issue of capillary blood flow was still at the time a contentious one,

> Many physiologists, believing that the power of the heart is not sufficient to propel the blood through the capillary system, have imagined the existence of other auxiliary forces, such as contractions of the capillaries themselves, or a spontaneous motion of the blood, – neither of which has been demonstrated by direct observation. On the contrary, it is irrefragably proved that the motion of the blood through the capillaries is effected *solely by the action* of the heart.[143]

Hall's final reference to the 1831 episode was in 1848 following yet another rejection the previous year, of his final paper to the Royal Society.

Over the years Hall tried desperately to enlist the support and patronage of commanding figures within the Royal Society. He felt he would get nowhere without such support. In 1848 he was writing his letter to the President-Elect himself, the Earl of Rosse who replaced the Marquis of Northampton following his resignation as President. The resignation was precipitated by a controversy involving Roget a nd Robert Lee and the award of a Royal Medal. Hall criticized the cliquism and quasi-nepotism within the Society. He told the President that,

> In 1831, I read a paper before the Royal Society, ... This paper contained the discovery and description of the intermediate blood-channels situated between the minute arteries and minute veins, both in the pneumonic and systemic circulations. It was rejected, and replaced by what I venture to designate, after the lapse of 17 years, as utter error, – (by a paper) by a *member of the Council*![144]

Wilson Philip may have been a member of Council but unknown to Hall he had not been exactly a favourite with the officers of the Royal Society. Children, the Assistant Secretary, wrote to John W. Lubbock the Vice-President and Treasurer in June 1833,

> My dear Sir, I am very glad to see your answer to Dr W. Philip which I am much obliged to you for having sent me as it exactly corresponds with my own opinion – the worthy Dr is very thin skinned, and not a little pertinacious of his own views of things ....[145]

Having been refused a place in the *Philosophical Transactions* the substance of Hall's 1831 papers was published as a single work, *A Critical and Experimental Essay on the Circulation of the Blood.*[146] A shift of interest to the nervous system was already evident and this early work considered its influence on the circulatory system. Whytt, Haller and Spallanzani are cited as earlier workers in the field and Legallois, Wilson Philip, Clift and Flourens as contemporaries. While Flourens held that the blood circulatory system was entirely dependent on the influence of the brain for its maintenance, Wilson Philip's view was at the other extreme. Speaking of the influence of the brain on the heart Hall believed that, 'the opinion of M. Legallois, that it derives its power from that organ, ... must be regarded as erroneous'.[147] Hall rightly drew attention to the need for establishing criteria for judging whether or not the circulation in an animal had ceased. In his opinion while destruction of the brain and spinal cord may not be followed by immediate cessation of the circulation, such surgical intervention is eventually accompanied by a gradual but progressive failure therefore 'It cannot, ... by any means, be said that the circulation is independent of the brain and spinal marrow.'[148]

Hall Seeks Support within the Royal Society – Papers Published

In spite of his setback, Hall persisted with efforts to gain publication in *Philosophical Transactions.* A number of physiological investigations were planned. But first the support of one of the Society's officers was sought. In November 1831 a copy of *On the Circulation of the Blood* was sent to Lubbock together with a letter requesting an interview. Lubbock's patronage was being sought because Hall felt that it was needed if he were ever to secure proper recognition from the Society for his physiological investigations. Unwisely and in a tone of recrimination, reference was made to his own unjust treatment compared to that of Wilson Philip. With a rather desperate and somewhat pathetic promise of good behaviour which could never have been intended for eyes other than Lubbock's, Hall sought to distance himself from certain of the would-be reformers within the Royal Society,

> ... if you will kindly befriend me, I think you will find that if I am admitted I shall not prove a useless member of the Royal Society; far less would I rank amongst the troublesome.[149]

This letter with its criticism of the institution, of its practices and in particular of one of its council members cannot have evoked sympathy. Nevertheless, Hall's next two papers on the 'Theory of the Inverse Ratio which subsists between the Respiration and Irritability, in the Animal Kingdom' and 'On Hybernation' were read in February and March 1832 and were published in *Philosophical Transactions.*[150] Hall's Fellowship followed on 5 April of that year.[151]

The two latest papers were related in their content. The expression 'inverse ratio' was not being used in its strict mathematical sense but to establish the principle that 'in cases in which the quantity of respiration is great, the degree of irritability is low' and vice versa. The indicator for the quantity of respiration was the volume of gas consumed in respiration. To ascertain this, Hall constructed a piece of apparatus called a pneumatometer, of which the original drawing is in the archives of the Royal Society.[152] The second paper represents an application of the principle in the first paper. Hibernation and natural sleep, both life-preserving processes, are compared with torpor which is destructive. Respiration and the evolution of heat are reduced in sleep and suspended in hibernation.

Hibernation differs from sleep only in degree since in both, nervous sensibility and muscular mobility are unimpaired and irritability is augmented. The reduction in respiration induces in sleeping animals what is described as a 'less arterial' (deoxygenated) condition of the blood. Such blood is unfit for stimulating the heart, and the accumulation of blood in the pulmonary veins and left auricle combined with the enhanced irritability of the heart muscles, lead to a sense of oppression which causes the animal to draw a deep sigh or even to wake up altogether.

Hall was right about the stimulating effect of the less arterial blood but it is its effect on the respiratory centres of the brain not directly on the heart muscle which provokes the sighing or waking. In hibernating animals with their 'power of hibernation', the enhanced irritability of the muscles enables the heart to maintain the circulation of the blood albeit more slowly. The animal suffers no sense of oppression and is not roused. Severe cold, like other sources of pain, rouses the hibernating animal from its lethargy but if continued, induces the state of torpor which can be fatal. These generalizations are embodied in one of Hall's contributions, on the physiology of hibernation, in the Notes of a translation of Edwards' work.[153]

Hall's Early Work on Reflex Action, 1833, Published by the Royal Society

Encouraged by the success of his two papers, Hall pressed on with his research and concentrated more sharply on the nervous system. The earlier research on respiration and hibernation informed the new work on reflex action for it was found that cold-blooded creatures, in which respiration is low compared to that in warm-blooded animals, are among the most suitable specimens for investigations on reflex action. This had already been stressed by Legallois.[154]

In November 1832 and in 1834 Hall read short papers to the Zoological Society, of which he was a member. The first paper 'On a Particular Function of the Nervous System'[155] already embodied the essence of his theory of reflex action and was the paper constantly used as the basis of his claim to priority of discovery when other physiologists, including Müller, published experimental findings similar to his own. The second paper[156] was illustrated with experiments on decapitated turtles. Between these two papers to the Zoological Society, a memoir 'On the Reflex Function of the Medulla Oblongata and Spinalis, or the Principle of Tone in the Muscular System' was successfully submitted to the Royal Society in June 1833.[157] Its publication was Hall's third success in the Royal Society. Flourens described the memoir as being 'de l'ordre le plus élevé en physiologie ....'[158]

1837 Memoir Rejected – Events and Correspondence

Research on the nervous system continued amidst the duties of a demanding practice and lecturing in private medical schools. A second memoir 'On the True Spinal Marrow and the Excito-Motory System of Nerves' was submitted to the Royal Society in 1837.[159] It was refereed by Herbert Mayo and William Lawrence, two members of Council and of the Committee of Papers, who although interested in physiology were both essentially anatomists, surgeons and Fellows of the Royal College of Surgeons. Herbert Mayo had been a pupil of Sir Charles Bell with whom he was later engaged in controversy over the discovery of the distinction between the spinal sensory and motor nerve roots. Mayo, who became a member of the Physiological Committee, had published the first part of *Anatomical and Physiological Commentaries,* in 1822.[160]

William Lawrence was Lecturer in Surgery at St Bartholomew's Hospital, with which he was associated for more than 60 years. He wielded considerable influence within the hospital, the College of Surgeons where he was nominated Professor of Anatomy and Surgery and the Royal Society. A course of Lawrence's lectures delivered at the Royal College of Surgeons in March 1816 had been printed as *An Introduction to Comparative Anatomy and Physiology*[161] and were followed three years later by a series, *Lectures on Physiology, Zoology and the Natural History of Man,*[162] dedicated to Blumenbach. Therefore, in spite of his non-experimental approach to physiology, Lawrence would have been considered competent to judge Hall's paper, particularly since its most significant component, in Hall's view, concerned his speculation about an aspect of the anatomy of the nervous system. This was his proposal concerning the existence of a separate, anatomical sub-division of the nervous system devoted exclusively to reflex action. Hall designated this separate system the Excito-Motory System.

Lawrence read Hall's paper and on 1 May 1837 sent it on to Mayo with no detailed review but simply a brief note saying,

> Whatever may be thought of its merits as a physiological investigation, (it) can not be published in the Transactions of the Royal Society in its present state.[163]

Mayo read the paper and returned it to Children the Assistant Secretary, with Lawrence's note and his own more detailed comments.[164] These claimed that the paper consisted mainly of details already published, including experimental data from other authors. It was also alleged that the paper contained 'monstrous physiological mistakes' and that it 'abounds in hypothetical views which have nothing but their boldness to recommend them'. The final comment underlined the prevailing weak position of physiology in England at the time. It also confirms the high status of publication in *Philosophical Transactions* and explains Hall's deep concern about his rejected papers,

> It is important that the Society should not publish false physiology. It takes fifteen years to overturn an erroneous physiological theory that has had the fortune to obtain the sanction and currency which the papers of the Transactions impart to physiological speculations; – There is no science of which there are so few authorities or critics in England, as physiology: – Whatever is given on anything is adopted without further enquiry.[165]

It was acknowledged that the paper contained a few very good but not specified facts and experiments which, it was suggested, could be reduced to three or four pages and published as a supplement to the 1833 memoir.

Although Hall could be repetitious, Mayo was less than fair with his dismissive criticism and allegation of plagiarism. A number of significant issues which merited more than inclusion in a supplement had been raised. There was the innovatory proposal, already hinted at in the earlier memoir, concerning the existence of a separate and distinct anatomical sub-system of the nervous system associated exclusively with reflex action. This was certainly a bold but also a plausible speculation. More significantly there was further elaboration of an earlier contention that reflex actions are not associated with sensation and volition. Moreover there was Hall's new concept of a principle of reflex function, held to be identical with a property long associated with the nervous system under a range of terms used by other physiologists. These included the *vis nervosa* of Haller, *motorische Kraft* or *vis motoria* of Müller and *excitabilité* associated with Flourens.[166] These concepts had remained inadequately related to physiology. Finally there was Hall's questioning of the long-held belief in the uni-directional mode of transmission of the *vis nervosa* within the nervous branches or fibres, that is '*from* their course in the nervous system to their destination in the muscular system'. Hall had demonstrated experimentally that transmission in nerves and the spinal cord occurs in both directions.[167]

Over and above these specific speculations about the reflex function of the nervous system, Hall had emphasized its special feature of bringing together in a coherent manner seemingly disparate motions which had, hitherto, been designated instinctive, automatic or, in the language of the previous century, sympathetic. This group of excito-motory actions, as they were called, had been simplified into two basic groups – those concerned essentially with ingestion such as swallowing and those concerned with egestion such as urination, defecation and birth, all controlled by special sphincters.

Notwithstanding these novel aspects of the memoir and in spite of the implication in Lawrence's note, Hall was not invited to revise and re-submit the memoir. Roget prepared the published abstract whose content and tone provoked considerable displeasure.[168] Roget was accused of expressing opinions when he should have confined himself to making an abstract of the paper.[169] Later the same year this rejected memoir was published along with that of 1833 under the title *Memoirs on the Nervous System.*[170]

Background to the Rejection of Hall's Memoir

The immediate reason why Hall's paper had not been re-submitted to the Royal Society is to be found in a letter which he had addressed to the Council of the Royal Society, 8 June 1837. The letter began 'Gentlemen, At the suggestion of the Secretary, I have withdrawn my Paper on the "True Spinal Marrow ....".'[171] Thus, technically Hall's paper had not been rejected; it had been withdrawn. Events and correspondence relating to the previous six weeks, are revealed both in later unpublished correspondence with Sir John Herschel, to whom Hall turned for help in October 1839 and in the published letter of 1848 and addressed to the Earl of Rosse. Rosse had succeeded the Marquis of Northampton who had felt obliged to resign with Roget in 1847 following the affair involving the award of a Royal Medal to Snow Beck. All these sources indicate that Hall had no option but to withdraw the paper. In 1846 Hall had put at the disposal of Wakley what he described as all the documents in his possession relating to his treatment at the hands of the Royal Society. Wakley had used the documents, including correspondence between Hall and Lawrence, to prepare a series of editorials attacking both Roget, whose removal he personally sought, and the Society itself. These editorials of April 1846 refer to Roget's sneering abstract of Hall's paper and to Roget's involvement in blackballing the paper.[172]

The Rosse letter contains Hall's claim that he had been informed, in confidence (by Robert Grant?) that 'Dr Roget, the Senior Secretary of the Society, had expressed his intention of procuring the rejection of this paper!'[173] Resolving not to be unprepared, Hall had written to Lawrence in April 1837 expressing his concern. He may have turned to Lawrence because they both belonged to the Medico-Chirurgical Society of which Lawrence had been President in 1831 and they were both regularly mentioned and published in the *Lancet*. Lawrence's role as one of his referees may also have been suspected but in any case he knew that Lawrence was a forceful committee man and hence potentially valuable. Lawrence was urged to attend the forthcoming Committee of Papers meeting at the Royal Society where a decision on the paper was to be taken. Hall's willingness to abandon for a time his professional work in order to pursue experimental investigations was set forth in his letter to Lawrence.[174] In his reply of 28 April Lawrence offered to be the medium of any communication or explanation which Hall might wish to bring before Council, whether concerning his present paper or his researches generally. Lawrence claimed that Council would be interested to learn of Hall's proposal to devote himself for a period to research but added, somewhat obtusely, that he realized Hall did not in fact desire the Royal Society to do anything in this matter.[175] That this was quite contrary to Hall's expectations is clear from a comment in a specially written circular, dated 7 June and sent to members of Council. Hall complained, 'To the Royal Society I naturally looked for countenance, for reward ....' adding 'Have I not, then, an unusual claim to the consideration of a Society *established* for the express purpose of "promoting natural knowledge"?'[176]

In response to Hall's expression of concern about unjust treatment and his further appeal for support, especially from someone well placed, Lawrence had offered the bland assurance that,

> I have never seen in the Council any disposition to neglect or to behave unfairly towards those who have paid the Society the compliment of addressing communications to it; I cannot think, therefore, that you need a friend in the Council as a security for fair treatment.[177]

Lawrence's loyalty to his fellow Council members is understandable. But he well knew that Council was not beyond reproach from other quarters. For example Roget had recently been accused by Robert Grant of plagiarism in the preparation of his Bridgewater Treatise.[178] Further, the 1836 award, in which both Mayo and Lawrence had taken part, of a Royal Medal to George Newport, a former student at University College and a friend of Roget's, had provoked criticism of Council by other disappointed aspirants as well as Hall. Thus Lawrence's reassurance, especially since three days later on 1 May he had written his letter to Mayo suggesting rejection of Hall's paper in its present form, was nothing short of hypocrisy.

After the crucial Committee of Papers meeting, Lawrence had written to Hall on 23 May 1837 explaining that he had unfortunately been prevented from reaching the meeting until a resolution on Hall's paper had already been passed.[179] His lateness could have been genuine since he did have a Council meeting on the same day, or it could have been conveniently contrived. Clearly in view of his own report on Hall's paper which, to be fair to Lawrence had implied revision, the resolution was inevitable. But to this absence Hall attributed the decision relayed to him by Children, the Junior Secretary, writing from the British Museum where he worked, that '*The Council of the Royal Society have not thought it right to print your paper* in the *Philosophical Transactions* in its present form.'[180] In fact, at the meeting the paper had actually been postponed so Children's wording would seem to be decidedly premature. In any case the wording of his message would have suggested scope for revision and re-submission had Children not added 'It has therefore been postponed till the next meeting in June; ... that you may in the meantime, if you please, withdraw it.'[181]

This was clearly an invitation to withdraw the paper but on 12 May Hall replied that he would never do this. Not surprisingly he asked for clarification of the phrase 'in its present form'. He also asked Children if he could take a copy of the paper and added 'To the Royal Society shall belong all the honour or all the disgrace of *rejecting* that paper. *Time* will determine which of the two that may be.'[182] Children replied immediately and his impertinent opening deeply offended Hall,

> Dear Sir – Just as you please. I have done what I thought due, in delicacy to you, by suggesting the postponement of your paper till the next meeting of the Council. I shall lay your letter before that body in June, but can make no further communication respecting your paper till I receive the direction of the Committee of papers.[183]

He added that permission of Council would be needed before a copy could be made.[184] The postponement was clearly a device to give Hall the opportunity of withdrawing his paper rather than having it formally rejected which seems to have been Council's intention.

On receiving Children's first letter, Hall had again written to Lawrence, this time about the fate of his paper. Not surprisingly, Lawrence found it difficult to reply and excused himself on grounds of pressure of work but promised to give the letter detailed attention later. In his more detailed reply 23 May, Lawrence mendaciously claimed that due to his absence at the time of Council's decision he knew little more on the matter than Hall himself. However, some hope was offered that '*the resolution not to print that paper in its present form* clearly implied that it might be printed in another form and under other circumstances'. There was no explanation of what was meant by 'other circumstances' and the latter part of his letter did imply almost a different paper. Hall was advised to persevere with his intentions, which had already been conveyed to Council, of continuing his investigations to 'fill up the present outline or sketch'. This, it was felt, Council would wish to publish.[185]

Little comfort was derived from Lawrence's letter. As it stood the memoir was a significant contribution to knowledge on the physiology of the nervous system and Hall was convinced that Council had already decided to reject his paper. So, thanking Children for trying to soften the blow of rejection, his own previous decision was overturned and he withdrew the paper.[186] But he simply could not accept this situation and a fortnight later on 8 June 1837 his letter of appeal to members of Council was circulated.

The tone of this circular, 'A Note addressed to the Members of the Council of the Royal Society; by Marshall Hall, MD, F.R.S., etc.' marks it as one of the most reasonable and dignified statements of protest and appeal for fair treatment made by Hall; a far cry from the unremitting conceit, aggression and quarrelsome nature suggested by Clarke and Jacyna.[187] A readiness to make any alterations in his rejected paper 'that may render it more worthy of the Royal Society and of science' was emphasized. The appeal went on to request a Commission to witness his experiments step by step and to check the inferences drawn from them. This liberty had been taken because Hall was convinced that 'his experiments involved a new and important principle in physiology'. There followed a reference to Hall's original intention of full-time dedication to a programme of extensive research on the Anatomy, Physiology, Pathology and Therapeutics of what Hall called the excito-motory system and an expression of willingness to receive any suggestions about the present paper. A sense of profound disappointment in the Royal Society, established as it had been 'for the express purpose of promoting natural knowledge' but which had so far failed to provide the hoped-for support, permeates the letter. The Commission which might be set up was begged 'to look over my Paper with care before you finally condemn my labours'.

With a note of urgency about the potentially damaging consequences of delay in publication, plagiarism was hinted at. Mr Grainger 'who has announced the discovery of the anatomy of the excito-motory system' was referred to. Mayo, and others it was claimed, was 'putting forth unfounded claims to my discovery'. And, displaying yet again a serious lack of judgement and political skill in dealing with what he regarded as the opposition, as well as hinting at certain leaks concerning his referees and their criticisms, Hall queried, 'Surely *Mr Mayo* was not one of the *Referees* of my Paper!' He continued, 'I have been told that my Paper is too long; that certain parts are approved, others disapproved.'[188] Hall had rightly suspected Mayo of being one of his referees. But in seeking Lawrence's support, even if he was aware that he was the other referee, Hall seems not to have appreciated Lawrence's attitude to his work in experimental physiology.

Hall's letters and circulars demonstrate how desperate he was to get the paper published. This urgency will only have increased his justifiable sense of hurt and outrage when on receiving his returned paper, he found on it an unworthy and facetious remark. The writer was later identified as Children. Of his comment Hall wrote,

> I think I have some reason to complain of the treatment of my paper whilst it was before the Council. Having quoted from Whytt an experiment of Redi, on the movements of the *Tortoise* when deprived of its head, someone has written, 'will they live after they are made soup of?' Such an observation needs no remark from *me*. It is, rather, an indignity put upon the Royal Society itself.[189]

Children was described by Roget, in his Bridgewater Treatise as 'my friend and colleague'. He had helped Roget revise his Treatise while it was going through the press, following Robert Grant's withdrawal of help when he found that Roget was plagiarising some of his work. Hence Hall knew that he would receive no support from Roget for any criticism of his Assistant Secretary.

The adverse reaction to the possibility of Mayo having been one of his referees was understandable. It is natural that the Royal Society would turn to another specialist in the same broad field but in referring the paper to Mayo, Council had turned to someone whose interest was at the time too sharply focused on the same subject. And in seeking to suppress Hall's work Mayo could, as Hall had implied, have been attempting to further his own interests. At the Council Meeting on 22 June 1837 Hall's letter was read out and the following resolution, concerning his appeal for witnesses to his investigations, was passed, 'That the proposal contained in the above letter be not acceded to.'[190] This decision was conveyed to Hall by J. D. Roberton who had recently replaced Children as Assistant Secretary. He wrote,

> I am directed to acquaint you that the Council of the Royal Society cannot accede to the proposal contained in your letter addressed to them on the 8th June last.[191]

Controversy with George Newport

During 1838 Hall was once more involved in correspondence with the Royal Society, this time concerning a paper by George Newport on the physiology of insects. Newport had received the Royal Medal of the Royal Society in 1836 for his work on the physiology of insects and in a letter to Herschel in 1839, Hall referred to a rumour that Roget had promised the Royal Medal to Newport for helping him with the Bridgewater Treatise.[192] Newport had recently had a paper published in *Philosophical Transactions.*[193] To it was added a lengthy footnote referring to his 'supposed obligations to Dr Marshall Hall' and to 'certain misrepresentations which were made on a recent occasion respecting my communications with that gentleman'. The whole tenor of Newport's note was one of rejecting any obligation towards Hall. Not surprisingly such a declaration in print was found to be offensive and on 1 February 1838 Hall sought permission for a brief reply in *Philosophical Transactions.*[194]

This reply which was not published, explained that on the 'recent occasion' referred to by Newport he himself had forwarded, through a Member of Council of the Royal Society, a letter sent to him by Newport. It contained the following comment about Hall's earlier work on respiration and irritability,

> I am highly delighted and flattered at your instructing me to make the important inquiries respecting insect respiration. I began them immediately, and they are still in progress, and am happy to say that everything yet observed agreed with your ideas of a low degree of respiration with a high one of muscular irritability, and also with the degree of respiration in the larva, or young state being less than

> in the imago, a perfect one. The series of these observations I am noting very particularly, in a separate book, and which when you require it, will be entirely at your service. I cannot as yet see how we can discover the exact time when an insect has decomposed the whole of the atmosphere air in which it is confined except by means of your Pneumatometer.[195]

Newport clearly had received help from Hall and in fact had acknowledged it, drawing attention at the same time to the fact that, in the early 1830s relations between himself and Hall had been more cordial than they were towards the end of the decade,

> In the summer of 1832, at the suggestion of my friend Dr Marshall Hall, I was lead to inquire whether the quantity of respiration in insects bears any relation to the comparative irritability of the muscular fibre in the different genera ... of insects.[196]

Hall added to the extract from Newport's letter for which publication was sought, his own comment that it would be obvious and palpable to everyone that the fact of having sent Newport's letter, made impossible the misrepresentation insinuated against him by Newport.[197] The letter was read out at the Royal Society on 8 February 1838 but the Council decision went against him, and Roget wrote on his communication: 'Ordered by the Council not to be printed. PMR.'[198]

Another avenue for criticizing Newport presented itself. Thus, one of Hall's Webb Street lectures on the Theory and Practice of Medicine published in the *Lancet,* 3 February 1838, contained an oblique reference to the plagiarism of a doctrine first propounded by Professor Grant. The doctrine concerned the nervous system of the *Articulata* and, according to Hall, it had been 'very awkwardly, very ungraciously, and very ungratefully published by another pupil, without acknowledgment of the source from whence it was derived.'[199]

This letter provoked a heated discussion between Hall, Newport and Grant which took place through correspondence in the *Lancet* during February and March 1838. Grant supported Hall's charge against Newport of plagiarism. Writing to Herschel later that year, Hall claimed that Newport had 'appropriated a discovery of Dr Grant's', also some experiments which Newport had made at Sandwich in Kent at Hall's request with instruments which Hall had sent to him.

In a letter dated 6 February which appeared in the *Lancet* of 10 February 1838 Newport identified himself as the target against whom the charges levelled in the Webb Street lecture must have been directed since he was the only student of Grant who had since published a detailed account of distinct motor and sensory columns in the nervous system of invertebrates, the subject of the controversy.[200] The correspondence in the *Lancet* continued. Marshall Hall repeated his assertions regarding Grant's contribution and reinforced his charge of plagiarism against Newport. The footnote in Newport's recent paper in the *Philosophical Transactions* was reproduced as was the unpublished 1832 letter from Newport together with Hall's own communication concerning the footnote which had been rejected by the Royal Society.

Hall's extreme frustration was evident from the fact that he went on to reveal personal details concerning his earlier support of Newport with food, clothing, accommodation and even money, from July 1833 to May 1834. The impecunious Newport had been found a home from which he was eventually dismissed for prying into personal letters and for blasphemy against his host.[201] Despite the undoubted provocation, Hall himself can be censured for such indiscreet almost spiteful revelations in print.

Newport's reply to Grant and Hall was published in the *Lancet* on 3 March. It was lengthy and contained a somewhat self-pitying description of the unfortunate circumstances in which he had found himself from time to time. This was combined with a charge of exploitation of these circumstances by 'two distinguished investigators of science' whose objective was, it was claimed, that of 'crushing their victim with one fell swoop'. Through a chronicle of dates and events, there was an attempt to justify the claim that his discovery about the invertebrate nervous system had been made before he became a pupil of Grant's and that his first paper to the Royal Society in May 1832 had preceded his joining the class.[202]

Nevertheless, there was an acknowledgement of meetings with Grant in and after January 1832 when his work, including his preparations and drawings, was discussed. Furthermore, the reproduction of notes taken by Bailey, a fellow student during Grant's lecture of about 23 November 1832, shows that while Newport's own notes do not contain it, Grant did indeed make two references to the issue in question. Newport claimed that in these references Grant did not deliver the information as a doctrine. But it is clear that the lectures and personal contributions during private discussions between Newport and Grant, had contained concepts which Newport claimed as his own.

Newport wrote to the President of the Council of the Royal Society and referred to the *Lancet* correspondence. He re-rehearsed the events of his association with Hall and Grant in an attempt to dissociate himself from the charge of having received help from either.[203] His letter was accompanied by a set of unpublished notes containing experiments on the respiration of insects made, it was claimed, three years before he met Marshall Hall. Two further letters from a Mr Weekes and a surgeon Mr William Gurnier, resident of Sandwich who had helped with the experiments as early as 1829 were added. Indicating that he was well aware of the gossip concerning the award of his Royal Medal in 1836, it seems that Newport had already sent to Roget copies of the notes in his communication, more than a year earlier on 20 January 1837 – that is about three months after the award of his medal. In his letter to the President and Council, Newport reproduced an extract of the letter giving reasons for the earlier action. Roget had been told,

> Herewith I take the liberty of forwarding to you copies of notes etc., respecting my observations on the Physiology of Insects, which may be useful to me in establishing my right to certain points contained in the papers published in the 'Transactions' but which certain parties have thought proper to dispute.[204]

The correspondence continued in the *Lancet* and Hall made a further reference to the whole episode in his next Webb Street Lecture. Sections of Newport's own papers were reproduced in the *Lancet* and they clearly indicated that both Sir Charles Bell and Professor Grant had certainly played a part in suggesting to Newport the ideas he claimed as his own. By this time the nature of the correspondence in the *Lancet* had irritated Grant for he himself then re-entered the controversy with a letter, in the introduction of which he delivered what might at first sight be construed as a brief admonition to Hall with, '... the question of priority lately agitated by Dr Marshall Hall, is likely to be better known to me than to the latter gentleman'.[205] However, the main objective of Grant's letter was still to defend Hall's stance in attributing priority to himself over Newport. But he preferred a more objective compilation of the information without the emotive criticism which undoubtedly marked some of the contributions of both Hall and Newport.

Not surprisingly Grant's careful analysis of all the unpublished and published material together with his correction of certain statements made by Newport concerning their discussions, provoked a further reply. A final attempt to defend his own position came from Newport who called for a cessation to the controversy, from which he undoubtedly emerged in a less favourable light than his one-time mentor, Grant.[206] Hall who had initiated the confrontation on behalf of his friend Grant and to embarrass Newport cannot have enhanced his own reputation or acceptability by his involvement. And there the episode ended.

Hall again Seeks Support within the Royal Society – Herschel Correspondence

Despite the bitter controversies with Newport and George during the late 1830s, Hall had still hoped for a change of heart on the part of the Royal Society. Thus, in October 1839, Sir John F. W. Herschel an early supporter of the reform group and someone whose patronage within the Royal Society Hall so earnestly sought, was approached. The correspondence lasted from October 1839 to April 1840.[207] On Hall's part there were expressions of bitterness and disappointment regarding the past, together with appeals for future support. On Herschel's part there were clear indications of courteous withdrawal from any attempt to involve him in Hall's affairs. No benefit to Hall was derived from the exchange.

Once again Hall had injudiciously rushed in with a personal, but this time sneaky attack, going straight for his main target – Roget. In his first letter of 1839 seriously incriminating evidence against Roget was referred to alongside claims that in spite of Roget's treatment towards him he still was not using the opportunity he had of exposing Roget. He wrote,

> If my object were further to criminate the Senior Secretary of the Royal Society, I might mention that it was stated amongst at least six persons (and in confidence I will give the names of Mr Mayo and Mr Kiernan) that he had been guilty of an act which, if knew, would render it impossible that he should retain his office another hour.[208]

Hall who complained bitterly of intrigues within the Royal Society, failed on more than one occasion to see that with his own silent gossiping through this kind of correspondence, he was himself contributing to such intrigues.

It was asserted that Hall was not alone in resenting the award of a Royal Medal to Newport in 1836, 'Mr Kiernan thought himself entitled to the Royal Medal, for his paper "On the Anatomy and Physiology of the Liver".'[209] Kiernan had invited Hall to accompany him to Dr Grant's house 'to ascertain by comparison with previous works on the same subject, the real merits of Mr Newport's papers'. Following these discussions, it seems that Mayo, acting on behalf of Kiernan, had re-summoned the Committee to reconsider their award. In a footnote Hall added, regretfully, that his own paper had not even been deemed worthy of being named as a competitor for the Royal Medal. Hall then alleged that to appease Kiernan the Committee had awarded him the Copley Medal.

The Journal Book of the Royal Society records that such an award was made to Francis Kiernan Esq., for his discoveries relative to the structure of the liver as detailed in his paper communicated to the Royal Society and published in *Philosophical Transactions* for 1833.[210] The paper had been highly regarded and the referee's review included,

> Hitherto, however eminent the English have been in physiology (and the most eminent of physiologists Harvey, was an Englishman), they have been behind the Germans and the Italians in Anatomy. The discovery which Mr Kiernan has made exceeds in originality and importance (and) is scarcely inferior to any single anatomical discovery on record.[211]

And this from a scientific fraternity which had actually supported and encouraged anatomy at the expense of furthering the interests of physiology. The tone of praise would suggest that Kiernan had good grounds for his personal disappointment. Of this award to Kiernan, Hall noted that one of the Vice-Presidents of the Society has observed that the Committee had 'attempted to correct one blunder by committing another'.

Towards the end of his letter to Herschel, Hall revealed his real reason for writing. Herschel was asked if, in view of the treatment Hall had so far received at the hands of the Royal Society, he could possibly submit another paper to it. Showing that he still believed in the merit of his rejected *Memoir* (1837) but spoiling yet again, any chances he might have had of Herschel's cooperation Hall asked in a confrontational manner,

> Can I have the remotest hope of justice being done to me in the approaching award of the Royal Medal? For I must conclude this letter by stating that my second Memoir, my rejected paper, comes within the following rule in regard to the next award of that medal: ....[212]

Hall quoted the advertisement from *Philosophical Transactions* setting out the conditions for the award of the Royal Medal. He did not despatch the letter immediately, for on 5 October a postscript was added revealing Hall's awareness that in fact he had no hope of gaining the present Royal Medal. He had realized, from consulting the current *Philosophical Transactions* that 'the *Royal* Medals are given for such Papers only as have been presented to the Royal Society and inserted in their Transactions'. Hall had clearly reflected on this predicament and, now having in mind the memoir on which he was currently working for submission to the Royal Society, the postscript was completed with a comment of resignation which nevertheless contained some hope about possible future events. The comment read 'so ... my Paper cannot now be a competitor for *this* Medal'.[213]

In adding the postscript Hall indicated that he felt it was still worth sending this lengthy epistle to Herschel with the hope of enlisting support. However even Hall must have entertained doubts about the wisdom of his proposed action. Perhaps he realized his lack of discretion in having named Mayo and Kiernan as critics of Roget, because there followed a further ten days before the letter was finally despatched, together with copies of his memoirs and a hurriedly written covering letter requesting Herschel to peruse the letter. Herschel was assured that a reply was not needed since 'I shall feel content to know that you know the facts which I have detailed'.

On the contrary, a reply was expected. Unable to tolerate Herschel's silence any longer, Hall wrote again on 21 November 1839 with the excuse that since sending his earlier communication he had developed doubts about its arrival. With this second, brief letter were sent several papers which Herschel was asked to examine in order that he might realize how highly esteemed the rejected *Memoir* (1837) had been on the Continent. Feigning further indifference about receiving any correspondence from Herschel, Hall asked only that the wrapping paper should be returned as evidence of the arrival of the two packets. In revealing that he now had a third memoir just ready for publication which, if he could submit it to the Royal Society with Herschel's sanction, he would despatch, Hall betrayed his urgent expectation of a reply from Herschel.[214]

A reply followed. The appearance of the draft of this letter with its numerous amendments indicates how difficult Herschel found its preparation. The letter began with thanks to Hall for sending the correspondence which was returned, from Flourens, Van Deen, Louis and Müller. Thus, with his second letter to Herschel, Hall must have sent the testimonials he had used to support his unsuccessful application for the Chair of Medicine at University College earlier that year. The receipt of Hall's earlier letter and books was also noted with the explanation that there had been no reply partly because this had been Hall's expressed wish but partly on account of the nature of the communication itself. This referred to discussions of a personal and professional nature which took place during Herschel's absence from England and which in any case involved subjects which he did not profess to understand.[215]

Herschel elegantly but firmly pointed out that the insinuations of partiality and oblique motives were most injurious to the Royal Society of which Herschel was a Fellow and senior office holder. He described as invidious and incorrect Hall's mode of addressing him as the 'acknowledged head of British Science', and he asked Hall to pardon him 'if I decline to appropriate it to myself in any capacity public or private'. Hall was left in no doubt that little would be gained by forwarding copies of his work. Herschel even offered to return Hall's memoirs since even if he retained them, his unfamiliarity with their subject would preclude the possibility of their being studied with advantage. Regarding the enquiry about re-submitting the *Memoir* (1837) Hall was again disabused of any expectation that support would be forthcoming from Herschel.

The letter ended with advice which could only have increased Hall's sense of desperation and rejection by the Royal Society. Herschel declared that as a Fellow of the Royal Society, Hall needed no other sanction for introducing his latest work to the Society. Nevertheless, he had especial difficulty in composing his final comment,

> If I might presume to advise on this occasion it would be to the effect that relying on your own consciousness of their originality and importance and putting behind you at once and for ever all derogatory ideas of partial bias on the part of the Council you should ... communicate your paper simply as from yourself.[216]

But Herschel was well aware of the problems of others including his influential friend Babbage in getting even his voice heard within the Royal Society.

While Herschel must have hoped that his letter would close the correspondence, Hall quickly wrote back expressing extreme disappointment at its tone and contents. The mode of addressing Herschel had been used with perfect sincerity and if any error had been committed it was one of judgement and not of motive or intention. Nevertheless, Hall realized little if anything was being achieved by appealing to Herschel who was informed that,

> When I resolved to address you ... I also resolved that that should be my last effort towards obtaining any Rights at the Royal Society; and I should feel humiliated by being again, unbefriended, a candidate for its publications or its honours.[217]

This letter from Hall created some unease and sense of guilt in Herschel, who wrote back attempting to assuage Hall's disappointment and bitterness. The draft of this letter shows that Herschel reiterated, and attempted to justify his reasons for not being disposed to offer the special patronage being sought. In looking for further excuses to enable him to preserve his detachment, he explained understandably that since he had, through pressure of work, very little time to devote to the Royal Society, he had made it his duty to confine this time to offering advice in areas within his competence. Supporting Hall's work on physiology would be a departure from this principle. Anticipating the disputatious proceedings which the rejection of yet another of Hall's papers would provoke, Herschel was fortunate in being able to take refuge in the apparently reasonable stand that,

> By making myself in some sort a party to any discussions which might follow, I would involve myself in the consideration of thorny and difficult questions of professional priority and originality at the obvious disadvantage of our understanding nothing whatever of the merits of the case.[218]

In fact there was a degree of hypocrisy in Herschel's claim of ignorance of the field. In a series of lectures in 1831 he had shown positive interest in and support for the notion of the brain as a source of animal electricity. The idea, proposed by J. C. Reil and others in the early nineteenth century was, in England, further supported by Todd and Bowman. According to Herschel, in seeking the origin of muscular motion, one could,

> ... look to the brain, a wonderfully constituted organ, for which no mode of action possessing the least plausibility had ever been devised, as the source of the required electrical power.[219]

Nevertheless, Hall was advised that if he really felt he needed additional support he should seek it from 'one of the many eminent, venerable and respected names in your own Profession – (not upon the Council) – ....' One wonders how Hall interpreted this piece of advice. The remainder of the letter can only have seemed even more of a *volte face.* Having in a previous letter declared that Hall needed no such support Herschel now said that support from such a prestigious individual 'must naturally carry some degree of weight ....'[220] Even though he did not wish to provide opportunities for blaming him for any further disappointment at the hands of the Royal Society Herschel tried to persuade Hall to reconsider the decision not to submit any further work to the Society.

Hall was touched by this letter. When, after some months delay, he eventually replied on 3 April 1840, it was described as obliging and kind. His own delay in replying was due to indecision about how to act in the face of Herschel's advice, his doubts about the Royal Society in general and about Roget in particular,

> Many times I resolved to adopt your kind suggestion. As often my heart recoiled from the idea of again submitting investigations to the same tribunal by which, (as time and the concurrent testimony of the ablest physiologists had determined) my former labours, the foundation of my present ones, were so unjustly handled; especially when I saw the name of Mr Owen, (with that of Dr Roget) one of my former judges, on the Council.[221]

Hall's state of despair was complete. He had finally decided to write a series of letters which Professor Müller of Berlin had agreed to publish in his *Archiv.* The first of this series had already recently been sent in the diplomatic bag through the Prussian Ambassador.[222]

Hall's Final Submission to the Royal Society – Rejection

Over the following seven years Hall was actively engaged in lecturing, contributing to debates on medical reform and commenting on social issues in which he felt injustices were being perpetrated. In addition, he continued to write on a range of medical topics such as scarification of the gums in dentition in the young[223] and on the mechanism of vomiting.[224] He also published articles on the nervous system and on diseases associated with it.[225] Other aspects of his investigations included research on respiration in birds[226] and the blood circulation in the acardiac foetus.[227] His books included *Diseases and Derangements of the Nervous System* published in 1841[228] and the first and second editions of his *Practical Observations and Suggestions in Medicine* which appeared in 1845 and 1846.[229] In 1843 he had published the memoir[230] for which he had vainly sought the support of Herschel to gain publication in *Philosophical Transactions.*

Despite having told Herschel in November 1839, that his attempt to present his *New Memoir* (1843) to the Royal Society would be his last effort to obtain what he called his rights at the Royal Society, Hall's resolve weakened over the years. Almost a decade later, he once again determined to test the fate of his work within the Society. Although, albeit precariously, Roget still held the office of Secretary, Hall presented another paper in January 1847.

The latest work, presented in two parts, was entitled 'Researches into the Effects of Certain Physical and Chemical Agents on the Nervous System.'[231] In carrying out a new series of experiments Mr Henry Smith, of Torrington Square, had again been an invaluable assistant, 'whom I cannot sufficiently thank for the untiring attention, as well as generous devotion, to this enquiry'. Michael Faraday, who was also investigating electricity, had spent two evenings witnessing some of Hall's experiments.

This paper represented a new research direction, associated as it was with an empirical study of nerve function. The paper represented only part of the broader study suggested by the title. It was concerned with investigating agents, in this case electricity, which can activate the nerves and lead, for instance, to muscle contraction. In the early stages of nerve muscle investigations, research on muscle contraction and nerve function were inseparable.

The Fate of Hall's Final Paper Within the Royal Society

Hall's paper was read before the Society on 4 March and 6 May 1847. At the Physiological Committee meeting of 2 June the first part was reported on and it was noted that the second part had been referred.[232] However, as frequently happened with Hall, at some time before 7 July he sent to Mr Weld the current Assistant Secretary, 'a recent addition to his paper which has been put into the hands of the referees'.[233] The Physiological Committee recorded that this had caused their report on the second part of Hall's paper to be deferred. In fact this additional material was only a brief note explaining that the new term *electrogenic,* referred to a phenomenon in nerves which he felt had been fully identified and described for the first time in his paper. It had been derived from the Greek.

That such a slight addition could have been held to delay the second report and final decision on Hall's paper seems incredible. A more likely explanation is that this was a stratagem for delaying the decision on it since the official referees, who did not present their joint report until October, had needed more specialist help in evaluating the paper. It had been passed on to 'A fellow of the Society well versed in the physical questions involved in the above papers ....' This Fellow had written his report in August and it was added to that of the referees as a 'Postscript'.[234]

Despite Hall's careful efforts both to acknowledge what he perceived as an earlier but related contribution in the field by Matteucci and to distinguish this from his own convincing experimental work on muscle contraction, following the effect induced in motor nerves by the passage of a Voltaic current, both referees claimed that Hall had been preempted by Matteucci. Moreover, the specialist writer of the 'Postscript' claimed that a number of Italians including Volta and Valli as well as Ritter in Germany had written on the subject up to 50 years earlier. This reviewer scathingly claimed that the only novelty in Hall's paper was the term *electrogenic* which he translated as meaning the act of generation of electricity. This power, he claimed, nerves do not possess. The official referees concluded their written report with the recommendation that Hall's paper 'be not inserted in the *Philosophical Transactions*'.[235]

It is the case that, through this new area of interest in his physiological investigations, Hall was attempting to enter a field hitherto almost exclusively developed by Continental workers with their more widely available laboratory facilities. He was certainly a 'Continent watcher' especially in areas of his own interests. And the implication that Hall was ignorant of other work was unjust. The failure of the Royal Society referees to witness Hall's experiments as invited by him, was unforgivable. An important and original aspect of Hall's work was that he perceived the phenomenon of the induced current in (motor) nerves as being associated with reflex action. He saw it as dependent on the reflex principle on which he had, ten years earlier, placed such emphasis. Hall's attempt to bring at least some attention to bear on the issue of nerve force was over-peremptorily dismissed. It is unfortunate that the term electrogenic was chosen since it provided the referees with the opportunity of implying that Hall's conception belonged to a bygone era, in which it was held that animals possess in their nerves (and muscles) a subtle fluid analogous to current electricity.

The official referees delayed the submission of their final report until after the 1847 summer period. On 26 October under the chairmanship of Sir Benjamin Brodie, present for once, the Physiological Committee with Rees, Bowman, Christie, Bishop, Rompel and Thomas Bell together with a trio Hall would have called his adversaries – Roget, Todd and Lawrence, 'It was resolved that Dr Marshall Hall's paper ... be not recommended to be printed in the Transactions.'[236] At the same meeting it was reported that a paper on the ganglia and nerves of the heart by Hall's associate Robert Lee was also rejected as was a paper on the Triton submitted by Hall's brother-in-law Mr Higginbottom.

There must have been further delay in conveying the decision and returning the paper to Hall, for on 5 November he wrote asking permission for an artist to go to the Royal Society to make a copy of his paper.[237] In due course, though, Hall was informed of the fate of his paper. His extreme disappointment was enhanced yet again by the tone of Roget's subsequent abstract of the paper. It began 'The professed object of the author, in the present paper is ....'[238] It was with some justification that Hall's wife, recalling this episode, observed that 'The "Philosophical Transactions" were still hermetically sealed to his contributions.'[239]

Hall's treatment at the hands of the Royal Society was in marked contrast to that of Erasmus Wilson. A paper which Wilson submitted in 1843 was rejected at a meeting of the Physiological Committee of the Society in June of that year. This information is contained in a letter written by R. B. Todd. Writing to Roget, Todd explained that he could not be present at the next Committee meeting and asked for a reconsideration of Wilson's paper. He even suggested that 'the paper be referred to Mr Wilson and myself to curtail and alter in such a way as may make it more suitable to the Society's transactions and that afterwards it may be submitted to the Physiological Committee again for their judgement.'[240] To this inappropriate appeal on Wilson's behalf, Todd added that Wilson had spent a great deal of time on the paper and that it seemed hard upon him to reject his paper without giving him more opportunity of adapting it better for publication especially since, as Todd put it 'it contains original matter'. Small wonder that Hall had felt he needed the positive support of a member of the Royal Society's establishment if he were to gain fair judgement of his work.

Todd's first intervention on Wilson's behalf appears not to have reversed the Physiological Committee's decision. But on 12 October of the same year, Todd was again writing to Roget on Wilson's behalf. This time Todd wrote a letter of introduction for the aggrieved Wilson to present personally to Roget. Todd explained that he was introducing his friend Wilson,

> ... in the hope that you will kindly hear him state his case, and advise him what he ought to do under circumstances which are certainly very trying and discouraging.[241]

One wonders just how Hall would have fared had he been able to call on the support of such an influential friend within the institution. Wilson's paper was printed in *Philosophical Transactions* the following year[242] and he became a Fellow in 1845.

New Elections at the Royal Society

Although Hall did see the results of his electrophysiology research in print in another journal[243] he might have been wiser to delay the submission of his latest memoir to the Royal Society, for in November 1847 only days after Hall had been informed of the fate of his latest paper Roget had announced his intention to resign the following year. He had been under great pressure in connection with his less than professionally honest behaviour towards Robert Lee over his work on the nerves of the pregnant uterus. The President, the Marquis of Northampton, had also felt obliged to resign over the same issue. At the close of 1848 new elections were held and the turn-out to vote was high, which the *Lancet* attributed to 'the present condition of the

Royal Society'.[244] The vote of thanks to the retiring President and Dr Roget 'were passed in a cold and formal manner'. The Earl of Rosse was elected President and Council proposed W. R. Grove 'a Chemist and Electrician', winner of the Royal Medal in 1847 and one time opponent of Roget, as his successor for Secretary.[245] The Physiological Committee at its meeting in July 1848 had drawn attention to the 'larger proportion of Papers communicated to the Royal Society on subjects of Anatomy, Physiology, and other branches of Natural History'. It had expressed its concern at the forthcoming loss of a Secretary (Roget) 'eminent for his acquaintance with those sciences' and it sent a resolution to Council urging it to replace Roget with a Secretary similarly 'conversant with those branches of knowledge'.[246] Thomas Bell was their choice of candidate.

According to the *Lancet,* Bell's supporters had accused Grove of intending to oust anatomy and physiology from the Society. Grove denied the charge. Over the years Hall complained more than once of intrigues within the Society but, as already hinted, Hall was clearly not above a little intrigue involvement himself. Correspondence between Hall and Grove reveals that there was undoubtedly some campaign against Grove. Hall told him,

> It has been again stated to me, and the author is given, that you *have* expressed yourself as desirous of removing physiology and physiologists from the Royal Society. If this be incorrect, it is a shocking calumny; and I think it right to give you the opportunity of contradicting it ....[247]

Grove was urged to allow his friends to mount a campaign on his behalf and he was given permission to use Hall's letters. Hall warned Grove that he had been informed in confidence – some confidence! – by Brodie that two Fellows were coming in from as far away as Birmingham and Tamworth to vote against him.[248]

It seems bizarre, until his personal motive is revealed, that Hall who had fought so hard to promote the experimental life sciences in the Royal Society should be supporting the opposition. In a further letter to Grove on 28 November 1848 it is clear that Hall was seizing yet another opportunity to denounce George Newport as the initiator and perpetrator of the calumny against Grove. In a thoroughly gossipy and mischievous letter, Grove was told,

> I do not think that even now you are acquainted with the name of the person who first and principally spread the report of your hostility to physiology. You will remember that I said his name did not begin with P. It was in fact Mr Newport. This was the name first

> mentioned to me by Mr Bishop who has thought fit since to change it for Mr Pratt's. Mr Newport added that he thought himself a proper person for the office of Secretary to the Royal Society.[249]

And no doubt Roget would have agreed and might well have encouraged Newport in his ambition. At the same time the *Lancet* found Grove preferable to Bell despite its wish that one of the Secretaries should be a physiologist. Bell, it insisted could speak only for Zoology and was no better suited than Grove to promote physiology. It referred to the Physiological Committee, of which Bell was Secretary, as 'that very suspicious locality, from which all the injustices and illegalities which have so damaged the Royal Society have proceeded'. It urged that 'Mr Bell should not be translated from the Secretaryship of the Physiological Committee to the Secretaryship of the Royal Society.'[250] In the event Bell was elected and Grove attributed his defeat to the calumny against him. He did later though become President of the Society. The *Lancet* considered that the stand taken by the natural science party had placed it in direct opposition to the physical party and that one benefit would be that 'with lynx-eyed watching from opponents, none of the misdoings of recent years will be repeated by either party'.[251] Another fight at the next election was predicted and Mr Christie's seat as Assistant Secretary was the target.

Hall's Letter to the Earl of Rosse

Hall's rejected 1847 paper was not published in memoir format. Already that year Hall had, in the second of two papers to the *Comptes Rendus,* compared the tetanic effect produced in muscles, on the one hand by the electrogenic condition and on the other by substances such as strychnine.[252] In the following year the rejected Royal Society paper was published in two parts in the *Edinburgh New Philosophical Journal.*[253] But, as in 1837 Hall could not let the matter rest.

In any case there was new leadership at the Royal Society and so on 15 November 1848 Hall decided to wade in once more, this time with the lengthy letter of complaint to the new President, the Earl of Rosse.[254] The letter was privately printed and one of the recipients, with a personal covering letter, was Herschel. Unequivocally did Hall blame Herschel's lack of support a decade earlier, for his continuing misfortune at the hands of the Royal Society,

> Ten years ago I wrote to you to explain the transactions of that year at the Royal Society. Had you then kindly listened to me ten years of unhappiness and so important to me the dishonourable proceedings

> at the Royal Society would have been prevented and I venture to add Natural Science would have been promoted.[255]

Herschel was reminded that 'the Royal Society and its Fellows have their duties as well as their Rights'.

The Rosse letter itself began with a somewhat ingratiating remark congratulating the Earl on his accession to the Presidential Chair of the Royal Society. Less ingratiating though was the onus immediately placed on the Earl for ensuring that in the future 'in that hall of British Science', the wrongs about to be detailed would not continue. There followed a summary of what Hall claimed as his unique contributions to the physiology of the nervous system together with the justifiable complaint that his experiments were rejected without anyone having troubled to witness them. According to Hall 'one incompetent referee has since confessed that he misunderstood my paper, even as to its principle term, that of the electrogenic state'. The Earl was then allied to Grove as another individual to whom Hall was, again indiscreetly, prepared to disclose confidential information about attempts to remove physiology from the Royal Society. The Earl was assured that this intelligence was only for 'your Lordship's private ear'. Rosse was informed that physicians had enjoyed a high profile among the original founders of the Royal Society and that it was only recently that physicians were bringing discredit on their profession and on their science.

In the final comment Hall somewhat uncharacteristically expressed regret that his letter was so egotistical. It was too late for reparation and he averred that had not a life subscription been paid to the Royal Society he would have taken the almost unprecedented step of resigning his Fellowship. The letter was a means of getting his own grievances on record to provide a deterrent to 'future Secretaries and future Councils and Committees of the Royal Society from such evil doings'.

Hall added a Postscript.[256] It was highly pertinent and contained a number of positive proposals for restoring peace and honour to the Royal Society. The first, already made by Babbage and Smith, sought the elimination of the secret and anonymous system of refereeing papers. The second, very significant for experimentalists like Hall, emphasized that dissections and experiments contained in papers simply had to be witnessed by these referees. The issue of unwitnessed dissections had been at the centre of the long drawn-out dispute involving Lee and Roget over the award of a Royal Medal to Snow Beck rather than to Lee in 1845. There followed a call for more mobility in the membership of Council and the Committees within which the present cliquism hindered innovation and progress in the Sciences. In all his 17 years as a Fellow, Hall had never been placed on Council, although it seems he had been proposed and blackballed, possibly at the instigation of Todd, in 1847. Furthermore, he had never been appointed to read one of the prestigious Croonian Lectures nor even nominated a referee of papers. Finally, Hall appealed for accessibility of the President and Council for complainants who should always be treated with civility and consideration.

Changes Following the Rosse Letter

Hall's letter must have been widely discussed within Council and in the Physiological Committee of the Royal Society during 1849. Whatever the real impact of the letter within the Society it is the case that Hall's status changed during 1850. In that year he was proposed and appointed to Council,[257] he was nominated Croonian Lecturer and gave three lectures summarizing his researches on the nervous system.[258] He was also nominated a referee to prepare a report on a paper by Lockhart Clarke on the spinal marrow.[259]

The report was both detailed and generous towards the author and Hall practised what he preached. Mr Clarke had been given,

> ... the full opportunity of showing his preparations both preserved and recent, in his own microscope assisted by Mr Browne, a gentleman long skilled in the use of that instrument, and who has watched the progress of Clarke's researches.[260]

Three evenings had been set aside to observe Clarke's work and the emphasis on these direct observation suggests that such practice still had not yet become compulsory for referees. Clarke's work was described as careful. It involved a technique which did not extend and distort the tissues as did those of Stilling. His pioneer work had begun in 1842 and his early preparations of the medulla oblongata, as mentioned by Hall in his Report, had been examined when Stilling visited Hall during a stay in London. As well as confirming and in some instances correcting certain views of Stilling and Charles Bell, Clarke had made contributions of his own relating to the anatomy of the spinal marrow. But Hall made no claims that Clarke's work corroborated his own earlier speculations about a separate excito-motor system which had involved special anatomical components in the spinal marrow. The final recommendation on Clarke's paper was, 'I cannot hesitate to state that I think Mr Clarke's paper very valuable and worthy of a place in the *Philosophical Transactions.*'[261] The paper was published.[262]

The selection of Hall as Clarke's referee may have been part of an attempt to make some reparation to him. But he was a fitting referee and despite his own disappointments had retained his interest in the wider research developments on the nervous system taking place in other parts of Europe. On 31 October 1850 he was writing to Dr Munk at the Royal College of Physicians who was soon to visit Holland.[263] Hall wanted to check if a translator of his work had actually received a parcel from Hall. Kölbe and Van Deen also seem not to have acknowledged receiving copies of Hall's first Croonian Lecture on the *Synopsis of the Diastaltic System.* Fastidious himself about such details Hall will have found this Dutch trio lacking in straightforward professional courtesy. Nevertheless, he was still interested in knowing about their work and had indeed asked about the latest Dutch medical and physiological research in a letter written the previous day,

> I wish you would kindly ascertain for me what Van Deen, Van der Kölbe, etc, are now doing, what research work they have published and at the same time the degree of notice the spinal system is obtaining. Indeed I should like to know all about the state of medicine and physiology in Holland at the present time. Are there any new laboratories in our science? Any new works? Any new subjects taught? ....[264]

Hall retained this curiosity about new work in his own field until his death.

Notes

1. See note 151 *Infra.*
2. Review of Jardine, G., *Outlines of Philosophical Education,* in *West. Rev.,* 4, 1825, p.175.
3. Desmond, A., *Archetypes and Ancestors,* 1982, p.16.
4. Roget, P. M., 'Electricity'; 'Galvanism'; 'Magnetism'; 'Electro-Magnetism', *Library of Useful Knowledge,* 2, 1832, pp.1–64; pp.1–32; pp.1–96; pp.1–100. A final 'Note' to the Electro-Magnetism treatise gives the results of Faraday's latest findings in the field, made since the 4 treatises went to press. In a 'Postscript' to the treatises, Roget's name appears for the first time.
5. Berman, M., *Social Change,* 1978, pp.xvii–xxv.
6. Babbage, C., *Decline of Science,* 1830, pp.56–7.
7. Sadler, T. (ed.), *Henry Crabb Robinson,* 3 vols, 1869, 3, p.400.
8. 'Abuses in the Royal Society', Letter from FRS, dated 3 April 1830, *Lancet,* vol.ii, 1830, p.16.

9. [Granville, A. B.], *Science Without a Head,* 1830, p.1. This work is unattributed but the British Library copy bears the MS addition 'by 'A Head without Science' – Sir J. South – .'
10. Barrow, Sir John, *Sketches of the Royal Society,* 1949, pp.8–9.
11. Banks announced 'his determination to watch over the applications for admission to the balloting list'. *Ibid.,* p.33. Banks made a number of enemies as a result of this action.
12. The Professor was Charles Hutton who, in 1784, had challenged Banks' dominance within the Royal Society. He had been elected Foreign Secretary in 1779 but Banks obliged him to resign on the grounds that, living outside London, he could not discharge his duties properly, despite the fact that his official duties at the Royal Academy in Woolwich required his presence there. *Ibid.,* pp.33–4. One of Hutton's friends took up his cause. *An Appeal to the Fellows of the Royal Society, Concerning the Measures taken by Sir Joseph Banks, their President, to Compel Dr Hutton to Resign the Office of Secretary to the Society for their Foreign Correspondence,* By a Friend to Dr Hutton, 1784.
13. 'On the Reciprocal Influence of Science and Education', Babbage, *Decline of Science,* pp.3–8.
14. *Ibid.,* p.3.
15. *Ibid.,* pp.29–30.
16. *Ibid.,* p.15; p.18.
17. South, Sir James, *Charges Against the Royal Society,* 1830. South also lambasted the Society for using '£2,000 on mace gilding, picture cleaning and other frivolities whilst they purchased not a single book to add to their imperfect library ....' *Ibid.,* p.15.
18. *Ibid.,* p.17.
19. South's work was reissued a fortnight later on 25th November because, according to South, Gilbert (the departing President) and Roget, had conspired to refuse a copy of the first edition for the Society's Library. In the Preface to the second edition, South claimed that his present of the book had not been mentioned, contrary to custom, and Roget had ordered its removal from the meeting room. South also claimed that several meetings had been held without Minutes being taken. South, Sir James, *Charges Against the Royal Society,* 2nd edn, 1830, p.v.
20. In the form of a Letter, *A Lover of Science, The Knighthood of Sir J. Smith,* 1830.
21. *Ibid.,* p.19.
22. For details of this episode, concerning the election of the Duke of Sussex, by the small majority of 8 votes, see, *Statement Connected with the Election for the Presidency of the Royal Society,* 1831. (This was published by a group of Herschel's supporters.)
23. [Granville], *Science Without a Head,* pp.1–7.
24. [Cuvier, Baron Georges,] 'Éloge Historique de Sir Joseph Banks, Lu le 2 Avril 1821' in Suttor, G., *Memoirs of Sir Joseph Banks.,* 1855,

pp.53–80. See also, Foot, G. A., 'The Place of Science in the British Reform Movement 1830–1850', *Isis,* 42, Part 3, 1951, pp.192–208.

25. Berman, *Social Change,* p.xx.
26. Babbage, C., *The Exposition of 1851,* 1851.
27. *Ibid.,* p.171.
28. *Lithographed Signatures of the Members of the British Association Who Met at Cambridge,* 1833, p.58, contains Hall's signature. At the 1833 Cambridge meeting, Hall was already a member of the Provisional Committee for the *Anatomy and Medicine Section.*
29. See Chapter 5 *Infra.*
30. For a detailed scholarly view of the origins of the British Association see Morrell, J. and Thackray, A., *Gentlemen of Science,* 1981.
31. Cannon, S. F., *Science in Culture,* 1978. See Chapters 5 and 6, 'Professionalisation', pp.137–165; 'Professionalisation and the B.A.A.S.', pp.167–200.
32. Macleod, R. M., 'Whigs and Savants: Reflections on the Reform Movement in the Royal Society, 1830–48', in Inkster, I. and Morrell, J., *Metropolis and Province,* 1983, p.56. See also Macleod, R, 'Retrospect: The British Association and its Historians' and 'Introduction on the Advancement of Science' in Macleod, R., and Collins, P. (eds), *The Parliament of Science,* 1981, pp.1–16; pp.17–42.
33. See Chapter 2 of Morrell & Thackray, *Gentlemen of Science,* pp.35–94. Babbage's work is described as 'a growling, complaining, but wide-ranging book, ...', *ibid.,* p.47.
34. For each annual meeting, lists of members resident in and near the venue, non-residents and those classified under 'Supplementary', (late payers?) together with programme details, were published. The British Library possesses, bound in one volume, such lists covering most of the meetings 1832 to 1845 (plus meeting for 1854). The title page is absent but the spine reads: *British Association for the Advancement of Science, List of Members, Journal, etc, 1832–54.* Over the years Hall rarely attended the meetings while Roget, although, not making the trek North to Newcastle, did reach places such as Plymouth and Manchester as well as the nearer Birmingham. But he occupied a place on Council for several years as well as sitting on Medical Committees. By 1835, although Roget was a life member, Marshall Hall was not, neither had he paid an annual subscription for that year. *Report of the Fifth Meeting of the British Association, Held at Dublin in 1835,* 1836, pp.1–37.
35. See *Lithographed Signatures of the Members of the British Association who met at Cambridge.* Not surprisingly, this meeting attracted a high turn-out from Oxford, Cambridge and London.
36. Cannon, *Science in Culture,* p.187.
37. W. Whewell to Rev. W. Vernon Harcourt, Letter, dated 1 September 1831, Todhunter, I., *William Whewell, D.D., An Account of his Writings,* 2 vols, 1876, 2, p.126. Todhunter points out, at the end of

the letter, that it had cost Whewell two shillings and sixpence (about 12p) to have the letter conveyed by post from Cambridge to York.

38. Cannon, *Science in Culture,* p.209.
39. *Ibid.,* p.192.
40. *Ibid.,* pp.193–4.
41. *Ibid.,* p.211.
42. *Ibid.* This is the committee in which Hall would have been most interested.
43. Letter, Whewell to Vernon Harcourt dated 1 September, 1831, Todhunter, *William Whewell, D.D., An Account of his Writings, 2,* p.128.
44. *Ibid.*
45. *Meeting of the British Association at Newcastle,* 1839, pp.1–44.
46. *Ibid.,* p.15.
47. Whewell, W., *Address Delivered [at] the Third General Meeting of the British Association,* 1833, pp.1–23.
48. *Ibid.,* p.21.
49. *Ibid.,* p.22
50. Morrell & Thackray, *Gentlemen of Science,* p.289.
51. The 'Preface' to *Report of the Third Meeting of the British Association Held at Cambridge,* 1834, p.iii, merely noted 'and to the present Report, by Dr CHARLES HENRY, on one branch of *Animal Physiology,* a more general review of the progress of that science will be added by the Rev. Dr CLARK'.
52. *Ibid.,* pp.59–91. Hall is not even mentioned in the Report.
53. Clark, W., 'Report on Animal Physiology, Comprising a Review of the Progress and Present State of Theory, and of our Information respecting the Blood and the Powers which Circulate it', *Report of the Fourth Meeting of the British Association Held at Edinburgh,* 1835, pp.95–142. Clark sought to justify his own delay in producing the Report with his opening remark, 'That physiology should have been a science slow and uncertain in its progress is scarcely surprising, when we consider how extensive are its objects.'
54. *British Association. General Meeting,* Cambridge, 24 June 1833, Provisional Committees of the Sections, 'V. Anatomy, Medicine, etc.'
55. It also reflects the lack of enthusiasm (esteem) for the medics on the part of the 'real' scientists, who marginalized them within the British Association as soon as they could dispense with their support. Physiology, by being associated with medicine, suffered as a result.
56. *Report of the First and Second Meetings of the British Association [Held] at York and at Oxford,* 1833, p.47, which refers to a subcommittee for 'Zoology and Botany', p.54, and refers to the deliberations of a 'Botanical Committee.' The nomenclature of the subject groupings was very flexible in the early years. The early subcommittees gave way to Sections and Committees and in 1835 Sectional Committees and Sections. There was also a certain amount of fragmentation within some of the groups and certainly the 'medical subjects' were

somewhat shunted around. See Morrell & Thackray, *Gentlemen of Science*, pp.453–4.

57. *Ibid.*, p.114.
58. *Report of the Third Meeting of the British Association Held at Cambridge*, p.xi. *Report of the Fourth Meeting of the British Association Held at Edinburgh*, pp.xxix–xxx.
59. *Report of the Seventeenth Meeting of the British Association Held at Oxford*, 1848, p.xv. The named section membership is presented and includes 'Section D – Zoology and Botany ... Section E – Physiology'. A note at the foot of the page reads, 'By direction of the General Committee at Oxford, Sections D and E, will be in future incorporated under the name of the Section of Zoology and Botany, including Physiology.' Owen and Newport were Vice-Presidents on the Physiology Committee.
60. *British Association, Eleventh Meeting, Journal of Sectional Proceedings*, No. 4, Plymouth, 1841. For Friday 6 August, there is a blank space under 'E Medical Science' even though a venue, the Hospital in Princess Street, had been booked for the papers. *Ibid.*, p.4.
61. The Transactions of the 1834 Edinburgh meeting contain a paper, under the sectional heading, by Sir Charles Bell, 'Observations on the proper method of studying the Nervous System', *Report of the Fourth Meeting of the British Association Held at Edinburgh*, pp.667–70, and at the Dublin Meeting in 1835, a committee had been set up to investigate the 'Pathology of the Brain and Nervous System.' *Report of the Sixth Meeting of the British Association Held at Bristol*, 1837, pp.283–4.
62. Little, E. M., *British Medical Association*, [1932], pp.16f.
63. A 'British Medical Association' had been formed in 1836 among the practitioners of Camberwell and Southwark. Its permanent President was Hall's friend, George Webster of Dulwich and Hall as one of its Vice-Presidents delivered the Annual Oration in 1840. Wakley and the University College Professors Grant and Liston, were on its Council. Webster, who was also a member of the Provincial Association opposed an amendment that the renamed Provincial Association should bear the longer title British Medical and Surgical Association. The present day British Medical Association was launched in Birmingham in 1856, Little, *British Medical Association* pp.31–7; 73–8.
64. 'Recommendations of the Committees', *Third Meeting of the British Association Held at Cambridge*, p.483.
65. Hall, M., and Broughton, S. D., 'Report of Progress made in an Experimental Inquiry regarding the Sensibilities of the Cerebral Nerves, Recommended at the last Meeting of the Association', *Report of the Fourth Meeting of the British Association Held at Edinburgh*, pp.676–80.
66. Morrell & Thackray, *Gentlemen of Science*, pp.312–24. In the early days, there was considerable resentment that a disproportionate share of funds was allocated to the physical sciences – especially astronomy.

67. Hall, M. Boas, *All Scientists Now,* 1984, pp.18f.
68. *Ibid.,* p.23.
69. Macleod, 'Reform Movement', p.61.
70. *Ibid.,* pp.66–70.
71. *The Record of the Royal Society,* 1897, p.114.
72. 'Address of His Royal Highness the President, Delivered at the Anniversary Meeting, November 30, 1831', in *Addresses Delivered at the Royal Society, 1831 and 1832,* 1833, pp.3–11.
73. *Ibid.,* p.4.
74. Andrade, E. N. da C., *Brief History,* 1960, p.14.
75. 'Address of His Royal Highness the President, Delivered at the Anniversary Meeting, November 30, 1832', *Addresses Delivered at the Royal Society, 1831 and 1832,* p.17.
76. *Ibid.,* pp.15–16.
77. Andrade, *Brief History,* pp.12–13.
78. 'Address...1832', *Addresses Delivered at the Royal Society, 1831 and 1832,* p.14.
79. *Statutes of the Royal Society,* 1831, p.16.
80. 'Address ... 1832', *Addresses Delivered at the Royal Society, 1831 and 1832,* p.15.
81. It is interesting that Hall's dispute with George Newport, within the Royal Society, in 1838 contributed to the setting up of the new committees following Council's initial resolve to set up a Committee to consider papers on physiology. Hall, M. Boas, *All Scientists Now,* p.68.
82. 'Address ... 1832', *Addresses Delivered at the Royal Society, 1831 and 1832,* pp.14–15.
83. *Address Delivered at the Royal Society, 1836,* 1836, p.7.
84. *Ibid.*
85. MS *Minutes of Council Meetings,* Royal Society, London.
86. Front page attached to MS *Minutes of Committee of Zoology and Animal Physiology,* Royal Society, beginning 23 November 1838. Hereafter, these Minutes will be referred to as MS *Minutes of Committee of Physiology.*
87. Owen was frequently exalted as the British Cuvier, although the title was actually first bestowed on Hall's friend, Robert Grant. Desmond, *Archetypes and Ancestors,* pp.117–18.
88. MS *Minutes of Committee of Physiology.* See note 86 *Supra.*
89. *Ibid.,* 23 November 1838. The physician, John Bostock, took the Chair at this Meeting.
90. *Ibid.,* 29 May 1841. The loyal members who had turned up were Wilson Philip, Roget (who after all worked in the Royal Society), Thomas Bell and Sir Benjamin Brodie who clearly arrived late, since he was listed last and Wilson Philip was already in the Chair. Nevertheless, Brodie signed the Minutes at the following meeting, to which he had arrived in time to take the Chair.
91. Thomas Bell, a dental surgeon at Guys (1817–61), was very

interested in Zoology and became Professor of Zoology at King's College London in 1836. He served the Royal Society well as one of its Secretaries (1848–53) and became President of the Linnaean Society in 1853.

92. MS *Minutes of Committee of Physiology,* 28 October 1840.
93. *Ibid.,* 27 November 1845. That this meeting was perceived as an important one is reflected in the large turn-out of Sir Benjamin Brodie, in the Chair, Lawrence, Gray, Kiernan, Professor Owen, Bostock, Bowman, Roget and Professor Bell (Thomas) plus Todd and Sharpey who had prepared the Report on Snow Beck's paper.
94. The dispute between Lee and Snow Beck, and their respective supporters and detractors, had begun some six years earlier and is dealt with in more detail in Manuel, Diana E, *Marshall Hall (1790–1857): His Life and Work with Special Reference to his Investigations on the Nervous System,* Unpublished Ph.D. thesis, University of London, 1979. Lee's earlier work on the nervous system of the uterus had already been published in *Philosophical Transactions.* Lee, R, 'On the Nervous Ganglia of the Uterus', *Phil. Trans.,* 131, Part 2, 1841, pp.269–75; Lee, R., 'An Appendix to a Paper on the Nervous Ganglia of the Uterus, with a Further Account of the Nervous Structures of that Organ', *Ibid.,* 132, Part 2, 1842, pp.173–9. Lee's and Snow Beck's 1845 papers were subsequently published in the same volume with a note claiming that while Lee's Supplement to his earlier papers was received by the Royal Society on 14 June 1845 and Read on 19 June 1845, Snow Beck's paper was alleged to have been received on 29 May 1845 and Read on the same day as Lee's. In fact, only the title of Snow Beck's paper was read on 19 June 1845, since the paper was incomplete and while Brodie had refereed Lee's Appendix, no Report on Snow Beck's paper existed. Lee, R., 'Supplement to a Paper, "On the Nervous Ganglia of the Uterus", *ibid.,* 136, Part 2, 1846, p.211. Snow Beck, T., 'On the Nerves of the Uterus', *ibid.,* pp.213–35. A significant and lengthy letter in this saga was that of Lee 'To the President and Council of the Royal Society of London', MS *RR.1.150,* Royal Society London. Lee set out all the events and the President was left in no doubt that 'the Statutes of the Society ... have been violated by the Council'. Lee then published his version of events which involved the erasure of Minutes from the Journal Book of the Council of the Royal Society in Lee, R., *Memoirs of the Ganglia and Nerves of the Uterus,* 1849.
95. Acting on behalf of a group of Fellows, Mr Wharton Jones, submitted a requisition for a Special General Meeting. MS *Journal Book of Royal Society,* 49, 1843–59, 7 January 1847. A special leaflet was produced announcing, above the name of Charles Richard Weld, Assistant Secretary of the Royal Society, and in response to a request by several Fellows, including Marshall Hall, 'a *Special General Meeting,* to be held on Thursday the 11th February [1847] at Two

o'clock, for the purpose of considering and determining on the legality of the circumstances under which the award of the Royal Medal in Physiology for 1845 was first recommended by the Physiological Committee....' The leaflet contained a printed version of the events leading up to the award. MS marginal notes, question certain facts and the same hand has added, 'Mr Kiernan told me it was *Dr Roget's* statement of facts (Kiernan was on Council).' Also attached to the leaflet are two sets of newspaper cuttings – one set supporting the Royal Society in its trouble over the Award, the other providing in greater detail the irregularities of the Award, *Royal Society of London, Special General Meeting,* 25 January 1847. Lee also implicated Roget. Lee, *Memoirs of the Ganglia and Nerves of the Uterus,* pp.27–8f.

96. MS *Minutes of the Committee of Physiology,* 4 July 1848. A special Committee 'to promote the Election of Professor Thomas Bell' was formed, and its deliberations set out in *Statement of Facts Concerning the Proposed Election of Officers and Council of the Royal Society for the Ensuing Year [London, 1848].* Three versions of the *Statement* leaflet were produced, the third reminding Fellows 'NB. The Election is on Thursday, Nov. 30, at 4pm.' Each successive list has a longer list of Committee members. Hall's name does not appear.
97. The charge about Grove's intentions was published in the *Lancet.* Details of the charge against Grove will be dealt with later.
98. For descriptions and illustrations of the Royal Society Medals, see: *Record of the Royal Society of London,* 2nd edn, 1901, pp.123–32. Granville, A. B., *The Royal Society in the XIXth Century,* 1836, p.60.
99. Hall, M., *Letter to the Earl of Rosse,* 2nd edn, 1848, p.9.
100. MS *Minutes of Committee of Physiology,* 2 November 1842.
101. Hall, *Letter to the Earl of Rosse,* p.9.
102. Granville, *The Royal Society in the XIXth Century,* p.60. Recipients of the award received both medals and it was envisaged that they might convert the gold one into money and retain the identical silver one.
103. *Ibid.* Granville provides a Table of the recipients – and their work – of the Royal Society Medals over the years, 1800–29. *Ibid.,* pp.62–5.
104. Hall, *Letter to the Earl of Rosse,* p.9.
105. Newport's first paper was published in *Philosophical Transactions* in 1832, p.383. His second paper was, Newport, G., 'On the Nervous System of the Sphinx ligustri, LINN., (Part II.), during the later stages of its Pupa and its Imago state; and on the Means by which its Development is effected', *Phil. Trans.,* 124, Part 2, 1834, pp.389–423. The announcement of Newport's award was published in *Phil. Trans.,* 126, Part 2, 1836, pp.v–vi.
106. MS *Minutes of Committee of Physiology,* 14 November 1849.
107. *Ibid.,* 21 November 1849.
108. *Ibid.,* 26 October 1841.
109. MS *Minutes of Committee of Physiology,* 6 November 1845. The

Committee immediately decided to recommend Professor Schwann of Louvain instead, for the Copley Medal. It was at this same meeting, that Lee's Supplement to his earlier papers on the nerves of the uterus, having been referred at the previous meeting, 27 October 1845, was in fact recommended not to be published.

110. MS *Minutes of Committee of Physiology*, 29 October 1846. The recommendation read, 'It was resolved that the Council be recommended to award the Copley Medal to Professor Owen, for his researches on the generation of the Marsupialia, his work on Odontography, and various other important researches and discoveries in Comparative Anatomy and Physiology.'
111. *Ibid.*, 17 November 1846.
112. *Ibid.*, 26 October 1847.
113. *Ibid.*, 1 November 1848.
114. Hall, M., *Note to Council of the Royal Society*, 8 June 1837, pp.1–4. Hall wrote this letter following the rejection of his 1837 Memoir on the Nervous System. He requested, in vain, that a Commission be set up to witness his experiments step by step to check the inferences which he had drawn from them.
115. *Record of the Royal Society*, 2nd edn, p.122. Over the years, a number of contributions were made to the Donation Fund which, in 1901, yielded an annual income of £390. There were other funds available at the time. *Ibid.*, pp.123–7.
116. Hall, M. Boas, *All Scientists Now*, p.40.
117. MS *Minutes of Committee of Physiology*, 1 February 1843.
118. Hall, M. Boas, *All Scientists Now*, pp.146–51; pp.163–4.
119. *Record of the Royal Society*, 2nd edn, p.156.
120. Letter, Whewell to Murchison dated 17 March 1850, Todhunter, *William Whewell, D.D. An Account of his Writings*, 2, p.357.
121. Hall, M. Boas, *All Scientists Now*, p.150.
122. In *Record of the Royal Society*, 2nd edn, p.104 it is claimed that the earliest mention of a paper being technically referred is on 25 May 1780, the next being a paper by Davy referred to Faraday in 1831. Thereafter the practice of referring became common and while initially at this time, those to whom the paper was referred were named, this practice was soon dropped. The Physiological Committee seems to have taken some time to learn the new practice even though the Minutes of 23 November 1838 refer to Mr Hilton's paper being referred to 'two members of the Committee'. *Minutes of Committee of Physiology*, 23 November 1838.
123. MS *Minutes of Council Meetings*, Royal Society, January 1849.
124. See Taylor, D. W., 'The Life and Teaching of William Sharpey (1802–1880) "Father of Modern Physiology" in Britain: Part I', *Med. Hist.*, 15, 1971, pp.126–53; Part II, *ibid.*, pp.241–59.
125. See p.198 and note 172 *Infra*, for further reference to the editorials. Concerning the abolition of the Committees, the *Lancet*, 8 December

1849 was lamenting the fact that 'the Committees – those seats of every variety of jobbing and intrigue – still remain unreformed'. On 29 December, it triumphantly announced, 'Another vast change has recently occurred at the Royal Society. The Committees, of which we have so often complained are now altogether abolished!', *Lancet,* vol. ii, 1849, p.614; p.696.

126. Maulitz, R. C., '"Physician Versus Bacteriologist": The Ideology of Science in Clinical Medicine', in Vogel, M. J. and Rosenberg, C. E. (eds), *The Therapeutic Revolution,* 1979, pp.92–4.
127. The bibliographical details of the first volume are: *Zeitschrift für Klinische Medizin,* Herausgegeben von Dr Fr. Th. Frerichs, und Dr E. von Leyden, Ester Band, Berlin, Verlag von August Hirschwald, 1880. Frerichs was described as 'Professor der medicinischen Klinik' and Leyden as 'Professor der med.-propädent. Klinik.' In 1883 Frerichs and Leyden were joined by two editors from Vienna, Dr H. v. Bamberger and Dr Northnagel both described as Professors 'der medicinischen Klinik.'
128. Manuel, D. E., 'Marshall Hall (1790–1857): Vivisection and the Development of Experimental Physiology', in Rupke, N. A. (ed.), *Vivisection in Historical Perspective,* 1987, pp.78–104.
129. MS Hall, M., *On the Anatomy and Physiology of the Minute and Capillary Vessels.* Communicated by J. G. Children, Sec. R.S. Read 28 April 1831. *Archives 1831, AP.15,* 32, 6, Royal Society London. Throughout the paper, Charlotte Hall had written 'bung' for 'lung' and 'capillaries' as 'cappillaries'. According to Barrow, the papers were read by one of the Secretaries and he himself 'often found them dull enough in themselves, and not always improved by the monotony of an official reader, sometimes weary of his task'. Barrow, *Sketches of the Royal Society,* p.5.
130. According to Hall's wife, 'the animals required for all these investigations, formed a little menagerie in a room in our house, devoted to that special purpose. There were assembled mice, hedgehogs, bats, birds, fishes, frogs, toads, tritons, snakes, etc.' Hall, C., *Memoirs,* p.82.
131. MS Hall, *Minute and Capillary Vessels.*
132. *Ibid.*
133. MS Hall, M., *On the Effect of Water Raised to Temperatures Moderately Higher than that of the Atmosphere, upon Batrachian Reptiles.* Communicated by J. G. Children, Sec. R.S., Read 1 May 1831; *Archives 1831, AP.15,* 32, 6, Royal Society London. The term 'batrachian reptiles' referred to amphibia such as frogs, toads and salamanders.
134. Edwards, W. F., *De l'Influence des Agens Physiques,* 1824.
135. *Ibid.,* p.v.
136. [Edwards, W. F.], *On the Influence of Physical Agents,* 1832.
137. Edwards, *De l'Influence des Agens Physiques,* p.vii.
138. *Ibid.,* p.vi.
139. *Ibid.,* p.viii.
140. This 1847 paper was to be Hall's final submission to the Royal Society.

141. Philip, A. P. W., 'On the Sources and Nature of the Powers on which the Circulation of the Blood Depends.' Read 16 June 1831. *Phil. Trans.*, 121, Part 2, 1831, pp.489–96. An abstract appeared in *Abs. Phil. Trans. 1830–1837,* 3, 1837, p.64.
142. Philip, 'On the Sources', *ibid.*, p.496.
143. Hall, M., *Gulstonian Lectures,* 1842, p.19.
144. Hall, *Letter to the Earl of Rosse,* p.5.
145. MS Letter, J. G. Children to J. Lubbock, 12 June 1833, *LUB. C. 177,* Royal Society London.
146. Hall, M., *A Critical and Experimental Essay on the Circulation of the Blood,* 1831. Hall dedicated this work to the surgeon, Henry Smith, who had helped in so many of his experiments.
147. *Ibid.*, p.112.
148. *Ibid.*, pp.140–1.
149. MS Letter, M. Hall to J. Lubbock, 10 November 1831. *LUB. H. 17.* Royal Society London.
150. Hall, M., 'Theory of the Inverse Ratio, which Subsists between Respiration and Irritability, in the Animal Kingdom', *Phil. Trans.*, 122, Part 2, London, 1832, pp.321–34. Hall, M., 'On Hybernation', *ibid.*, pp.335–60. An Abstract of the papers appears in *Abs. Phil. Trans. 1830–1837,* 3, 1837, pp.104–6.
151. Hall's Election, which bears Roget's initials, was on 5 April 1832. The statement about his candidacy reads:

    > Marshall Hall., M.D., of Manchester Square, and desirous of becoming a Fellow of the Royal Society, we, whose names are hereunto subscribed, do, of our personal knowledge, recommend him as deserving of that honour, and as likely to prove a valuable and useful Member.

    Hall's supporters included Michael Faraday, J. C. Brodie, William T. Brande, Astley Cooper, Herbert Mayo, Joshua Brookes, Robert Lee, Charles M. Clarke, Thomas Bell and Edward Hawkins. MS *Election Certificate* 47, VIII, 1830–1840, Royal Society, London.
152. MS *PT. 74. 15,* Royal Society, London.
153. 'Notes to Chapter I, II, III, IV, V and VI' in [Edwards], *On the Influence of Physical Agents,* pp.469–72.
154. According to Legallois, 'Les animaux à sang froid fournissent une preuve évidente de ce que j'avance ici.' Legallois, J. J. C., *Expériences sur le Principe de la Vie,* 1812, p.iv.
155. Hall, M., 'A Brief Account of a Particular Function of the Nervous System', *Proc. Comm. and Corresp. of Zoo. Soc. Lond.,* 1830–4, Part 2, pp.190–1.
156. Hall, M., 'Dr Marshall Hall Showed Some Experiments', Read 12 August 1834, *ibid.,* 1833–5, Part 3, pp.95–112.
157. MS Hall, M., *On the Reflex Function of the Medulla Oblongata and Spinalis, or the Principle of Tone in the Muscular System,* Royal Society,

London, Archives, 1832, *AP.17.5*. It was published as Hall, M., 'On the Reflex Function of the Medulla Oblongata and Spinalis, or the Principle of Tone in the Muscular System', *Phil. Trans.*, 123, Part 2, 1833, pp.635–65. An Abstract appeared in *Abs. Phil. Trans. 1830–1837*, 3, 1837, pp.210–11. Hereafter in the text, this memoir will be referred to as *Memoir* (1833). It was published with a second rejected Memoir on the nervous system as Hall, M., *Memoirs* on the Nervous System, 1837. In the text, the rejected memoir will be referred to as *Memoir* (1837). In the notes, references to both the 1833 and the 1837 Memoirs will be taken from *Memoirs on the Nervous System*, 1837 and referred to as *Memoirs* (1837).

158. Flourens' tribute to Hall, when announcing his death to the Academy of Sciences, was on the same generous lines, *Comptes Rendus*, 45, 1857, p.281.
159. The title of the memoir was *On the True Spinal Marrow, and the Excito-Motory System of Nerves* and it was read before the Royal Society on 16, 23 February and 2 March 1837.
160. Mayo, H., *Anatomical and Physiological Commentaries*, 1822.
161. Lawrence, W., *An Introduction to Comparative Anatomy and Physiology*, 1816. The work was dedicated to J. E. Farre, physician to the London Infirmary for diseases of the eye, who hailed from the West Indies. Lawrence also worked at the Infirmary, which played an important part in treating the epidemic of eye diseases which spread across England in the early nineteenth century. It originated among the troops returning from Egypt and quickly spread to the civilian population.
162. Lawrence, W., *Lectures on Physiology, Zoology, and the Natural History of Man*, 1819. Lawrence began the lectures by replying to the recent Charges made against him by Abernethy, who had accused him of propagating, as a sceptic and in conspiracy with a bunch of French physiologists, opinions detrimental to society.
163. MS Letter, W. Lawrence to H. Mayo, 1 May 1837, *RR.1.86–88*, Royal Society, London.
164. MS Letter, H. Mayo to J. G. Children, 8 May 1837, *ibid.*
165. *Ibid.*
166. Hall, *Memoirs* (1837), p.xiv. The *vis nervosa*, the property through which nerves activate muscles to respond, was distinguished from the *vis insita*, the property of irritability of the muscles. See Fearing, F., *Reflex Action*, 1930, p.71; pp.90–3.
167. Hall, *Memoirs* (1837), pp.4–5.
168. *Abs. Phil. Trans. 1830–1837*, 3, 1837, pp.463–4.
169. In the Rosse letter, p.7, Roget is accused of 'misrepresentation and insult'.
170. See note 157 *Supra*.
171. MS Letter, M. Hall to Members of the Council, dated 8 June 1837, Royal Society London, *MC.2.255*. The letter was accompanied by 'A Note Addressed to the Members of the Council of the Royal Society',

by Marshall Hall, 7 June 1837. A copy of the Letter and the Note were contained in MS *Minutes of the Council Meetings,* 22 June 1837, Royal Society, London. They had been read out to Council and the following resolution taken, 'That the proposal contained in the above letter be not acceded to.'

172. Editorials: 4 April 1846, *Lancet,* vol.i, 1846, pp.390–3; 11 April 1846, *ibid.,* pp.418–19, 25 April 1846, *ibid.,* pp.469–74. It was to the third of these – an unusually long editorial – that Wakley added copies of the correspondence that Hall had lent him, together with Hall's covering letter. This included the comment, 'I have long resolved never again to enter into controversy, and I have publicly and privately expressed this determination....' The change of mind was in order to spare any other investigator from occupying such a position as his own within the Royal Society. *Ibid.,* p.469.
173. Hall, *Letter to the Earl of Rosse,* p.6.
174. The gist of Hall's letter is inferred from Lawrence's reply of 28 April 1837, and which Wakley published in full, *Lancet,* vol.i, 1846, pp.471–2.
175. Letter, W. Lawrence to M. Hall, *ibid.*
176. MS Copy of Letter, Marshall Hall to the Council of the Royal Society, dated 8 June 1837, *Minutes of Council Meetings of Royal Society,* 22 June 1837, Vol.1, 1832–46, Royal Society, London. A longer version of this letter entitled *A Note Addressed to the Members of the Council of the Royal Society; By Marshall Hall,* dated 7 June 1836, pp.1–4, *MC.2.255,* Royal Society, London, was distributed to Members of the Council.
177. Letter, W. Lawrence to M. Hall, 28 April 1837, *Lancet,* vol.i, 1846, pp.471–2.
178. Roget, P.M., *Animal and Vegetable Physiology,* 2 vols, 1834. This was the fifth of the *Bridgewater Treatises on the Power, Wisdom and Goodness of God as Manifested in the Creation.* Marie Boas Hall looks briefly at this issue in, Hall, M. Boas, *All Scientists Now,* p.85, Note 30 pp.234–5. Marshall Hall also referred to the matter in the first of a series of letters which he wrote to Sir John F. W. Herschel, a Vice-President at the Royal Society. MS Letter, M. Hall to J. F. W. Herschel, 1 October 1839, *HS.9.198,* Royal Society London.
179. Letter, W. Lawrence to M. Hall, 23 May 1837, *Lancet,* vol.i, 1846, pp.471–2.
180. Letter, J. G. Children to M. Hall, 11 May 1837, *ibid.,* p.471.
181. *Ibid.*
182. Letter, M. Hall to J. G. Children, 12 May 1837, *ibid.*
183. Letter, J. G. Children to M. Hall, 12 May 1837, *ibid.*
184. According to the Statutes of the Royal Society, 'IV. The original copy of every Paper read at the Society shall be considered the property of the Society, if there be no previous engagement with its Author to the contrary; but any Author may have a copy of his own papers by leave of the Council.' *Statutes of the Royal Society,* 1831, p.19.

185. Letter, W. Lawrence to M. Hall, 23 May 1837, *Lancet*, vol.i, 1846, p.472.
186. Letter, M. Hall to J. G. Children, 15 May 1837, *ibid.*, p.471. Hall's tone in the letter could be construed as sarcastic.
187. Clarke, E., and Jacyna, L. S., *Nineteenth Century Origins of Neuroscientific Concepts*, 1987, pp.114–15.
188. Hall, *Note to Council*, p.4. It is not known which mole was responsible for this leak, nor whether it was a deliberate disclosure to increase the chances of Hall withdrawing the paper of his own accord.
189. *Ibid.*, pp.3–4. Hall referred to Children's lampoon in his lectures on the Theory and Practice of Medicine, delivered at Webb Street during 1837–8, and published in the *Lancet*, vol.i, 1837–8, p.729.
190. MS *Minutes Council Meetings*, 22 June 1837.
191. Letter, J. D. Roberton to M. Hall, n.d., *Lancet*, vol.i, 1846, p.472.
192. MS Letter, M. Hall to J. F. W. Herschel, 1 October 1839, *HS.99. 198–205.*
193. Newport, G., 'On the Respiration of Insects', *Phil. Trans.*, 126, Part 2, 1836, p.551.
194. MS Letter, M. Hall to the Council of the Royal Society, 1 February 1838, *MC. 272*, Royal Society London. Hall included the reply which he was requesting Council to publish. *MC.273. Ibid.*
195. *Ibid.*
196. Newport, 'Respiration of Insects', p.551.
197. MS Hall, Letter, *MC.273*, Royal Society, London.
198. *Ibid.* Roget had exerted pressure on his pen, as he wrote this decision, indicating considerable emotional response on his part.
199. 'Lectures on the Theory and Practice of Medicine, by Marshall Hall', *Lancet*, vol.i, 1837–8, p.650.
200. 'Reclamation of Mr Newport', *ibid.*, p.715.
201. 'Letter from Dr Marshall Hall', *ibid.*, pp.748–9.
202. *Ibid.*, pp.812–17.
203. MS Letter, George Newport to the President and Council of the Royal Society, 7 March 1838, *MC.2.279*, Royal Society, London.
204. *Ibid.*
205. 'Further Observations on Dr Hall's Statement Regarding the Motor Nerves of Articulata', *Lancet*, vol.i., 1837–8, pp.897–900.
206. 'Mr Newport's Second Reply to Dr Marshall Hall', *Ibid.*, pp.950–2.
207. The correspondence, which has already been briefly referred to, contains the following items:

    MS Letter, M. Hall to J. F. W. Herschel, 14 October 1839, *HS.9.199*

    MS Letter, J. F. W. Herschel to M. Hall, 16 October 1839, *HS.9.200*

    MS Letter, J. F. W. Herschel to M. Hall, October 1839, *HS.9.201*

MS Letter, M. Hall to J. F. W. Herschel, 21 November 1839, *HS.9.202*
MS Letter, M. Hall to J. F. W. Herschel, 27 November 1839, *HS.9.203*
MS Letter, M. Hall to J. F. W. Herschel, 3 April 1840, *HS.9.204*

The appearance of Herschel's letters to Hall, with their numerous deletions, insertions, etc., indicate that they were drafts from which Herschel prepared his final copies. In December 1840, Hall sent to Robert Grant an autographed copy of the printing of his first letter to Herschel. It was later deposited in the Library, University College London.

208. MS Letter, M. Hall to J. F. W. Herschel, 1 October 1839, *HS.9.198.*
209. *Ibid.*
210. Kiernan, K., 'On the Anatomy and Physiology of the Liver', *Phil. Trans.,* 123, part 2, 1833, pp.711–70. The intention to award one of the Royal Medals in 1836 for work in Animal Physiology was announced in 1834, *ibid.,* 124, part 2, 1834, p.iv.
211. MS *Journal Book,* 30 November 1839, 8, 1836–43, Royal Society, London, p.56.
212. MS Letter, M. Hall to J. F. W. Herschel, 1 October 1839, *HS.9.198.*
213. This rule about publication was not adhered to later when Snow Beck was awarded a Royal Medal in 1845 which, Robert Lee felt, should have been awarded to him.
214. MS Letter, M. Hall to J. F. W. Herschel, 21 November 1839, *HS.9.202.*
215. MS Letters, J. F. W. Herschel to M. Hall, 16 October 1839; October 1839, *HS.9.200, 201.* These documents are clearly Herschel's draft attempts to reply to Hall's first letter. One contains a reference to Hall's letter of 2 November 1839. Hence the dates in October 1839 assigned to these letters are inaccurate as is their position within the series *HS.9.* 198–205.
216. MS, *ibid., HS.9.200.*
217. MS Letter, M. Hall to J. F. W. Herschel, 27 November 1839, *HS.9.203.*
218. MS Letter, J. F. W. Herschel to M. Hall, October 1839, *HS.9.201.* This letter should bear a date post 27 November 1839.
219. Herschel, J. F. W., *Discourse on the Study of Natural Philosophy,* 1831, p.342.
220. MS Letter, *HS.9.201.*
221. MS Letter, M. Hall to J. F. W. Herschel, 3 April 1840, *HS.9.204.*
222. Hall, M., 'Briefe über das Nervensystem an Professor J. Müller'; 'Erster Brief Ueber die Vis Nervosa Hallers', *Archiv für Anat. Physiol.,* 1840, pp.451–66.
223. Hall, M., 'On the Motive for the Scarification of the Gums during

Dentition', *Lancet,* vol.i, 1844, pp.244–5.

224. 'Dr Marshall Hall on the Mechanism of Vomiting, in a Letter to Dr Anderson', *Lond. and Edin. Mon. Journ. of Med. Sci.,* n.s. 4, 1844, pp.182–3. See also *ibid.,* pp.813–14. In holding that vomiting was a reflex action, Hall was engaged in dispute on the topic.
225. Hall, M., 'Memoirs on Some Principles of the Pathology of the Nervous System', Memoir I, *Med.-Chir. Trans.,* 22, 1839, pp.191–217; Memoir II, *ibid.,* 23, 1840, pp.120–67; Memoir III, *ibid.,* pp.168–91.

    'On the Difficulties in the Study of the Nervous System Set Forth by Legallois', *Lond. and Edin. Mon. Journ. of Med. Sci.,* n.s. 1, 1841, pp.853–8; 'Note on the Use of Setons Especially in the Treatment of Paraplegia', *ibid.,* pp.803–5; 'On the Irritability of the Muscular Fibre in Paralytic Limbs', n.s. 4, 1844, pp.655–60.

    'Remarks on the Nature and Treatment of Stridulous Convulsion in Infants', *Lancet,* vol. ii, 1841–2, pp.505–8; 'Notes on Neuritic Sciatica', *ibid.,* p.508.
226. Hall, M., 'Note on the Respiration of Birds', (Review of part of new edition of work by Flourens), *Lancet,* vol.ii, 1841–2, pp.271–2.
227. Hall, M., 'On the Circulation in the Acardiac Foetus', *Lond. and Edin. Mon. Journ. of Med. Sci.,* n.s. 3, 1843, pp.541–7; *ibid.,* n.s. 4, pp.775–6.
228. Hall, M., *Diseases and Derangements of the Nervous System,* 1841.
229. Hall, M., *Practical Observations and Suggestions in Medicine,* 1845. *Idem, Practical Observations and Suggestions in Medicine,* Second Series, 1846.
230. Hall, M., *New Memoir,* 1843.
231. MS Hall, M., *Researches into the Effects of Certain Physical and Chemical Agents on the Nervous System, Archives 1847, AP.29.* 5–6, Royal Society, London.
232. MS *Minutes of Committee of Physiology,* 17 March 1847, Royal Society, London.
233. MS *AP.29.*5–6, *ibid.*
234. MS Report on Dr M. Hall's two papers received 28 January, 15 April, and the Appendix 26 June 1847 entitled *Researches into the Effects of Certain Physical and Chemical Agents on the Nervous System,* with the 'Postscript' by 'A Fellow of the Society Well Versed in the Physical Questions Involved ....' The main report is dated October 1847; the 'Postscript' 17 August 1847. *RR.1.86–89,* Royal Society, London. The content, status and fate of this paper will be dealt with in greater detail in Chapter 5 *Infra.*
235. *Ibid.*
236. MS *Minutes of Committee of Physiology,* 26 October 1847, Royal Society, London.
237. Hall was aware that the Council meeting was on the following Monday and to save Weld the trouble of writing back to him, Hall

explained that he proposed to send an artist around to the Royal Society to make the copy on Friday unless he heard to the contrary. MS Letter, M. Hall to C. R. Weld, 5 November 1847, *MS.4*.189, Royal Society, London.

238. *Abs. Phil. Trans. 1843–1850,* 5, 1850, pp.667–8; 674–5.
239. Hall, C., *Memoirs,* pp.169–70.
240. MS Letter, R. B. Todd to P. M. Roget, 28 June 1843, *RR.1.265,* Royal Society, London. The subject of the paper is not mentioned in the letter although a note in another writer's hand added to the back of a later letter from Todd on the same matter refers to 'Mr Erasmus Wilson's Paper on the Structure of Muscular Fibres.' The same note refers to the need for a referee 'well versed in the use of the microscope' and finishes with the words 'Dr Todd for example'. What partial concern!
241. MS Letter, R. B. Todd to P. M. Roget, 12 October 1843, *RR.1.266,* Royal Society, London.
242. Wilson, E., 'Researches into the Structure and Development of a Newly Discovered Parasitic Animalcule of the Human Skin – the Entozoon folliculorum', Communicated by R. B. Todd, *Phil. Trans.,* 134, Part 2, 1844, pp.305–19. The paper had been received 23 November 1843 and read 14 December 1843.
243. Hall, M., 'Researches into the Effects of Certain Physical and Chemical Agents on the Nervous System', *Edin. New Phil. Journ.,* 45, 1848, pp.252–67; Hall, 'On the Electrogenic Condition of the Spinal Marrow, and of the Incident Spinal Nerves', *ibid.,* 46, 1849, pp.27–8.
244. Editorial, *Lancet,* vol.ii, 1848, pp.591–2.
245. *Ibid.,* p.646.
246. MS *Minutes of Committee of Physiology,* 4 July 1848, Royal Society, London.
247. MS Letter, M. Hall to W. R. Grove, 13 November 1848, Library of Royal Institution London.
248. MS Letter, M. Hall to W. R. Grove, 26 November 1848, Library of Royal Institution London.
249. MS Letter, M. Hall to W. R. Grove, 28 November 1848 , Library of Royal Institution London.
250. Editorial, *Lancet,* vol.ii, 1848, p.591.
251. *Ibid.,* p.646.
252. Hall, M., 'Comparaison Entre les Éffets Tétanoides des États Électrogéniques, et Ceux de la Strychnine, de la Narcotine, etc.', *Comptes Rendus,* 24, 1847, p.1054.
253. See note 243 *Supra.*
254. Hall, *Letter to the Earl of Rosse.* See note 99 *Supra.*
255. MS Letter, M. Hall to J. F. W. Herschel, 23 November 1848, Herschel Papers, *HS.9.205,* Royal Society, London.
256. Hall, *Letter to the Earl of Rosse,* p.14.

257. Hall was present at the March meeting of Council. MS *Minutes of Council Meetings,* 20 March 1851, Royal Society London. He was re-elected for 1852 on the same day as George Newport's recent award of another Gold Medal was announced. Hall attended no further meetings.
258. Hall delivered the Croonian Lectures at the Royal College of Physicians in 1850, 1851 and 1852. The first lecture was published as *Synopsis of the Diastaltic Nervous System,* 2nd edn, [1850]. In the Preface, Hall explained that he had printed the lecture in its present format at his own expense because 'in the present condition of the profession ... "Monographs do not sell" ....' The other two Croonian Lectures were: *Synopsis of Cerebral and Spinal Seizures,* [1851]; *Synopsis of Apoplexy and Epilepsy,* [1852].
259. MS *Referee's Report on a Paper by J. Lockhart Clarke on the Structure of the Spinal Cord.* By Marshall Hall, 18 February 1851, RR.2.44, Royal Society, London.
260. *Ibid.*
261. *Ibid.*
262. Lockhart Clarke, J., 'Researches into the Structure of the Spinal Cord', *Phil. Trans.,* 141, Part 2, 1851, pp.607–21. An abstract of the paper appeared in *Abs. Phil. Trans. 1850–1854,* 6, 1854, pp.1–3.
263. MS Letter, M. Hall to Dr Munk, 31 October 1850, Royal College of Physicians, London.
264. MS Letter, M. Hall to W. Munk, 30 October 1850, Royal College of Physicians, London.

5

Hall's Work on the Nervous System

Background

Hall's work on the nervous system, and in particular on the concept of reflex action, began around 1830, when there was considerable interest both in England and on the Continent in the properties and functions of the nervous system. The main period of animal experimentation on a large scale, involving investigation of the effects on muscle activity and hence on animal motion, of such procedures as decapitation, section through or destruction of the cord and isolation of limbs, had not begun much before the middle of the seventeenth century. Significant contributions to the development of a general theory of reflex action, that is involuntary motion, the phenomenon to which Hall devoted so much of his research, were made by a number of individuals during the seventeenth and eighteenth centuries. They included Descartes, Willis, Perrault, Duverny, Chirac, Preston, Mayow and Whytt.

Other developments in the wider field of nerve-muscle physiology over the same period concerned attempts to distinguish, through experiment, the special properties of the two types of tissue. This work on nerves and muscles contributed to the gradual refinement of the theory of reflex action. It included the speculation by Glisson, confirmed almost a century later by Haller and others, concerning the inherent capacity of muscle to contract; this property of muscle tissue came to be called irritability. Glisson in his major work *De Ventriculo*[1] had, through his experimental work, undermined the Galenical theory of animal spirits within the nervous system, as an explanation for muscle contraction. Glisson's quantitative experiments indicated that muscles did not increase in bulk during contraction as was required by the theory of animal spirits.

In the eighteenth century, the Scottish neurophysiologist Robert Whytt contributed through his extensive repertoire of experiments, to the maintenance of interest in the study of nerve-muscle phenomena. Section 13 of his *Essay on Vital Motions*[2] is devoted to 'Concerning the Motions Observed in the Muscles of Animals after Death, or their Separation from the Body.' Whytt referred to the experimental work of his predecessors including Robert Boyle's investigation on the female silk worm moth which not only continues to admit the male after losing its head but also lays eggs. Redi's work on decapitated tortoises, one of which survived the operation 23 days and another from November to the following May was also cited, as was Bacon's account of the events following the excision of a malefactor's heart which was tossed up several times to a considerable height when thrown into a fire.[3] Whytt detailed a range of his own experiments on animals such as eels, frogs, pigeons, cats, vipers, a cock and a new-born chick. His procedures included decapitation, removal of the heart or other viscera, and examination of the effects of these procedures on the viscera themselves and on the behaviour of the eviscerated or decapitated animals. Whytt also investigated the effect of injecting opium into the gut. He seems not to have been the subject of censure, as was Marshall Hall, for carrying out certain procedures some of which must have inflicted pain on the animals involved.

Contributions to the Concept of Reflex Action

According to the Preface of *On the Diseases and Derangements of the Nervous System*,[4] it was during his research into the circulation of the blood that Hall had noted what he regarded as a remarkable phenomenon and one which led him to concentrate on reflex action. He noticed that the separated tail of the eft (newt) moved on being irritated by the point of a scalpel. Hall described the incident in his *Memoir (*1833),[5]

> I was first struck with the phenomena of the reflex function of the spinal marrow in the separated tail of an eft. On being excited by the point of a needle passed lightly over its surface, it contracted and moved as if it still formed a part of an entire animal.[6]

It was this movement in an isolated tail following irritation which was considered remarkable, because the occurrence of movement following stimulation in a decapitated creature such as a frog was quite familiar.

A survey of the literature on the physiology of the nervous system revealed similar observations by predecessors including

Redi (1626–98), Mayow (1641–79), Whytt (1714–66) and Prochaska (1749–1820). But these writers, it was claimed, had simply made observations which, lacking interpretation, had remained sterile. Whytt had already spoken of vital (reflex) actions in connection with such isolated phenomena as that of limb withdrawal following stimulation of a foot in a decapitated creature. But he and others had failed to see them as but one example of the wide range of movements which Hall came to embrace within a general theoretical framework of reflex action. Nevertheless, many basic isolated facts concerning the functioning of the reflex mechanism were known by the end of the eighteenth century.

Hall was outspoken in stating his claim to be the first person really to have understood the true nature of reflex action in all its aspects. Others have since bestowed this credit on a variety of individuals, all of whom have come to be recognized as playing some part in the evolution of the theory of reflex action. These writers criticized the choices of their predecessors and contemporaries.

The recognition of reflex actions as a special class to which physiologists turned their attention certainly began in the seventeenth century with the speculations of Descartes and Willis,[7] both of whom had their later advocates. Descartes had the enthusiastic support of Huxley in 1874[8] and also that of Lhermitte[9] writing at the beginning of the present century. While one may be reluctant to echo Huxley's enthusiasm about Descartes as a physiologist, Fearing, in 1930, did credit Descartes with having recognized the principle of reflex action three centuries earlier.[10]

Descartes's writing on automatism, in which he presented his model of the body as a machine, included what seems to be an enunciation of the principle of reflex action. But though Descartes is frequently credited with being the originator of the doctrine of reflex action, this opinion is based largely on his chance use of an analogy in support of his theory of the human body as a material machine directed by a rational soul. He did not, says Fearing, elaborate the implications of his generalization or offer experimental proof of the fact itself. Whytt is generally regarded as having made the most significant contribution to elucidating the doctrine of reflex action up to the middle of the eighteenth century. But although he described a number of reflex actions he never used the word 'reflex'.

According to Richet, in his *Physiologie des Muscles et des Nerfs,* 1882, one of the best French histories of research on the subject, the concept of reflex action developed slowly and depended on the contributions of several people.

> Il nous parait très simple aujourd'hui de considerer le système nerveux comme l'appareil directeur et harmonisateur de l'organisme, tenant sous sa dependence la sensibilité, le mouvement et l'intelligence. Mais autrefois cette notion n'existait pas; l'idée banale qui fait du système nerveux l'appareil de la pensée n'est devenue banale que depuis un siècle à peine. Il a fallu une longue série d'erreurs pour que cette verité soit acquise.[11]

An interesting comparison was made between the rate of development and the number of individuals involved in the origin of this concept compared to that of other biological concepts. According to Richet,

> La découverte des actions réflexes n'est pas, comme celle de la circulation du sang ou des phénomènes chimiques de la respiration, due au genie d'un seul homme. Elle s'est faite graduellement, insensiblement, pour ainsi dire; et l'enfantement de la théorie actuelle a duré une très longue période de temps, c'est à dire près de deux siècles (1630 à 1830).[12]

That Hall was making his contribution to knowledge of the physiology of the nervous system in a period marked by more intense activity than that referred to by Richet, is reflected in the *British and Foreign Medical Review* of 1840 which noted 'It is not a little remarkable that most of what is definitely known regarding the special functions of the nervous system has been ascertained within the last thirty years.'[13] Despite his tributes to Descartes and Willis, Richet attributed to George Prochaska from Moravia, the credit for having first formulated a general theory of reflex action, thanks to his 'esprit synthetique qui avait permit à Prochaska de systematiser les actions reflexes'.[14] Others, including Beaunis,[15] agreed with Richet. Hall became involved in a serious controversy over Prochaska's contribution. Ironically in spite of its sometimes harsh criticism of Hall, it was precisely the compliment that he had been the first to formulate, through a synthesis of several existing ideas, a general theory of reflex action, which the *British and Foreign Medical Review* paid to Hall.

Eckhard, on the other hand, in his highly informed study written a year earlier than Richet's work, was one of the earliest people to recognize the significance of Willis's work. In Eckhard's opinion, Willis's observations and conception of reflex action approached current ideas of the 1880s. Eckhard noted that Willis identified a location for the seat of reflection and placed it in the encephalon of the brain. Eckhard criticized Prochaska for not having brought to the investigation of reflex action any experimental contributions comparable to those of Whytt, much admired by Eckhard. Richet had assigned no particular contribution to Hall and Eckhard felt that the significance of his work had been exaggerated.[16]

The decapitation experiments of Whytt's predecessors had demonstrated that the power of responding to stimuli did not reside wholly in the brain. Whytt himself in the first half of the eighteenth century carried out the crucial experiment, which demonstrated that the capacity of the muscles of a decapitated creature to respond to stimulation of the skin depended on the presence of the spinal cord. Destruction of the cord led to cessation of responses. Whytt further refined this observation when he found that it was even possible to provoke responses in isolated portions of the body, provided that the segment of the spinal cord to which the nerves of the isolated portion were connected, remained intact. As Whytt observed, '... when the body of a frog is divided in two, both the anterior and posterior extremities preserve life and a power of motion for a considerable time'.[17] As in a decapitated creature, removal of the corresponding segment of cord was accompanied by removal of the capacity to respond. In another work, Whytt summarized his observations,

> Further, it ought to be observed that when, after decollation, the spinal marrow of a frog is destroyed with a red hot wire, no visible motion is produced in its limbs or body, by pricking, cutting, or otherwise hurting them: ....[18]

As a result of his experimental findings, Whytt was sceptical of the contemporary theory of peripheral nerve connections as an explanation of the phenomenon of sympathy, in which stimulation of one part of the body provokes response in an anatomically unconnected part. Already for Whytt it was most likely that these instances of sympathy were due to some connection between the corresponding nerve endings within the spinal marrow. In writing on muscle response in a decapitated frog following destruction of the spinal cord, Whytt proposed that,

> ... there was no sympathy between the different muscles, or other parts of the body, as was observed while the spinal marrow was entire: from whence it seems to follow, that the nerves distributed to the several parts of the body have no communication but at their termination in the brain or spinal marrow; and that to this, perhaps alone, is owing the consent or sympathy observed between them.[19]

Whytt's work on vital or reflex actions thus drew attention to the necessary presence of the spinal cord for the execution of reflex actions in decerebrate animals, and for at least the corresponding portions of the cord in isolated segments of the body. He also saw the necessity for the existence of nerves connecting the sites both of stimulation and response to the spinal cord. But Whytt's explanation of the functioning of the neuro-muscular system depended on the existence of a non-material sentient principle or soul.

Nevertheless, his work led to the formation of models of reflex arcs in which the spinal cord, in reflecting ingoing influences from sense organs into outgoing influences passing to muscles and glands, was compared to a mirror with its reflecting properties. In the 1820s and 1830s attention was focused, as for instance in the work of Bell and Magendie, on distinguishing between the ingoing or sensory and the outgoing, or motor nerve pathways involved in reflex actions. Other researchers such as Hall's friend, R. D. Grainger, concentrated on the structure of the spinal cord itself[20] and on investigating its role in sensation. Hall even attempted to establish the existence of a distinct set of afferent and efferent nerve fibres associated with reflex action and connected to a particular region of the spinal cord.[21]

Legallois (1770 –1814) in his work *Expériences sur le Principe de la Vie,* published in 1812,[22] was primarily interested in locating the source of movements of the heart but claimed that the principle which he had been investigating had much wider application with regard to involuntary responses generally. In the *avant-propos* to his work he explained,

> Ce que j'y ai dit du coeur, pouvant s'appliquer aux autres organes des fonctions involuntaires, la question peut être considerée plus générale-ment comme la determination du siège du principe qui preside à cet ordre de fonctions.[23]

Hall's *Memoir* (1833) almost opens with a quotation from a posthumously published work of Legallois and although Hall contradicted him on the crucial issue of the role of sensation in reflex action, he was greatly influenced by Legallois' work and that of other French physiologists.

During his research on the principle controlling involuntary motions Legallois, while repeating experiments similar to those of Whytt, extended Whytt's repertoire in experimental detail, observation and interpretation. He noticed, and Hall was to take this up later in support of his contention that the will plays no part in reflex action, that decapitated creatures show no spontaneous motion. They move little compared to the intact animal and only when stimulated.[24] According to Legallois,

> Si on les place sur le dos, elles s'agitent parfois pour changer de situation; mais elles y restent, parce qu'elles ne savent plus faire

> les mouvemens convenables pour se remettre sur le ventre. Tous ces animaux font en general peu de mouvemens, 'à moins qu'on ne les touche, et l'on conçoit que cela doit être, puisque de tous les sens il n'y a plus que le toucher qui puisse leur transmettre des impressions.[25]

In his search for an explanation of the movements in decapitated animals, Legallois was impressed by the fact that movements in 'un tronc vivant sans tête, semblent assez souvent provoqués par une sorte d'instinct ou de volonté'.[26] It was his opinion 'que le principe de la vie du tronc réside dans la moëlle épinière'.[27] The spinal cord was therefore seen as the sole source of the vital properties associated with the trunk but, contrary to Hall's view, it was also associated with sensation and voluntary movements of the trunk. According to Legallois,

> ... il est facile de démontrer que cette prerogative de la moëlle épinière, d'être la source du sentiment et de tous les mouvemens volontaires du tronc, lui appartient exclusivement à tout autre organe.[28]

Legallois was further led to the opinion that the spinal medulla was built up of several 'centres de sensations'[29] which were, through sectioning of the cord, capable of being rendered separate and distinct, although in the intact animal they were linked in such a way that they formed a single interacting system. Hall pursued this notion of separate centres as, at a later date, did Sherrington who constructed his model of an integrated nervous system on the basis of a number of linked units.

Legallois had emphasized the greater suitability, for experimental work, of cold-blooded animals over warm blooded in which haemorrhage and the interruption of respiration more quickly render the preparations useless. Legallois' recommendation was taken up in Hall's own choice of animals for experimentation both for their advantages and, probably, to deflect anti-vivisection pressures from himself. Much of Legallois' work was demonstrated to the Faculty of Medicine in Paris during 1809 and reported on by a group of Commissioners in the same year. It was also investigated in September 1811 by Humboldt and others on behalf of the National Institute of France which, speaking highly of Legallois, recommended firstly publication of the work in its *Transactions* and secondly remuneration for Legallois to defray the expenses occasioned by his experiments.[30] This was the kind of financial support and encouragement which, in vain, Hall sought from the Royal Society.

Another French physiologist was admired by Hall when he spoke of 'the admirable work of M. Flourens'. Flourens' experimental work of 1822–4[31] on pigeons had led him to conclude, contrary to Legallois, that the will to perform voluntary, locomotor movements emanated from the cerebral hemispheres, their coordination from the cerebellum and their arousal from the spinal cord and its nerves. Thus for Flourens – and it figured significantly in Marshall Hall's explanation – sensation and volition were associated with the cerebrum but not with the spinal cord.

Issues of Priority

Thus, the cohort of individuals working on the neuro-muscular system during the 1830s and 1840s included Bell, Grainger, Mayo and Carpenter as well as Hall in England, Volkmann and Müller in Germany and Flourens and Magendie in France. Not surprisingly with such an array of investigators concentrating simultaneously on this field of study, similar questions were raised and sometimes similar conclusions arrived at. Inevitably, there arose disputes concerning priority of discovery. Some commentators on work on the nervous system had as an aim not only that of explaining the nature of recent advances in knowledge, but also that of considering just this issue of priority of discovery. As one reviewer explained, 'The object we have in view is to explain not merely *what* has been done, but to *whom* the credit of each discovery is attributable.'[32] The *British and Foreign Medical Review* was somewhat ambivalent. In claiming that few controversies are more bitter than those which relate to priority of discovery, it asserted that, 'The knowledge is that which the world wants; of what practical importance is it by whom that knowledge has been first attained?'[33] But it added,

> We do not think however, that the world should be left in ignorance on the subject; because if the real claimants do not make known their pretensions there will always be found upstarts quite ready to seize the reward, without the slightest title to it.[34]

Bell and Magendie each felt that he had been the first to elucidate the separate sensory and motor nature of the anterior and posterior spinal nerve roots. And each had his supporters, in the sometimes heated controversy involving the medical journals, lectures and the like, which followed the publication of their work on the topic.[35] Bell was accused of falsifying earlier versions of his work. Decades after Bell's work and in Schafer's *Textbook of Physiology* (1900), Charles Sherrington spoke of 'The Bell Experiment'.[36] The *Lancet* in 1912 published, over a number of weeks, articles and a heated correspondence between Bell and Magendie supporters under the titles 'The Bell-Magendie Controversy' and the 'Bell-Magendie Question'.[37] More recently, in 1944 Olmsted, in a biography of Magendie, referred to 'The designation, "Bell-Magendie rule" ....'[38] Both were given recognition when the term 'Bell-Magendie Law' was applied to their findings. One earlier commentator on their dispute had declared,

> We apprehend, therefore, that if the opinion as to the function of the different parts of the spine and nerves prove correct, although the honour of perfecting it belongs exclusively to M. Magendie, yet the merit of having first made it must be surrendered by him, at least in part, to Mr C. Bell. And it is with pleasure we observe, that in a late Number of his Journal, the French physiologist has done ample justice to the claims of our countryman.[39]

The effects of the 'encroachment of one another's territories', as it was put, was thought to result from a change of practice. A touch of the xenophobic, anti-French attitude existing at that time claimed that,

> This concomitance of pursuits among so many persons is not, we apprehend, the effect of mere chance. At one time it seems to have been an understanding among the scientific in general, that, when a man was lucky enough to strike out a new path of inquiry, had favoured the world with some of his discoveries, and pledged himself to prosecute them, he was to be allowed to pursue his course without disturbance or encroachment. And such, we believe, is still the understanding in Britain. Not so in France: There, as soon as anyone has opened a new field of discovery, and gathered up the produce of a part of it, his brethren on all hands conceive themselves entitled to enter it also, and to partake in the harvest along with him.[40]

Observing that while this newer practice might be unfair to the original discoverer, it could not be doubted that it must benefit science for two reasons. Firstly, the field of enquiry is exhausted much sooner and much more thoroughly and secondly the process of comparing and contrasting the range of relevant research should lead to more valid results.

One of Hall's potential, contemporary competitors in his work on reflex action was Johannes Müller (1801–58) who, working on reflex action, came to some of the same conclusions as Hall. In 1831 Müller, in an article 'Nouvelles Expériences sur l'Éffet que Produit l'Irritation Mécanique et Galvanique sur les Racines des Nerfs Spinaux' published a number of physiological details which did not accord with those presented by Hall in his paper to the Zoological Society and in his *Memoir* (1833).[41]

Müller admired Hall's work and in the first edition of his *Handbuch der Physiologie*[42] referred to it in a section entitled 'Von der Reflexion in den Bewegungen nach Empfindungen'.[43] This section was retained with few changes in subsequent editions. Hall translated and published the earlier editions of Müller's discussion together with his own comments, in the *London and Edinburgh Philosophical Magazine.*[44] A German translation of Hall's *Memoir* (1833) was published in Müller's journal *Archiv für Anatomie und Physiologie*,[45] together with correspondence on the nervous system between the two men.[46]

While Hall felt that Müller's complimentary attitude towards his contributions to reflex function, underlined and enhanced his prestige and helped to keep alive interest in these enquiries, Hall, nevertheless, felt that Müller's work represented a threat to his own position as a 'discoverer'. In a Postscript to *Memoir* (1837) Hall declared,

> ... there is now a danger lest the discovery itself should, inadvertently, be partly assigned to the distinguished physiologist whom I have mentioned. I think it right, therefore, to state very distinctly, that whatever credit does attach to this discovery, ... is certainly and entirely due to me.[47]

In his translation and commentary on Müller's account, Hall claimed that they had independently engaged in research on the nervous system and had come to nearly similar conclusions concerning the role of the spinal marrow in reflex action. In expressing satisfaction that his own work should be confirmed by someone of Müller's stature, Hall was at pains to establish that the claim to priority of publication of their mutual ideas was in his favour. This was based on the fact that even though his first Royal Society Memoir and the first volume of Müller's *Handbuch der Physiologie* were both published at roughly the same time in 1833 he himself had really put forward his ideas at a published meeting of the Zoological Society in November 1832. Müller acknowledged this. But from Hall's translation it is clear that Müller was not entirely indifferent to the question of establishing priority. He said,

> The phenomena now to be examined were observed at nearly the same time by Dr Marshall Hall and myself. As the greatest part of the 'Nervenphysik', as here given, was completely prepared several years since, so also this Chapter on the reflected motions after perceptions, was written down almost exactly as here given several years ago.[48]

While writers including Fearing claim that Müller, in referring to Hall's preliminary report to the Zoological Society, acknowledged priority to Hall, it may be that with his reference to 'several years ago' Müller was trying to establish a date earlier than 1832 when his own work was actually conceived and written down. Summarizing his own interpretation of reflex action and dismissing older notions of peripheral nerve connections as an explanation of the phenomenon Müller believed,

> It results from this that the general twitchings which take place in animals on touching particular parts, do not result from communication of sensorial and motor nervous filaments, but that the spinal marrow is the connecting medium between the sensorial-centripetal and the motor-centrifugal excitement.[49]

Elsewhere in the same section he observed,

> The phenomena which we have now described, first from our own observations and then from those of Dr Marshall Hall, have all this in common with one another, that the spinal marrow is the connecting link between a sensorial and a motorial motion of the nervous principle.[50]

Thus, like Hall, Müller emphasized the crucial role of the cord as the connecting link between the centripetal and centrifugal elements of the reflex arc. But while Hall confined the reflex function exclusively to the medulla and cord, Müller rightly pointed out that certain cerebral nerves, especially the optic and auditory, can also be associated with reflex phenomena. He was less dogmatic than Hall regarding the mechanical aspects of the reflex act and in Fearing's view, his writing indicates an early reaction against the mechanical theory of reflex action associated with Hall's name.[51] Müller recognized sooner than Hall the possibility of psychical activity as a modifying factor in reflex action. More significantly, Müller associated certain movements, which Hall undoubtedly classified as reflex, with sensation and he did not accept Hall's view of a distinct excito-motory system.

Hall's Work on Reflex Action

Hall took up the subject of nerve cord physiology where Whytt and Legallois had left it. Most of the major theoretical aspects of the theory of reflex action were known by the end of the eighteenth century. Hall's theory of reflex action was one proposed mainly to explain or systematize existing information. Theorizing and system building were seen as the hallmark of good scientific work and Hall firmly dissociated himself from any claims to have established all the facts or phenomena associated with reflex action. Recognition for 'the establishment of the *principle* and of the special *organic seat,* and the development of the

Theory, or System' of reflex phenomena was his goal.[52] Hall coveted the role of clarifier and reformer of the concept of reflex action, which would facilitate identification of the principle, make possible a systematic classification of an extensive series of disparate experimental facts, including physiological, pathological and therapeutic phenomena which were 'previously entirely unarranged'.'[53]

A typical physiological indicator of activity in the nervous system as a whole and in the spinal cord in particular is some form of muscle response. Four types of muscle action in animals were initially described by Hall. There are, firstly, voluntary actions which involve the will and originate in the cerebrum, from which the influence generated passes along the spinal marrow and out via the spinal nerves of the responding muscles. Secondly, are the motions of respiration which like voluntary motions originate within the central nervous system. In the third type of muscle action, termed involuntary, there is an immediate application of a stimulus to the muscle itself. The heart, directly stimulated by the blood, is one example. Its muscle responds due to its inherent property of irritability. To these three types of muscle response, Hall added his fourth group of motions. These persist, even following the removal of the cerebrum and medulla oblongata, and after the voluntary and respiratory motions have ceased. This fourth type of motion, exemplified by a very wide group of involuntary motions, was the group of reflex actions on which Hall was now concentrating. These actions, as the newt tail demonstrated, are associated with the spinal medulla and cease only when this is destroyed.

The essence of Hall's observation on the newt tail must, as he claimed, have been included in his first paper to the Zoological Society in 1832, for as well as detailing experiments on salamanders, frogs and turtles the Reporter added Hall's account of a newt experiment in which 'the tail, entirely separated from the body, moved as in a living animal on being excited by the point of a needle passed lightly over its surface.'[54] Hall summarized his interpretation of his observations and claimed,

> 1. that the nerves of sensibility are impressible in portions of an animal separated from the rest; in the head, in the upper part of the trunk, in the lower part of the trunk:
> 2. that motions similar to voluntary motions follow these impressions made upon the sentient nerves: and
> 3. that the presence of the spinal marrow is essential as the central and cementing link between the sentient and motor nerves.[55]

A special feature of this fourth type of muscular motion, it was emphasized, was that its motive influence did not originate in any central part of the nervous system, but at a distance from that centre. Thus, this motion which Hall considered to be neither spontaneous in its action nor direct in its cause he saw as being excited by the application of appropriate stimuli to certain membranous parts from which 'the impression is carried to the medulla, *reflected,* and reconducted to the part impressed, or conducted to a part remote from it, in which muscular contraction is effected'.[56]

Having been enthused by his own observation on the newt tail Hall embarked on a programme of investigations of this type of phenomenon. His first planned investigation was performed on a turtle. Details of the procedures and observations on the severed head and torso of the creature included,

> This animal was decapitated in the manner usual with cooks, by means of a knife, which divided the second or third vertebra. The head being placed upon the table for observation, it was first remarked that the mouth opened and shut, and that the submaxillary integuments descended and ascended, alternately, from time to time replacing the acts of respiration. I now touched the eye or eyelid with a probe. It was immediately closed .... I then touched the nostril with the probe. The mouth was immediately opened widely, ...[57]

The procedure was repeated and the same observations made. Finally the medulla and brain were gently withdrawn and it was observed that 'All the phenomena ceased from that moment!'[58]

Other parts of the decapitated animal including the limbs, tail and anal sphincter were stimulated as were the same parts when the trunk region was cut in half, again 'in the manner usual with cooks', to produce three – and when the tail region, too, was cut off four – distinct regions of the animal's body. Hall's similar experiment on a newt is shown in **Plate V.** Despite the fragmentation and separation from the body the limbs moved rapidly, the anal sphincter was perfectly circular and closed and the limbs and tail 'possessed a certain degree of firmness or *tone,* ....'[59] When the spinal marrow was withdrawn, as with the head region when the brain was removed, all the associated spinal phenomena ceased. The limbs and tail no longer responded, but became flaccid and lost their resilience; the sphincter lost its circular form and contracted state, becoming lax, flaccid and shapeless.

There followed another research programme of experiments on frogs which had been made to ingest a watery solution of opium. This was found to enhance reflex function. The animals became affected with symptoms very similar to those of tetanus in that the limbs became rigidly extended. In addition the properties of the cutaneous and motor nerves were so enhanced that the slightest stimulus, for example even a breath of air, induced spasmodic movement of the body and limbs. Having decapitated and divided the body of one of the opium-treated frogs below the third vertebra, it was found that the eyes remained tightly shut and no motion could be detected on irritating the eye, eyelid or skin. Both anterior and posterior parts of the divided body region remained tetanic as before and the enhanced condition of the nerves persisted. However, on removal of the brain and the corresponding sections of the spinal marrow these phenomena changed. The eyes were immovable but no longer retracted; the muscles of the limbs were flaccid and the sensory nerve showed no excitability.

Plate V: Responses in sectioned newt

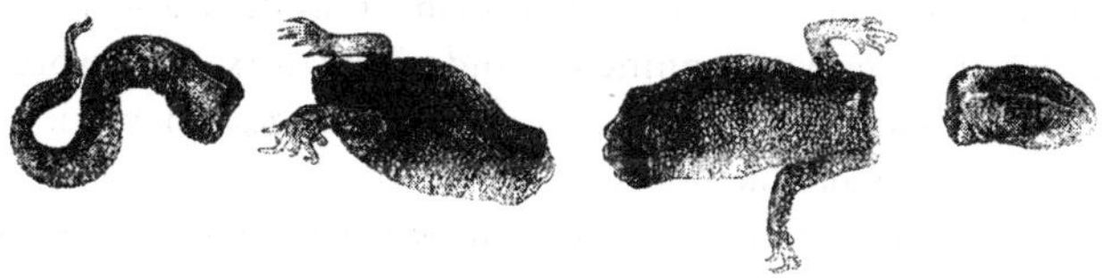

The separated parts continued to respond provided the brain and spinal cord were intact.

Hall, M., *Synopsis of the Diastaltic Nervous System* [2nd edn], 1850. Wellcome Institute Library, London.

Any explanation or interpretation of such observations as Hall had made on his range of treated animals and parts of animals needed, according to him, at least two components. First the phenomena needed to be related to a physiological principle of action. For Hall, this was the special reflex function of the spinal medulla. Secondly, these phenomena needed to be related to some function in the animal economy. Mindful of the aim to link his research to practical medical applications, Hall outlined the role of this reflex function in the healthy organism and went on to relate it to conditions of disease.[60]

It was the combined task of elucidating the principle and its functions which Hall took up, elaborated and extended over much of the next 25 years. And it was precisely the light which he did throw on the subject wherein lay his discoveries, his unique contribution to knowledge of the nervous system. On one occasion he declared that he had spent 25,000 hours on his researches.[61] But speaking of this declaration and of the attitude to it of some of his critics, Canguilhem cuttingly observed, 'On a vu George, Jeitteles, Pflüger reprocher à Marshall Hall d'avoir passé vingt-quatre mille heures de travail a enforcer, selon eux, des portes déjà ouvertes.'[62]

Concerning the role of reflex action it was noted that while three of the four types of motion outlined by Hall were known only by actual movements or muscular contractions, one important manifestation of the fourth or reflex action was that it existed also as a continuous muscular action, 'as a power presiding over organs not actually in a state of motion'.[63] Here was a recognition of the phenomenon of muscle tone which pervades the whole muscular system and maintains postural equilibrium in a healthy body. The reflex function is also responsible for a wide range of hitherto disparate and apparently unconnected movements or reactions. These include, as well as the familiar limb movements following the unexpected pinching of a hand or foot, such movements as those associated with the larynx, pharynx, sphincters and ejaculators. In a word it is associated with the processes of ingestion and egestion and it guards the body's entrances and exits.[64]

Respiration although singled out as the second type of muscle action was later held also to be fundamentally a reflex action although to some extent controllable by volition. In the unhealthy body the reflex principle could be morbidly affected, such change being responsible for a range of disorders including hydrophobia, chorea, *paralysis agitans,* certain forms of epilepsy, of tremor and of asthma. Furthermore, it could be augmented, sometimes fatally, by certain poisons such as opium.

Plate VI: Diagrammatic Illustration of the True Spinal or Excito-Motory System

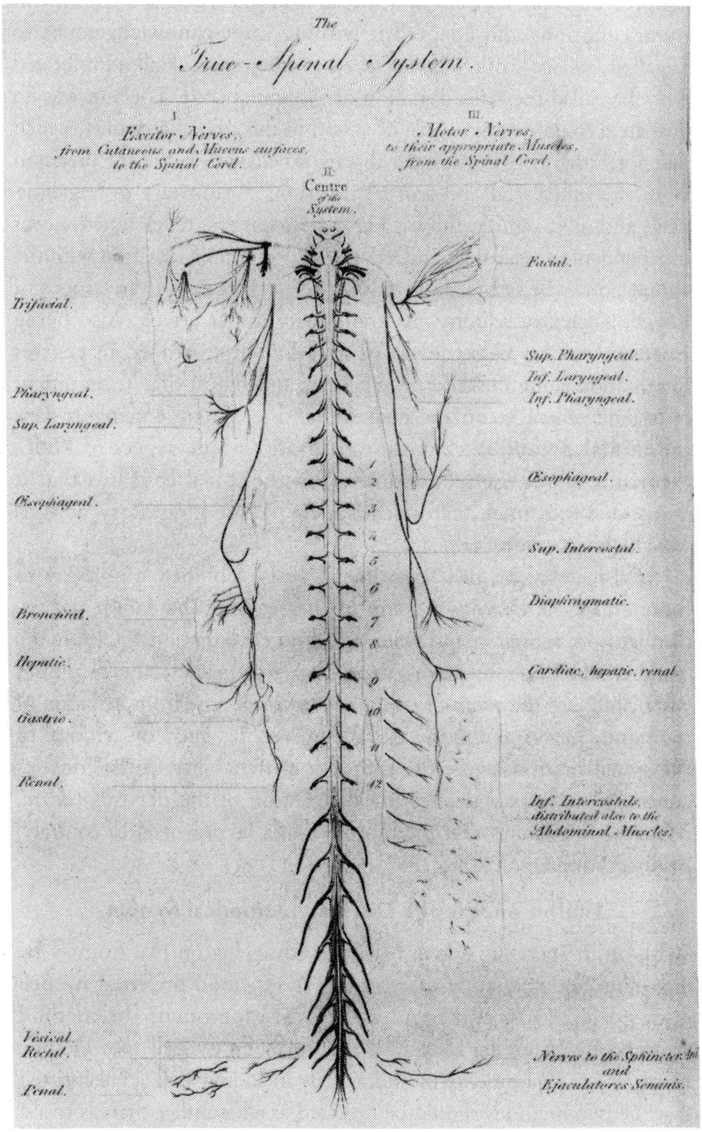

Hall, M., *On the Diseases and Derangements of the Nervous System*, 1841. Wellcome Institute Library, London.

Searching for some explanation of the nature of the principle underlying the phenomena noticed throughout his experiments, Hall adopted Legallois' concept of a special principle associated with involuntary motions and adapted it without any acknowledgement to Legallois, whose work was described as interesting. Hall adumbrated what he called the reflex function of the spinal cord. The aim was 'to give an account of a principle of action in the animal economy, which has not hitherto, I think been distinguished with sufficient precision from the other vital and animal functions'.[65] Contrary to Legallois' view, and most significant, was Hall's assertion that reflex function was independent of sensation. Indeed, the association of sensation with the spinal cord, the 'siège de la puissance nerveuse', was an aspect of Legallois' theory which was strenuously under attack. Also being targeted was his belief, even in decapitated creatures, in centres of sensation along the cord; by extension this meant that 'le sentiment et le mouvement volontaire peuvent subsister et être entretenus dans un animal décapité, ...'[66] An even more unique aspect of Hall's theorizing about the reflex function of the spinal cord was that it demanded a distinct anatomical system. These innovatory notions were highly controversial.

Hall felt that his own experiments had established 'a property or function of the nervous system, of the sentient and motor nerves, distinct from sensation and voluntary instinctive motion'.[67] Clearly the contraction of the muscles in a decerebrate frog made tetanic by opium was plainly not the result of volition. Nor did it arise from irritation of the motor nerves or the muscles themselves. It ceased on removal of the spinal marrow while the property of irritability in the muscles continued unimpaired after the destruction of the nervous centre, because it was found that the muscles could be provoked to contract, by direct stimulation.

Hall's Concept of a Distinct Anatomical System

Work on the nervous system had been advancing on two fronts – by dissection and the study of anatomy, the method preferred by Bell during the 1830s, and by physiological experiment the method favoured by Flourens, Magendie and formerly by Bell too. Despite strictures on experiments by individuals including Bell, Hall insisted that the two methods should be regarded as allies rather than as rivals. He personally placed greater reliance on experiment, 'The scalpel may confirm what experiment had determined but it can do no more. It may suggest experiment; but experiment alone can definitely resolve the problem.'[68]

The most unique aspect of Hall's theory of reflex action and embodied in the title of his first rejected Royal Society memoir was the assertion that the distinct reflex function of the spinal cord demanded a distinct anatomical system, **Plate VI,** a series of afferent and efferent nerves associated exclusively with reflex action and linked to a region of the central spinal system similarly associated specifically with reflex function. The nervous system as a whole, **Plate VII,** therefore, consists of,

I. The Cerebral, or the Sentient and Voluntary;
II. The True Spinal, or the Excito Motory; and
III. The Ganglionic, or the Nutrient, the Secretory, etc.[69]

Speaking of his proposed system Hall argued,

> Suppose the cerebrum and cerebellum, the *centre* of the first sub-division of the nervous system, and the ganglionic, or the second sub-division of this system, removed, *this remains.*[70]

A separate system was, therefore, seen as a logical necessity. As Sherrington put it, Hall was trying to establish the existence of a great sub-system for unconscious reflex action as distinct from the other main sub-system associated with sensation and volition.[71] Lack of concrete anatomical evidence of the existence of either the system of nerves or the essential central connections was admitted. The sub-system was not identifiable with the unaided eye since it 'blended'[72] into the rest of the nervous system. Notwithstanding these limitations a schema and a diagrammatic illustration of the proposed system was produced.

One well-known supporter of this view on a distinct reflex system was R. D. Grainger currently working on the spinal cord. A reviewer of *Observations on the Structure and Functions of the Spinal Cord,* drew attention to this support of Hall,

> Firmly convinced of the correctness of Dr Marshall Hall's views concerning the reflex function, Mr Grainger naturally concluded that if a separate function existed in the nervous system, it must correspond with a separate order of nerves.[73]

Like Hall, Grainger lacked faith in the capacity of the microscope to reveal structural connections between the fibres of the nerves and the grey matter of the cord but believed that his macroscopic examination of preserved tissue did reveal such a connection. George Newport, with whom Hall had tussles in the Royal Society, claimed to have seen two sets of nerves in the *Articulata.* Another supporter of Hall was the physiologist William Carpenter who as editor of the *British and Foreign Medical Review* was at times harshly criticized by him as the author of reviews antipathetic to his work. R. B. Todd, by contrast, strongly refuted the idea of Hall's special anatomical system.

Plate VII: Table showing Hall's Three Divisions of the Nervous System

TABLE OF THE NERVOUS SYSTEM.

- I. *The entire Nervous System is divisible into*
  - I. *The Cerebral.*
  - II. *The True-Spinal.*
  - III. *The Ganglionic.*
- II. *The Cerebral System is*
  - I. *The Seat of the ψυχη, or soul, and*
  - II. *The system of*
    - 1. *Sensation and of the Senses.*
    - 2. *Volition and of Spontaneous Motion.*
    - 3. *Sleep and Fatigue.*
- III. *The True Spinal System.*
  - I. *The Principle of Action is the Vis Nervosa.*
  - II. *Its Modes of Action are excited and reflex or direct.*
  - III. *The Reflex Functions are those of*
    - 1. *Ingestion and Retention.*
    - 2. *Egestion and Exclusion.*
  - IV. *The Direct Functions are*
    - 1. *The Tone* } *of the Muscular Fibre.*
    - 2. *The Irritability* } *of the Muscular Fibre.*
- IV. *The Ganglionic System.*
  - I. *The Principle of Action is the Vis Nervosa.*
  - II. *The Mode of Action excited, direct or immediate.*
  - III. *The Functions those of*
    - 1. *The Internal Muscular Organs,*
      - 1. *The Heart and Arteries.*
      - 2. *The Stomach and Intestines.*
    - 2. *Nutrition.*
    - 3. *Secretion, &c.*

Mixed Functions. (bracketing the Cerebral System, items 1–3, and the True Spinal System, items I–II)

Mixed, in the Œsophagus; the Rectum? (bracketing the True Spinal System, items III–IV, and the Ganglionic System, items I–III.1)

Hall, M., On the Diseases and Derangements of the Nervous System, 1841. Wellcome Institute Library, London.

At the time of Hall's birth the prevailing view on the structure of the spinal cord was that it consisted of a bundle of nerves running between the brain, from which all nerves were held to originate or in which they terminated, and the periphery. Both the German Johann August Unzer and, according to Hall, his pupil Prochaska, had shared this view. Hall distinguished between two regions of the spinal cord. On the one hand was outlined that portion, the white matter of the cord, described as the intervertebral cord of the sentient and voluntary nerves linking the periphery with the cerebrum and cerebellum. On the other hand was a quite separate entity, the true spinal marrow or grey matter which was continuous with the medulla oblongata of the hind brain and which was associated with the proposed system of excito-motory nerves and hence with reflex action. Speaking of the spinal cord in vertebrates Hall summed up his description,

> The entire medulla spinalis in the *Vertebrata* consists then of *two* portions, so intimately blended together, indeed as not to be easily separated by the anatomist, and, perhaps, only to be distinguished by physiological experiment and pathological observations.[74]

Anatomy, experiment and observation of pathological conditions in patients were, therefore, seen as sources of evidence to support Hall's speculation.

In his small memoir *Idea of a New Anatomy of the Brain* (1811), Bell gave as one aim that of refuting the notion of individual nerves carrying impressions associated with both sensation and movement. It was first necessary to show that nerves are composed of filaments,

> ... the nerves which we trace in the body are not single nerves possessing various powers, but bundles of different nerves, whose filaments are united for the convenience of distribution, but which are distinct in office, as they are in origin from the brain.[75]

In 1824, *An Exposition of the Natural System of the Nerves of the Human Body* included a drawing of a nerve A made up of a bundle of distinct filaments B, with one of these filaments dissected out.[76] Bell did not yet know that the filaments B were themselves composed of many thread-like axons associated with neurones. The different roles of the nerve threads were then outlined. Even when the filamentous nature of nerves was accepted, until Bell's observations, a nerve such as the one he had described was thought to have all its threads alike and all equally capable of exciting a muscle or conveying a sensation.

> It is imagined that impressions, thus differing in kind, are carried along the nerves to the sensorium, and presented to the mind; and

> that the mind, by the same nerves which receive sensation, sends out the mandate of the will to the moving parts of the body.[77]

On the contrary, Bell claimed, each filament or tract of nervous matter has its peculiar endowment, independently of the others which are bound up along with it. Thus each filament retains throughout its whole length either the property of conveying sensation, when it is described as afferent, or of stimulating motion or secretion, when it is described as efferent.

Bell had also outlined 'a set of nerves, called vital nerves', described in 1811 as 'nerves, controlling secretly the operation of the body; ....'[78] These vital nerves, whose ganglia were referred to as 'knots', were,

> ... less strictly connected with the sensorium, or which have upon them knots, cutting off the source of sensation, and thereby excluding the vital motions from the government of the will.[79]

The work of Bell, whose notions of afferent and efferent nerves were crucial to his own concept of reflex action, was known to Hall. But there was not even the slightest hint of the genesis from Bell's research of his own idea of a discreet anatomical system, independent of the cerebrum, within the main nervous system. Nevertheless, his proposal could be seen as an extension of Bell's notion of separate nerve fibres possessing individual and different functions. Hall did not explicitly distinguish between nerves and filaments within nerves.

Hall's decidedly guarded recognition of Bell's stature was tied up with the different emphasis the two men paid to the pursuit and value of experimental physiology and anatomy, in revealing the structure and functioning of the nervous system. In stressing the importance of experimental work to his physiological discoveries Hall declared that it was the results of these investigations which led him to infer the existence of a separate system for reflex actions. By contrast, Bell in 1824 claimed that his conception of the functioning of the nervous system arose by inference from its anatomical structure. For him 'the few experiments which have been made, were directed only to the verification of the fundamental principles on which the system is founded'.[80] In a later edition of this same work, Bell underlined this with the even stronger assertion that if his older brother John, who had initiated Charles himself into a study of the nervous system, had started with experiments 'they would have misled him'.[81]

But earlier in his life Bell had definitely carried out experiments, even though he had by the late 1820s abandoned them. A paper to the Royal Society in 1821 actually refers to his experiments, 'On the

Nerves; giving an account of some experiments on their structure and functions, which lead to a new arrangement of the system.'[82] Procedural details include '... on cutting the respiratory nerve on one side of the face of a monkey ....' and 'The respiratory nerve being cut across in a terrier; the side of the face was deprived of all expression, ...' Bell also acknowledged that these experiments did involve suffering, with his observation '... although, as I have said, touching the nerve gave great pain'.[83]

Bell was as keen as Marshall Hall to stake his claim to priority of discovery on the nervous system. Striking out at his arch-rival Magendie whom he does not name, for his practice of vivisection and at what Bell considered the *ad hoc* nature of French experimental physiology, Bell bitterly complained that,

> In France, where an attempt has been made to deprive me of the originality of these discoveries, experiments without number and without mercy have been made on living animals; not under the direction of anatomical knowledge, or the guidance of just induction, but conducted with cruelty and indifference, in hope to catch at some of the accidental facts of a system, which it is evident, the experimenters did not fully comprehend.[84]

This concern of Bell about recognition for his work associated with discerning the nature of the sensory and motor nature of the spinal nerve roots was as intense as that of Hall towards his own work. It had driven Bell to publish his work earlier than he had intended. Later in his life he altered editions of earlier work. Even though he had already spent years of patient investigation on the subject, Bell still felt that the work was in an imperfect state. In 1824 he admitted, for instance, in his most recent series of lectures, that new facts and suggestions had arisen which he would have liked to incorporate in his latest volume.[85]

Hall may well have derived from Bell some impetus for his idea of a separate system associated with reflex action. His ideas on the structure of a distinct sub-system for reflex action evolved and were expressed in a sequence of research papers and other publications as was the new nomenclature which he used but which did not survive. By 1836 the terms *excitor* and *reflex motor* had been adopted for the afferent and efferent nerves associated with reflex action and linked to what was described as 'the true spinal marrow.'[86] The system thus consists of,

> ... a series of incident *excitor nerves* and of *reflex motor nerves*, which, with the true spinal marrow as their centre or axis, constitute the *true spinal system*, as distinguished from the cerebral.[87]

An alternative name to 'true spinal system' was 'excito-motory system'. The term 'excitor' was intended to emphasize the notion that reflex action did not depend upon sensation. In 1850, when Hall delivered the Croonian Lectures before the Royal College of Physicians, the audience was informed that the system had yet again been re-named; '... the spinal system may henceforth be designated – The Diastaltic Nervous System, – ....'[88] The term diastaltic, previously used in music, was suggested by a friend Mr Hoblyn 'as congeneric with *peristaltic*' and would, Hall hoped, underline the distinctiveness of the system.[89] According to the *OED,* Hall might have used the term diastaltic to indicate acting through or along but he was more likely to be using it in another sense to signify separate or distinct.

Despite his own firm belief in a separate system Hall still needed anatomical supporting evidence to convince his critics of its existence. It was acknowledged that most of the nerves associated with the spinal cord were bound in the same sheaths: 'Sentient and voluntary nerves are blended with the excitor and motor nerves ....'[90] The greater complexity of the nervous system in higher animals made it less likely that they as a group would provide clear unequivocal evidence. He wondered, though, if there existed a group of lower animals in which the functional components of the nervous system were both physiologically and anatomically simple and distinct,

> Is there, in any class of animals a distinct *Anatomy* of the excito-motory power? Are there excitor nerves distinct from nerves of sensation? Are there motor nerves distinct from nerves of volition?[91]

The *Articulata* (*Arthropoda*) in which the spinal cord is not tightly constrained within a rigid vertebral column as in vertebrates had been examined. In the lobster it was found that as in vertebrates the ganglionic nerves are incident and excitor, that is, associated with the many reflex actions of the animal; and the nervous columns or cords transmit influences in both direct and retrograde directions. At the same time, the aganglionic nerves are, as his friend Professor Grant of University College London had speculated, purely motor and direct in their mode of action. It would be in the lowest forms of the *Invertebrata,* 'in which sensation and volition are nearly extinct' by which Hall meant exist only at a low level, and where the 'animal lives a sort of excito-motory life only'[92] that one should expect to have the clearest indication of a distinct excito-motory system.

Even in the vertebrates themselves some anatomical evidence of a binary system was discerned since the olfactory, optic and acoustic (auditory) nerves 'are nerves of *sense* only and destitute of excito-motory power'.[93] Evidence of at least one nerve which is exclusively voluntary, that is a nerve which conveys the acts of the will but possesses no motor or excito-motory power, was sought. The nerve supplying the *levator palpebrae* muscles of the eyelid was considered to be just this nerve, based on the fact that all other skeletal muscles require *tone* which is maintained by the excito-motory system, even during sleep.[94] Hall was not right about the lack of tone during sleep in either the *levator palpebrae* or the four recti eye muscles whose nerves it was thought might also be purely voluntary. As for a corresponding purely excitor nerve, it was likely that one probably did not exist absolutely in health although the pneumogastric nerve was reckoned to be the least sentient and the most purely excitor of any nerves in vertebrates.[95]

Evidence from pathology had long been used in support of a range of speculations about the physiology of the nervous system. In certain diseases, such as those involving the brain and where sensibility of the face is destroyed, the sentient power has been annihilated while the excito-motory continues. In such cases, eyelashes and nostril may remain as susceptible as ever to stimuli. Deliberate experimental destruction of the cerebrum leads to cessation of the functions of the sentient and voluntary nerves while leaving the excito-motory functions intact. There was no speculation on the effect of deliberate destruction of the spinal cord while leaving the brain intact. Nor would there have been because an animal being thus treated would have been subjected to intense and cruel pain.

Whereas gross anatomical knowledge was readily available, the lack of fine anatomical or histological knowledge was a major problem. While he did not carry out extensive microscopic studies himself, Hall believed his other investigations and interpretations would stimulate further enquiry into the fine structures involved, in reflex action,

> It appears probable that the facts of this paper may lead to some important additions to our knowledge of anatomy by inducing an accurate inquiry into the origin, course, connection, and distribution of the subcutaneous, or submucous and muscular nerves which constitute the arcs of the reflex function.[96]

His early work on capillaries showed that Hall was not antipathetic to histology. He believed, for instance, that it was the height of folly to try to understand the diseases of the nervous system without an intimate knowledge of its anatomy as well as of its physiology and pathology. Neither microscopes nor histological techniques were really adequate until much later in the century for discerning and confirming the nature of the neurone and the synapses or connections between neurones within the brain, spinal cord and ganglia. These two concepts were crucial to the development of the contemporary model of a simple reflex arc. The findings of experimental work were definitely ahead of those provided by the microscope.

Although the histological nature of the central link within the spinal cord remained a mystery until just after his death, Hall had always emphasized the key role in reflex actions of these central anatomical structures,

> The anatomy of the reflex function must be particularly studied in the medulla oblongata in the portions of the spinal marrow which give origin to the nerves of the anterior and posterior extremities; in that of the intermediate space which supplies the trunk; in that which is the source of the nerves of the sphincters, the ejaculators, etc., and in that which supplies the tail ....[97]

Hall's frustration at the inadequacy of microscopes and histological techniques generally was echoed in the *British and Foreign Medical Review,* whose editor Carpenter came to support Hall's concept of a separate excito-motory system. He sympathized with Hall that, 'the microscope and the galvanometer have alike proved useless in the most skilled hands'.[98]

The new compound microscope greatly improved matters, especially through the work of Fontana, Fodera and Milne Edwards on the Continent. Dutrochet had made the first observation of nerve cells which he described as 'corpuscles globuleux' and 'corpuscles neurveux' in his work of 1824[99] and the cell theory was put forward in 1838, but Hall made no mention of it. By March 1858, more than two decades after Hall had started his investigations on the nervous system and less than a year after his death, Rudolf Virchow in a series of 20 lectures on cellular pathology delivered at the Pathological Institute of Berlin, included two sessions on the fine structure of the nervous system.[100] Referring to the earlier work of Wagner which showed that nerve fibres branched throughout their length, Virchow had appreciated the implications of this structure. Combining this knowledge with that of the uni-, bi- or multi-polar structure of neurones, some of which originated in the grey matter of the cord rather than exclusively in the brain as demanded by the old theory, Virchow expressed his opinion that: *'Every special function possesses its special elementary, cellular organs; every mode of conduction finds paths distinctly traced out for it.'*[101]

Much had happened in the few years after Hall left the field of neurophysiology in the early 1850s. With Virchow's knowledge at his disposal, Hall would have had to accept the redundancy of his concept of a distinct anatomical nervous system in favour of an alternative one which saw within the nervous system as a whole, distinct pathways as opposed to distinct structures associated exclusively with reflex actions.

The Role of Sensation in Reflex Action

An equally significant and highly contentious feature of Hall's concept of reflex action, was its complete exclusion of sensation and volition from playing any part. His *New Memoir* (1843) contains a restatement of an earlier objective '... to separate the reflex actions from any movements resulting from *sensation* and *volition*'.[102] That he had met with some success is reflected in the section on reflex action in George Henry Lewes' *The Physical Basis of Mind* (1878). This literary critic and author attributed to Hall the 'modern form'[103] of the reflex theory and claimed that 'The peculiarity of the Reflex Theory is its exclusion of Sensibility from the actions classed as reflex; ....'[104] Neuromuscular activities were held to be generically alike in all animals because of the fundamental similarity in the structures involved.

The task of analysing and evaluating Hall's assertion on this issue involves paying attention to the meaning Hall and others attached to such terms as sensation, feeling, perception and consciousness. It also requires information on events taking place within the central nervous system where ingoing fibres are connected with outgoing fibres. Insight into these issues could elucidate mid-nineteenth-century answers to such pertinent questions as: Does a decapitated frog really *feel* when its foot is pinched and it struggles to withdraw it? Does the decapitated frog *will* the movements it makes when it swims?

Another significant factor is that of recognizing the different organs and structures with which sensation has over time been associated. In *A Dissertation on the Sensible and Irritable Parts of Animals,*[105] Haller explained that certain parts of the body are endowed with sensibility or as Whytt expressed it, 'feeling'. Both men associated presence or absence of sensibility with presence or absence of nerves. But for Whytt the manifestation of reflex actions was also due to the presence of the sentient principle throughout the neuromuscular system. For Haller the *vis nervosa* acting through the nervous system was responsible for them. For Hall, reflex actions depended on a special power located within the spinal medulla and which acted through the agency of a 'principle of action' which Hall believed and showed in his introduction of *New Memoir* for example, to be the *vis nervosa* of Haller.

A useful glossary of the meanings of some of the significant terms which attempt to define certain abstract qualities of the nervous system appeared in Le Gros Clark's *The Practical Anatomy and Experimental Physiology of the Nervous System* (1836). Unexceptionally, volition is given as meaning the power of willing an action and it was applied to the control exercised by the will over muscles which are capable of being so influenced.[106] Such power ceases the moment the cerebrum is destroyed. *Sensation* is defined as *perception* through the medium of the senses. In adding that current ideas of sensation should not be confined to this definition in its common meaning, Clark revealed his affinity with Whytt's philosophy. He said of sensation,

> ... though diffused in a more or less perfect degree over the surface of the body, the sense of touch – by which is meant the capability of distinguishing between hot, cold, rough, smooth ..., we must include all impressions of which the *sensorium* is capable of being made cognizant; ....[107]

Hall's proposition regarding the absence of sensation from reflex actions first appeared as a simple, unsubstantiated statement in his 1832 paper to the Zoological Society, but it was one of the main issues in his *Memoir* (1833). Actions dependent on reflex function 'have been erroneously ascribed to sensation and volition or instinct'.[108] More assertive statements appeared in later publications. For example, in *Memoir* (1837) his first objective was 'the *distinction* of the excitomotory property from sensation, volition, instinct, and all the functions of the cerebrum, or of the ψνχη, or mind'.[109] Hall had been led to this view from his experiments on decapitated animals and isolated body segments. On being touched, the limbs of the decapitated animals moved as in the living animal. But this motion ceased on destruction of the spinal marrow tissue within the attached vertebrae.

The confusion between reflex function and sensation and volition was analogous to the confusion which had earlier surrounded the concept of irritability. The facts relative to irritability were well known to Harvey and others long before Glisson and Haller finally separated this principle of motion from sensation and volition with which, like reflex function, it had previously been confounded. And Hall reminded his readers that no one attempted to deprive Glisson and Haller of the merit of having first pointed out a distinct principle of muscular action. Two exceptions were allowed to the assertion that no one had dissociated reflex actions from sensation and volition. These were Blane, who in his contribution to *Select Dissertations on Several Subjects of Medical Science* (1822),[110] had certainly seemed to exclude sensation from the reflex actions and Van Deen who had stated that volition plays no part.[111]

If an animal can be placed in such circumstances that voluntary and respiratory movements will certainly not take place but yet the animal retains the power of moving then, Hall argued, it may be concluded that volition and the motive influence or respiration are annihilated. This condition is produced by removing the cerebrum and medulla oblongata. Such an experiment had already been carried out by Legallois and Flourens and repeated by Hall himself on a range of animals including a snake (*Coluber natrix*), viper (*Vipera berus*), the toad (*Bufo vulgaris*), frog (*Rana tempororia*) and the eft (*Triton cristetus*).

Observations on a lively snake were described in some detail. The animal had been divided between the second and third vertebrae and it immediately lay motionless except for occasional slight head movements including gasping. After a time the creature was stimulated; it began immediately to move vigorously and continued to do so for some time. These movements were explained on the basis of each change of position bringing some fresh part of the surface of the animal into contact with the table or other objects which thereby caused renewed stimulation of the creature. Eventually the animal quietened down and, being carefully protected from all external impressions, it died in the precise position and form which it had last assumed.[112]

Hall refined his technique and using a soft substance such as cotton wool, was able to arrest the animal's movements, induce a state of quiescence and record the outline of this state on paper. Before death, the slightest stimulus applied to the animal would renew excited movements in it. But unless externally stimulated these surgically treated animals would remain quiescent until they died.

Hall argued that his proposition that movement in these animals did not depend upon sensation, was supported by the fact that both the final position which the animal might be in at the onset of quiescence and the stimuli which might be applied, could be such as would be accompanied by extreme or continued pain if sensibility were undestroyed. Thus in one experiment it was found that an animal remained partially suspended over the acute edge of a table. In others the infliction of punctures and the application of a lighted taper did not prevent a potentially lively animal from passing into a state of complete and permanent quiescence.[113] Hall's implication was that in an intact animal such mutilation would be accompanied by prolonged signs of pain in the form of motion.

The absence of spontaneous motions in decapitated animals was stressed as proof of the absence of volition. This absence removed all evidence of sensibility in excited motion in such animals and indeed positively excluded its influence. Opponents such as Volkmann claimed that decapitated creatures are capable of manifesting spontaneous motions[114] and Hall repeated one of the experiments in which Volkmann claimed to have observed spontaneous motion. It involved the division of the spinal marrow in a frog which was followed by the immediate extension of the two hind legs. The position of the frog was soon seen to have changed such that the legs were drawn up to the normal sitting position. Volkmann admitted that no further movements took place. Hall explained these observations as attributable to the frog being initially in a state of shock, during which period the excito-motor power and reflex actions are suspended for a time and no change occurred. After a short interval the excito-motor power with its phenomena returned. The extended limbs were stimulated by the firm pressure against the table and an excited action, involving the retraction of these limbs to their normal sitting position, occurred.[115] But no further new motion took place.

Earlier and Contemporary Views on Sensation

Whytt had already attempted to explain how it was that decapitated animals could respond to stimuli. Like Hall he had observed that unless stimulated such creatures remain in a quiescent state. For Whytt the explanation lay in his concept of the sentient principle, by which some have claimed that Whytt meant the soul. The sentient principle was seen as extending beyond the brain into the spinal cord and nerves and into the muscles themselves. It was this principle within the nervous system which transmitted the effects of stimulation to muscle. Muscle for its part reacted through the agency of the same sentient principle. The sentient principle also enabled muscle, when stimulated directly, to react. Whytt was aware that response following stimulation of the skin of a decapitated animal depended on the presence of the spinal cord, and that even if the trunk were cut in half responses could be provoked in both sections provided the corresponding portion of the spinal cord remained intact.

Whytt's writings reveal his dilemma as to whether or not an animal deprived of its brain is aware of stimuli applied to its body, in the same way that an intact animal is,

> We have no other way to satisfy ourselves that an animal is alive, or endued with feeling, but by observing, whether it shows an uneasiness when anything hurts or tends to destroy any of its parts, and an endeavour to remove or avoid it.[116]

Since the bodies of a variety of decapitated creatures make the same movements, even if they suffer the further process of being eviscerated, as they would if intact, we are, Whytt believed, naturally led to conclude that they are still in some sense alive and endowed with feeling.[117] By feeling Whytt meant 'animated by a sentient principle'. The further fact that the muscular parts of these creatures move after being cut into pieces and are still sensible of punctures, means that they continue still to be animated. 'The motions of the separated parts of animals are owing to the soul or sentient principle still continuing to act in them.'[118] Whytt acknowledged the objection which was commonly levelled against this belief,

> It has been objected that, the separated members of animals, though they continue to move for some time, are nevertheless neither animated nor endued with any kind of sensibility, otherwise the animal to which those members belonged ought to feel pain when they are irritated.[119]

This objection was dealt with in Whytt's *Physiological Essays* and recapitulated in his *Essay on the Vital and Other Involuntary Motions.* It is clear that Whytt used the term 'feeling' in more than one sense and endowed with feeling all those parts in which, for him, the sentient principle existed. 'Simple sensation' which was possessed by simple organisms, did not involve the brain and was, furthermore, not attended by reflex consciousness which by implication Whytt associated with sensation in higher animals. According to Whytt,

> ... since the nerves, which are continuations of the medullary substance of the brain and spinal marrow, have been proved to retain their powers of feeling, and of putting the muscles in motion, for some time after their connection with the brain is cut off, we have reason to believe that the motions excited by *stimuli* in the muscles of animals, after they are separated from their bodies are owing to some kind of feeling or simple sensation (such as oysters or other animals of the lowest class, who have no brain, are endued with) in those muscles or their nerves, which, though not attended with any reflex consciousness, a power the soul only exercises in the brain ....[120]

It was just this simple sensation which remained in decerebrate creatures. Although Whytt insisted that muscle responses provoked by a stimulus simply could not be explained unless one acknowledged some kind of feeling, it is clear that he realized that much remained unexplained regarding the nature of feeling and the soul. This was due to our ignorance for,

> ... if feeling be not a property of matter, but owing to a superior principle, it must follow, that the motions of the heart, and other muscles of animals, after being separated from their bodies, are to be ascribed to this principle; and that any difficulties which may appear in this matter are owing to our ignorance of the nature of the soul, of the manner of the existence, and of its wonderful union with, and action upon the body.[121]

This theme of more than one kind of sensation – simple sensation unaccompanied by consciousness and sensation accompanied by consciousness – recurred after Whytt's time and throughout the nineteenth century. Thus it was in essence the position taken up by many including Müller, Claude Bernard writing at the time of Hall's death, and by Lewes who, a decade later, criticized the narrow definitions of sensation and consciousness which Hall had used. Hall dismissed contemporary discussions about sensation with and sensation without consciousness, claiming that the phrase was a contradiction in terms and that the participants in this discussion had mystified the subject.[122]

In 1824 Flourens had published the results of experimental investigations of the relation between the nervous system and the phenomena of sensation and movement. Flourens had worked on pigeons and had concentrated on investigating the effects of removal of regions of the brain. He had pondered the separate nature and the separate location within the brain of the two properties of *sentir* and *mouvoir,*

> Mes expériences montrent, de la manière la plus formelle, qu'il y a deux propriétes essentiallement diverses dans le système nerveux, l'une de *sentir,* l'autre de *mouvoir;* que ces deux propriétes different de siège comme d'éffet; et qu'une limite précise separé les organes de l'une des organes de l'autre.[123]

According to Le Gros Clark, it is to Flourens that we are indebted for having determined that the cerebrum is the source of sensation and volition, to which he added perception. These properties cease on destruction or removal of the cerebrum. However, others including Calmeil, had earlier granted that while the brain was, in the intact animal, the seat of sensation, in a decerebrate animal the cord was capable of taking over certain functions.[124]

In the fourth edition of his *Outlines of Human Physiology* (1837) Mayo added, following the appendix, a final and unnumbered page of text in which he referred to Marshall Hall. Hall was accused of investigating, under the name of 'reflex function of the spinal cord..., a principle, which was explicitly laid down in my *Anatomical Commentaries* published in 1823 ....'[125] Mayo based his claim on a passage which began,

> An influence may be propagated from the sentient nerves of a part to their correspondent nerves of motion, *through the intervention of that part alone of the nervous centre, to which they are actually attached.*[126]

It was further claimed that while Hall had added some new experimental evidence to what was already known about this principle, he was not entitled to claim in 1833 that 'the independent power of the spinal cord had never been accurately distinguished from sensation and volition'.[127] Mayo asserted that he himself had already done this in both the earlier and the present editions of *Outlines of Human Physiology.* It seems clear that through what could only have been a last minute addition to his current publication, Mayo's explicit statement against Hall had been provoked by his own recent reading, as a referee at the Royal Society, of Hall's (1837) unpublished *Memoir* which with Lawrence, he was jointly responsible for rejecting from *Philosophical Transactions.*

The fourth edition of Mayo's *Outlines of Human Physiology* must have appeared just before Hall himself published his *Memoir* (1837) in which he referred to Mayo's 'useful work on Physiology'. The note about himself required, he felt, a short reply, to the effect that Mayo must surely have a treacherous memory. Of Mayo's claim to have dissociated reflex functions from sensation and volition Hall asserted,

> He has not only not '*carefully distinguished* this view from the agency of sensation and volition', but he has, in several paragraphs, *explicitly* confounded them together.[128]

Mayo's confusion was illustrated by a quotation from the third and fourth editions of *Outlines of Human Physiology* which referred to the 'mechanism of sensation' possessed by each segment of the spinal cord, of sentient organs and nerves as well as of voluntary nerves.[129] Thus Hall's allegation about Mayo was justified. Alexander Walker in *Documents and Dates of Modern Discoveries in the Nervous System* (1839) also thought that Mayo was not at all clear about the subject and that his use of the term 'voluntary nerves' was probably an oversight. He found inexplicable Mayo's claim to have distinguished reflex function from sensation and volition.[130] At the same time, in a letter to the Council of the Royal Society,[131] Grainger, who ironically was also a target for Hall's criticism, went into print supporting Hall against Mayo.

While Mayo was undoubtedly an acknowledged authority on the nervous system he was in this instance, in Grainger's view, in error,

> The views advanced by Mr Mayo require a more detailed notice, because his authority on all questions relative to the nervous system is highly and justly esteemed; and particularly because in the last edition of his *Outlines of Physiology*, he has assumed the credit of having established the principle on which the phenomena above stated depend; while that work, as well as his *Anatomical Commentaries*, contains statements calculated, as I conceive, to perpetuate the errors which have so long and so generally prevailed.[132]

From careful analysis Grainger concluded that for Mayo the reflex action of the spinal cord was certainly dependent on sensation and volition. This, of course, was Hall's view of Mayo's position.

In April 1838 the *British and Foreign Medical Review* produced a lengthy review of Hall's *Memoirs* (1837), Grainger's *Observations on the Structure and Functions of the Spinal Cord* and Mayo's *Powers of the Roots of the Nerves, in Health and in Disease* all of which had been published in the previous year.[133] The review discussed the role of sensation in reflex action and the problems involved in defining the terms used. The difficulty of describing the properties of sensation as we do those of bile or urine, even though the production of all three is dependent on the functioning of the nervous system, was emphasized. Concerning sensation in lower forms of animal life the *British and Foreign Medical Review* reviewer believed that 'It is impossible to determine what degree of sensation and consciousness such beings possess.'[134] Regarding the role of sensation in the processes of ingestion and egestion in these creatures, he opined that such processes do not necessarily involve sensation as a condition of their performance.

Turning to higher vertebrates which have been deprived of their cerebrum and cerebellum, it was noted that certain aspects of the behaviour of animals thus treated had been interpreted as indicating the persistence of sensibility in the spinal cord. However, this conclusion had been greatly invalidated by other experiments – and Hall's work must have been included here – which led to an opposite inference,

> It must be recollected that the doctrine of the existence of sensibility in the spinal cord is founded only on vague inference from such experiments as those of Flourens, who was himself led by them to an opposite conclusion; and we think that the inference of its

> non-existence, founded upon the experiments we have just quoted, is at least as legitimate.[135]

Details of Grainger's own experiments together with his interpretation were presented. The data included the results of simple division of the spinal cord of a salamander. Stimulation of the feet and tail produced withdrawal reactions, all of which seemed to indicate the persistence of sensation and volition. However, Grainger had noted that in spite of their reaction to mechanical irritation the hind limbs and tail remained motionless under the impulse of the will. This inference concerning the will had been based on the fact that when the creature attempted to walk, which it frequently did, the hind legs and tail were dragged along completely paralysed. Grainger had concluded that the power of the will had been destroyed by the sectioning of the spinal cord. The fact that sensation too was absent, Grainger had claimed to demonstrate by his observation that if one proceeded so carefully that the animal did not see the approaching hand, it was possible to cut off the limb of a salamander without the creature moving or giving any other expression of suffering.[136]

Müller, like Hall, believed that the reflected motions following stimulation of the skin of a decerebrate animal, do not provide proof that the stimulus has excited true sensation in the spinal marrow, because the nervous principle can no longer be conducted to the brain, the organ of consciousness. However, between Hall and Müller there was an area of dispute on this issue. In Müller's opinion Marshall Hall went too far in claiming that in health 'every motion on true sensation is induced by the will, and that all excitation of sensitive parts in the reflected motions are without sensation'. As Müller pointed out, many reflected motions such as sneezing and coughing actually follow sensations,

> The view I take of the matter is the following:- irritation of sensitive fibres of a spinal nerve excites primarily a centripetal action of the nervous principle conveying the impression to the spinal cord; if the centripetal action can then be continued to the *sensorium commune,* a true sensation is the result; if, on account of division of the cord, it cannot be communicated to the *sensorium,* it still exerts its whole influence upon the cord; in both cases, a reflex motor action may be the result. In the first case the centripetal action excites, at the same time, sensation; in the latter case it does not, but is still adequate to the production of reflex motion, or centrifugal reflection.[137]

Hall clearly heeded Müller's criticism and by the time his *New Memoir* (1843) appeared, he had modified his views. It was now acknowledged that sensation and volition could be involved. Nevertheless, he insisted that they were not essential. Almost as if he had never expressed his earlier contrary view, he now wrote,

> I believe but few persons now require to be convinced that in the reflex actions, whether experimental or physiological, neither sensation nor volition bear any essential part. Sensation may accompany, and volition may modify, some of the physiological reflex acts; but, as I have stated, neither is essential.[138]

When in 1850 he came to give his first Croonian Lecture, his final major lecture on reflex action in London, Hall's thinking had undergone further significant modification. The role of emotion in the execution of reflex actions or as they were now called, diastaltic acts, was included. Under the heading '*Sensation; Volition; Emotion*', there appeared, 'Some diastaltic Acts, as that of *deglutition*, are entirely unattended by sensation. Others are accompanied by this psychical act. The association is not necessary.'[139] Having admitted that volition can modify many diastaltic acts such as respiration, Hall attributed an even stronger influence to the emotions,

> Every diastaltic Act admits of modification by emotion, as, indeed, do all peristaltic movements and secretory processes. This effect is seen in Respiration, in the action of the heart, in the secretion of saliva, etc. The influence of emotion is, indeed, both diffusive and extreme.[140]

There was one very persuasive piece of evidence claimed by Hall's opponents, including George Paton who was highly critical of Hall's mechanistic interpretation of reflex action,[141] to support their belief that reflex actions were accompanied by sensation, which could be associated, therefore, with the spinal cord. This was the assertion that no difference could be discerned in the movements provoked by stimulation of decerebrate animals compared to intact animals. Whytt had also made this observation in his essay on the *Vital and Other Involuntary Motions of Animals.* The implication was that movements in a decapitated animal were just as integrated and just as purposive as in an intact one.

In interpreting the behaviour of decapitated animals, the notion of purposiveness has been an important one. Its manifestation in an animal's behaviour was held to imply that the animal had been made aware of, had perceived, the stimuli reaching it and had responded appropriately. Paton was not alone in holding that no difference could be discerned in the nature of the movements carried out by stimulation of intact or decerebrate frogs. Mayo too in his *Anatomical Commentaries* (1822) claimed to have found that where the foot of an operated animal was stimulated, the movement of the limb was exactly similar to that which the animal would make if in undisputed possession of its sensation.[142] Grainger on the other hand disputed this claim, also on experimental grounds. He pricked the foot of a rabbit both before and following section of the spinal cord and found that the limb movements in the two cases were not at all identical.[143]

Hall's mechanistic explanation of reflex action included his conception of the spinal cord as being programmed to execute the motions necessary for the survival of the organism. In his *New Memoir* (1843) he claimed that,

> Every physiological act of the reflex excito-motor power is obviously designed; the act of deglutition, the act of inspiration, the closure of the eyelids, of the larynx, the action of the sphincters; - all is replete with obvious design.[144]

Hall attributed this design to an omniscient Creator.

The view of Nasse, Volkmann and others was that the appearance of design observed during some experiments on reflex action indicated a mental operation. This was dismissed by Hall.[145] Writing in 1836, Le Gros Clark indicated one reason for the delay by English physiology in undertaking serious investigation of the association between mental processes and the nervous system,

> The more purely metaphysical inquiry concerning the actual relation and connection of the material with the immaterial part of our nature is, of course, no business of the physiologist.[146]

Not only was there resistance at the time to developing experimental physiology in a more rigorously scientific manner, there was also resistance to opening up and developing the new field of psychology.

Evidence from Pathological Conditions of the Nervous System

In the 1838 review of the work of Hall, Grainger and Mayo the *British and Foreign Medical Review* had observed that interesting though Grainger's experiments had been, and probable though the conclusions concerning the exclusion of sensation and volition were, it was still felt that the experiments did not yield the really satisfactory evidence which could only be provided by the human subject with its power of speech. Positive evidence was available that reflex motions,

> ... may be excited without consciousness on the part of the individual, by the experiments which nature performs on the human subject; the effects of disease or injury replacing in these instances the scalpel of the vivisector.[147]

Delivering a rebuff at Hall, the reviewer quoted some of the most striking experiments from Grainger 'by whom they appear to have been more discerningly executed than the similar ones detailed by Dr M. Hall'.[148]

In *Essay on the Circulation of the Blood* (1831), Hall had drawn attention to the value of the various cases of monstrosity which a physician encounters. They were described '... as a sort of natural experiment'. Their real value lay in the fact that in general they presented less equivocal phenomena than any which result from an actual operation involving the infliction of violence or of pain. In addition the results of disease and accident also afford examples of undesigned experiments.[149]

The *British and Foreign Medical Review* alluded to a case of paraplegia in which irritation of the foot produced its retraction, without consciousness or will on the part of the patient.[150] The original citation was at the time a solitary instance but was followed by an even more striking incident in the *Edinburgh Medical and Surgical Journal*.[151] Such pathological conditions had been considered in Hall's *Memoir* (1837) when he posed the question,

> What is the condition of those muscles, in the human subject, from which volition is cut off, in disease of the cerebral lobes, or of the spinal marrow – in hemiplegia and paraplegia?[152]

Britain had already seen an increase in spinal injuries. These were due not only to an increase in road traffic with heavy carriages being responsible for many of them, but also to injuries in war and to the early nineteenth-century building boom and other industrial developments in urban centres such as London. Hall believed that he was one of the first to study the effects of such injuries. Such pathological conditions were considered in *Diseases and Derangements of the Nervous System* (1841),

> I believe we possessed no published facts of this kind before the attention of the profession was excited to them by my first publication on the subject ..., in 1832. Since that time, however, they have accumulated. Mr Barlow and Dr W. Budd each communicated to me a case of reflex actions in paraplegia, about the same time, in 1836. I was indebted for a third to Dr Elliot of Carlisle; a fourth I visited by the invitation of Mr Liston, at University College Hospital, and a fifth on board the Dreadnought ....[153]

Among the cases of paraplegia included in the work was one of priapism (abnormal erection of the penis) in a man who sustained injury of the spinal cord following a fall in a quarry. The victim had no consciousness of the condition of his penis unless 'he put his hand to the part, or looked at it'.[154] Details of the case had first been sent to Sir Benjamin Brodie by Dr Macartney of Dublin with whom Hall had checked the details of the case.

Hall included also a case cited by Dr Baly in his translation of Müller's *Elements of Physiology.* This concerned a woman of 50 in Baly's care at the St Pancras Infirmary. She had recently, through a stroke, been attacked with hemiplegia of the left side.[155] Nevertheless, pinching or even a slight touch on the sole of the foot or ankle of the paralysed left leg caused the limb to be retracted and the toes extended, the patient being unconscious both of the stimulation and of the movement. At the same time no effect was produced in the opposite leg although it was sensible to the slightest touch.[156] Hall's account of hemiplegia and paraplegia in *Memoir* (1837) noted that in these conditions, instead of the '*cerebro-spinal* life' with which the limbs were endowed before, a '*spinal* life' alone remained.[157] Hall also included the observation that.

> The paraplegic patient who, being asked whether he felt the puncture which made his limb move, replied – 'No, but my legs do!' – spoke a language which, however emphatic, was unphilosophical and unphysiological.[158]

Observations on the behaviour of anencephalous infants had already, in the previous century and earlier, been cited as evidence that involuntary actions were not dependent on the brain for their execution. Thus, Preston in 1697 had published in the *Philosophical Transactions* the case of an anencephalous infant which on delivery he had found to lack, as he described it, brain, cerebellum and medulla oblongata.[159] The infant survived for half an hour. Preston claimed that he had personal experience of three such cases. He reviewed the observations of Duverny and Chirac based on their experimental removal of cerebrum and cerebellum from pigeons and dogs. He noted that all their preparations had possessed an intact medulla spinalis. It was this part of the nervous system, Preston felt, which was essential for life. Fearing notes that Preston's paper can leave no doubt but that he fully appreciated the fundamental significance to vital function of the spinal cord. Thus, to some extent both Preston and Duverny partly anticipated Whytt. Of more immediate interest though, was Preston's statement that the pigeon on which Duverny had experimented 'had the use of sense ....'

Later accounts of anencephalous infants were provided by several writers including Lawrence, co-referee with Mayo of Hall's *Memoir*

(1837), Lallemand and Ollivier.[160] Lawrence's account appeared in the *Medico-Chirurgical Transactions*[161] and was, on the one hand, quoted by Marshall Hall and Le Gros Clark in support of the view that where the brain is absent, sensation is also absent and on the other hand by writers such as Mayo in support of the opposite view, that the brain is not necessary to maintain consciousness. Hall spoke of the anencephalous condition as 'displaying the interesting spectacle of the phenomena of the true spinal or excito-motory system, exclusively, in the human subject; ....'[162] He acknowledged that while the foetus may support an extra-uterine life for several hours, it is not viable since the influence of the brain is required in addition to that of the spinal marrow, for the continuance of life.

Lawrence's account was actually based partly on hearsay evidence. His friend Mr Armstrong of Islington sent to him the body of an anencephalous child which had been born on a Sunday and had lived until the following Thursday evening. The description of its behaviour was presumably relayed by Armstrong but it is not clear if the anatomical details were revealed as a result of Armstrong's dissection or if the specimen had been sent to Lawrence intact.

The brain and cranium were deficient but there existed a soft tumour about the size of the thumb end over the Foramen Magnum. The infant had apparently moved briskly at first but had remained quiet afterwards except when the tumour was pressed; this action produced general convulsions. The infant had breathed naturally and did not seem to lack body warmth until its powers declined. From a fear of alarming the mother, no attempt was made to test suckling and the child was given a little food by hand. There followed details in the paper which could only have been revealed by dissection. The tumour at the base of the skull was found to be the swollen upper end of the medulla spinalis, that is, the hind region of the medulla oblongata. According to Lawrence, '... brainless children are incapable of executing the functions necessary for the continuance of existence after birth.'[163] This accorded with Hall's view.

Hall speculated on the cause of death of the infant. He acknowledged that he 'really could not produce a sufficiently accurate history of its life to answer this question' and thought that it had probably died for want of food at the hands of the nurse who may have doubted whether any attempts ought to be made to prolong its life. Observations on similar neonates were reinforced by other cases in which difficult deliveries had necessitated rendering the infant anencephalous by removing the brain through the anterior fontanelle.

Hall's Research in Electrophysiology

Hall's final major research programme involving work on animals was that embodied in his memoir presented to and rejected by the Royal Society in 1847.[164] The main aim of this research was to throw more light on reflex function through working on the wider issue of nerve function but concentrating on one aspect of it, namely the nature of induced currents in nerves. This memoir could be seen as linked back to Hall's third paper submitted to the Royal Society 16 years earlier and dealing with the influence of increasing heat on physiological processes in 'batrachian reptiles'. In the present paper the agent under investigation is electricity and the physiological phenomenon, nerve function.

In 1780, a decade before Hall's birth, the work of Galvani was providing fresh impetus and throwing new light on the intriguing phenomenon of nerve function and the notion of animal electricity.[165] Repeating earlier experiments of individuals like Swammerdam as well as extending this repertoire with his own work, Galvani proposed a theory of animal electricity, or galvanism as it came to be called. He believed that nerve and muscle possessed an electric fluid probably secreted by the brain and distributed throughout the body via the nerves.

Galvani's findings were wrongly opposed by a fellow Italian, Volta who, working on the electric fish, denied the existence of animal electricity.[166] But ironically, aided by the findings of other contemporary experimentalists Volta's work actually led to consolidation rather than refutation of the concept of animal electricity or galvanism. It also resulted around 1800 in the invention by Volta of the artificial electric organ which he saw as analogous to the electric fish organ and which came to be called the Voltaic pile or cell. It was made up of alternating plates of unlike metals, such as silver and zinc separated by pieces of moistened cardboard. The Voltaic pile, which Hall used in his experiments, made available a stable flow of current electricity. The related Cruikshank trough was used when, as Hall put it, 'I wanted the more considerable power.'[167]

Faraday, who had in 1831 promoted research into electromagnetic induction, was among those early nineteenth-century scientists who also paid some attention to the biological aspects of electricity.[168] Not surprisingly, he too investigated the anatomy and physiology of the electric fish. However, along with Volta and also with Todd and Bowman in England, Faraday questioned whether the nervous agent was in fact identical with current electricity. He speculated that the responsible agent in the nervous system might be an unknown inorganic force. It is possible that it was Marshall Hall's belief in a hitherto unacknowledged and hence unknown force, the reflex function, which he was trying to relate to nerve function, which drew Faraday to his work, with the anticipation that he might find evidence in support of his own speculations.

In the early part of the nineteenth century, research in Italy had subsided for about three decades, until the appearance of work by Nobili (1784–1835) and Matteucci (1811–68) in the 1830s and 1840s.[169] During this period, while French physiologists were still preeminent in a number of fields their work in the biological domain of electrophysiology was greatly inferior to their physical science work on electricity. During the 1840s the focus of experimental physiology moved to Germany through the brilliant research of du Bois-Reymond. He was joined by Helmholtz who was then overtaken by his own pupil Hermann.[170]

Hall's work began against the background of Nobili's and Matteucci's work and the earlier work of Ritter,[171] an advocate of 'romantic biology' or Naturphilosophie. This philosophy related human physiology to the rest of the organic world and even to aspects of inorganic nature. Ritter believed that this interdependence was determined by a subterranean electric force. He was working on electrophysiology across the period preceding Galvani's death in 1798 to beyond Volta's invention of the pile in 1800. A number of significant findings were associated with Ritter but his advocacy of romantic biology cost him dearly in terms of serious recognition. Hall's respect for Ritter's work was applauded by du Bois-Reymond who also rediscovered him but who exaggerated his contribution in the field.[172]

Matteucci, whose research was reported in the French Academy of Sciences in 1842 was interested in the physical aspects of electricity and between the mid 1830s and 1850 was actively engaged in practical research on animal electricity using the electric fish, the common frog and, later, mammals. His numerous findings, mainly on muscle physiology, were also published in *Philosophical Transactions*[173] and were therefore available to Hall. They contained some research which was new and significant.

Matteucci had discovered a resting potential in muscle, that is a difference in potential between the inside and the surface of muscle. He had also demonstrated an injury current in cut muscle and had presented his observations and interpretation of induced muscle contraction, the phenomenon on which Hall was working. While being, like Hall, an excellent experimentalist he was handicapped by the lack of adequate instruments to detect and measure electric currents. But he was adept at instrument making and constructed apparatus including his preparation called the galvanoscopic frog for detecting electric currents in tissues. The galvanoscopic frog consisted of two separate nerve-muscle preparations which were connected by the muscle of the first preparation, whose nerve was stimulated by a voltaic cell, being cut and having inserted in it the nerve of the second preparation. When stimulation of the first nerve, from a Voltaic source for instance ceased, it was found that the second muscle contracted. This physiological effect came to be called 'induced contraction'. Hall was interested in the effect which must have been produced in the second nerve which itself had not been directly stimulated. This secondary effect was what he called the 'electrogenic' state of nerves.[174] Also, developing an original idea of Nobili, Matteucci constructed a biological battery composed of a 'pile' of frog nerve-muscle preparations.

But Matteucci, who paid generous tribute to fellow countryman Nobili, was curiously lacking in confidence about his own work and on more than one occasion retracted and denied his own observations and interpretations. By 1847, he had denied any association between muscle action and electricity leaving the way open for Hall to claim that while Matteucci had corroborated work on the effects of the Voltaic current on frog nerve and muscle tissue he himself was the first experimentalist to fully and accurately notice and describe the secondary, induced condition in nerve structures.

Hall had thus turned his attention to a phenomenon associated with nerve function even though he was not investigating the nature of nerve function *per se*. He claimed that he had tentatively begun his investigations 'many years ago' in connection with early work on the excito-motor property of muscular nerves and the property of irritability in muscles.[175] This early work was also associated with the clinical studies on 'various kinds of paralytic affections'. The investigations were, therefore, for Hall relevant to contemporary medicine. However, the present research was to concentrate on a study of muscular nerves; consideration of nerve function in the incident nerves and in the spinal marrow as well as explicit reference to the relevance of this research to pathological studies were to figure in later enquiries. Similarly investigations on other physical and chemical agents such as 'mechanical injury, heat or cold, strychnine or the hydrocyanic acid' would also have to wait until a later date.

Although nerve function was the main general issue, muscle contraction was used as the indicator of this function. The two physiological phenomena are of course closely related but Hall's critics in accusing him of failure to acknowledge the work of predecessors, such as Matteucci, did themselves on occasion fail to appreciate that these predecessors including Matteucci were concentrating on muscle rather than on nerve physiology. Indeed Matteucci failed to demonstrate electrical currents in nerves.[176]

Hall's paper gave meticulous details accompanied by equally detailed and elegant illustrations of an extensive series of investigations on a range of frog preparations in which frog nerve-muscle preparations were incorporated in electric circuits obtaining voltaic cells of varying strengths. There was explicit reference to working on animals in which the spinal cord had been divided near the cranium 'to annihilate sensation and to obviate all idea of the infliction of suffering ....'[177]

As well as variation in the strength and period of the stimulating current used, the metal electrodes were also varied – silver, copper, platinum – as was their position and their areas of contact on the nerves. Thus, it was possible to control the direction of the voltaic current within the nerves such that the current could be made to pass either in the direction towards (direct current) or away from (inverse current) the muscle. Hall was here questioning belief in the uni-directional transmission of the nerve force, an issue raised in his 1837 rejected memoir on the nervous system. In addition, the effect on muscle contraction of both the gradual desiccation of the nerve tissue and of wetting the nerves with water was observed. Further, the necessity of considering other variables such as 'The vigour, the sex of the animal, the season of the year etc.' was also appreciated. Hall added, 'So much more difficult are physiological than physical, experiments and investigations.'[178]

It was acknowledged that experiments investigating the electrogenic condition of nerves were extremely difficult and liable to error and that the reproducibility of the same phenomena in successive and apparently identical experiments was not always possible. The source of these discrepancies needed to be traced. As had already been noted by predecessors, it was remarked that on completing the electric current 'energetic muscular movements' ensued. Less widely noted was the observation that on breaking the current after 5, 10 or 15 minutes and on completing the circuit without the pile ('joining wires' as Hall expressed it), the effect instantly produced was of the 'most remarkable spasmodic contractions and tetanoid rigidly of the muscles of the limbs'.[179] These effects 'as instantly subsided on the restoration of the voltaic circuit and the reapplication of the Voltaic influence'.[180] Hall was clearly impressed by the difference in the effect on the muscles 'of making and breaking the voltaic circuit; and of connecting the wires before or after breaking that circuit (which was) quite extraordinary'.[181] Hall's own primary interest was in the effect produced on breaking the circuit whether he was using a single nerve-muscle preparation or Matteucci's galvanoscopic frog system.

Hall's work was an isolated instance of electrophysiological research on nerve function in England. But due to lack of support in the form of recognition and money, or more accurately, in the face of positive discouragement by the Royal Society, Hall lost heart and soon abandoned the work. Just as his efforts at confirmation or refutation of his concept of a separate system for reflex action were frustrated by inadequate technology in the form of poor microscopes and histological techniques, so in this new line of research Hall, an experimenter of great ingenuity and experimental skill, was also hampered by the lack of sufficiently sensitive instrumentation. This was absolutely essential for detecting and recording minute and brief electrical changes in nerve and muscle tissue.

Despite the important Italian work on nerve-muscle physiology in the early decades of the century and the interest of French physiologists, the centre of excellence of research in the field had, already in the 1840s, moved to Germany.[182] Emil du Bois-Reymond, a pupil of Müller, and his friend Helmholtz belonged to the 1847 group of materialistic physiologists who aimed to reduce all biological phenomena to the principles of physics and chemistry.[183] Du Bois-Reymond worked exclusively on electrophysiology and dismissed the work of his Italian predecessors including Matteucci. He introduced new equipment and especially a much needed, more sensitive galvanometer for measuring nerve currents. He claimed to have detected pre-existing currents in resting nerve and muscle tissue. However, du Bois-Reymond's work was overtaken by that of his pupil Ludimar Hermann who, to du Bois-Reymond's considerable annoyance, showed that the supposed resting current was in fact a current due to injury.[184] There was no such thing as a natural, pre-existing current in muscles and nerves if the tissues were perfectly uninjured. Although a British School of physiology rose to eminence in the final quarter of the nineteenth century[185] under Hall's intellectual successor Charles Sherrington, born in the year of Hall's death, it can hardly be claimed that Hall laid its foundation stone. But with Royal Society backing, Hall could have played a more significant role in its inception.

Controversies

During the years of his work on the nervous system, Hall became involved in a number of altercations of varying degrees of severity over his ideas on reflex action and in particular his views on sensation. Before he retired, Hall's work sparked off considerable discussion which culminated in a Continental debate between two German physiologists, Eduard Pflüger (1829–1910) and Rudolf Heinrich Lotze (1817–81), on whether or not a headless creature is also soulless. Hall had actually retired by the time the controversy really got under way. Also following his retirement and during travels in the United States of America, Hall became involved in a difference of opinion with two American physicians over the related issues of whether headless creatures possess sensation and behave purposefully. The combatants agreed to differ. The third but earlier controversy had been more significant, for it contained the serious charge of plagiarism against Marshall Hall.

Difference of Opinion in America

During his American tour and while at Charleston on 13 December 1853, Hall addressed a large audience of medical men, students and other scientific and literary gentlemen. Using alligators he conducted 'vivisections, with the object of demonstrating the truth of the theory of automatic or excito-motor actions, as distinguished from sensation and volition, after the division of the spinal cord'.[186] The reporter of the *Charleston Medical Journal and Review* produced details of Hall's experiments on a young alligator in which the cord had been divided at its junction with the brain. His demonstrations included one showing that the animal could be maintained in an uncomfortable position, which it would not have tolerated in the intact condition, because it had lost volition but not the power to move. This latter property he demonstrated by applying a lighted taper to the tail whereupon the animal writhed and moved from its former position.

Hall had travelled on to New Orleans to witness experiments on the physiology of crocodiles carried out by the two physicians, Cartwright and Dowler, who had already published accounts of their experiments in 1853. Their interests, Dowler's mainly in the nervous system and Cartwright's in the blood system, coincided with those of Hall. Their experiments included the resuscitation of a fairly extensively dissected crocodile, which had been allowed to sink to a condition of near death. The published account of their work indicates that the attitude to animal experimentation involving even large animals, in mid-nineteenth century America was more tolerant than that prevailing in England.

Cartwright began the first paper with, 'Four crocodiles were subjected to vivisection in the courtyard of my office, on the 1st and 6th of the present month. One was nearly ten feet long, another about six and a half feet and the other two of smaller size.'[187] Veritable dragons compared with the frogs and newts Hall usually worked on. Cartwright gave one of the crocodiles to Dowler to 'perform any experiment he might see proper'. The work of the two experimenters was watched by a large number of professional colleagues.

When the crocodiles were decapitated the beasts, according to Cartwright, continued to make 'intelligent' motions for some time afterwards and darted forth their headless necks at their persecutors. This proved 'that they are not only alive but retain the passion of anger without a head, about as well as with it'.[188] According to Cartwright, Dowler had repeatedly shown that,

> Sensibility, motion, the will and intelligence, continue to be manifested in the body after it has been cut off from the brain and spinal marrow; that pinching the distal portion of any divided nerve will cause motion and sensation in the part to which it is distributed ....[189]

Cartwright also believed that his own work refuted the hypothesis 'that the chief motive power of the blood is derived from the mechanical propulsion effected by the contraction of a muscle, called the heart'.[190]

The *Boston Medical and Surgical Journal* had thought that Hall supported their views. But neither proposition enunciated by Cartwright could have been further from Hall's own position on these two issues. Not surprisingly, Hall sent to the editor of the journal a copy of a letter he had sent to Dr Dowler on 5 October 1853, and the journal was left in no doubt as to the difference of opinion between Hall and the two Americans and that Hall's objective in going to New Orleans, at the invitation of Dowler, was very different from what they had first imagined. Hall wrote,

> The great question is, I believe, as to *spontaneous motion* after the removal of the cerebrum and the cerebellum. The *appearance of design* is, I believe, deceptive. In any case, if we do differ in opinion we must agree to differ.[191]

Acknowledging its new awareness of Hall's position, the journal published a synopsis of one of Hall's earlier lectures delivered in New York and published in the *New York Times*,

> The doctrine of the reflex action is Dr Hall's own .... He divided the spinal marrow of the frog. By this operation all sensibility was removed. According to the learned doctor's theory, it was no longer capable of feeling pain. It might squirm at will, or even cry out as a wounded frog may, but it can't be said to suffer any pain – it is incapable of being hurt.[192]

Cartwright's papers were nevertheless of special interest in that he appealed to the physiologists and psychologists of young America to rethink the philosophical basis of their subject. Cartwright appealed for a rejection of the prevailing Greek doctrine which supposed that life with all its attributes was located in the brain and nerves. He wanted this to be replaced by the Hebrew doctrine of Moses which held that life with its attributes of sensation, volition, mobility and intelligence, existed in the blood.[193] For Cartwright, the role of the nervous system was subordinate to that of the blood system which gained its life from the air without any assistance from the heart. He regretted the loss of certain clinical practices which followed from failure to appreciate this,

> ... the erroneous theory, that the nervous system is the fountain of life, has caused the old practice of bleeding, vomiting and purging for concussions of the brain and severe shocks of the body, to be abandoned, and the stimulating of the nervous system, to produce reaction, to be substituted for it.[194]

He was convinced that his own and Dowler's observations on the crocodile experiments were fully explicable on the basis of Mosaic physiology.

It could have been Cartwright's Mosaic interpretation of physiology together with his own religious interests, which re-kindled Marshall Hall's motivation to take up the study of Hebrew during a stay in Rome in the winter of 1854 to 1855. Of this pursuit which her husband took up so late in life, Charlotte Hall said,

> He had long had an earnest desire to read the Old Scriptures in the original, and, thinking that this was a good opportunity for study, engaged a Rabbi to give him a daily lesson. His Hebrew books and writing were now constantly beside him .... The Rabbi declared he had never ever had a pupil who made such rapid progress.[195]

The Pflüger-Lotze Controversy and its Background

Although Hall's final work on the nervous system appeared in 1850 his ideas on the role of sensation in reflex action figured in the discussion of later writers including Longet, Claude Bernard and Lewes. These views eventually contributed to the sparking of a controversy located mainly on the Continent and concerned with the issue of what was termed 'the Spinal Soul'. Writing three years after Hall's death, Longet observed,

> Si les physiologistes reconnaissent assez unanimement le cerveau proprement dit (*lobes cérébraux*) comme le siège exclusif des facultés intellectuelles, il y a dissidence entre eux sur la détermination de la portion encéphalique qu'on doit regarder comme le *siège de la faculté de perçevoir les impressions sensitives.*[196]

Flourens had claimed that in losing its cerebral lobes an animal loses the power of perception of all sensations. His former chief, Magendie, did not agree that the lobes were the unique site of perception. Longet did not concur with Flourens either and contended that many of the phenomena, in some cases those cited by Flourens himself, and observed in decerebrate animals with mid- and hind-brain intact, could only be explained if the animal had in fact *perceived* some sensations.[197]

Like others before and after him Longet based this assertion on his observations of the behaviour of a range of animals including young dogs, rabbits and birds from which he had removed the cerebral lobes. He noted that his preparations survived the operation for periods ranging from days to weeks. While they fell into a state of quiescence, which Hall would have attributed to shock following their operations, these preparations did at intervals, he claimed, rise from this condition either spontaneously which Hall would have denied, or in response to even light stimulation of the skin. Once aroused the animals could for a while manifest behaviour as in the intact state including attempts to move away from unpleasant stimuli such as ammonia vapour applied to the nostrils or pinching of the whiskers or of a foot.[198]

Longet referred to the 'acceptation rigoureusement metaphysique' of the term '*sensation*' as 'ne l'appliquant qu'à tous les cas d'exercice de la sensibilité *avec conscience*'.[199] He agreed that the seat of sensation in this sense was indeed as Flourens had claimed, the cerebral lobes which are also the seat of intelligence and volition. He further felt that perception of auditory, visual, olfactory and gustatory impressions probably required the presence of the cerebral lobes. He defined sensation as 'la réunion en un seul fait de trois faits élémentaires: l'impression, la transmission, la perception'.[200] But Longet was of the opinion that an animal deprived of its cerebral lobes is still endowed with 'la *perception* simple, en quelque sorte bête'.[201] He called this form of simple perception *sensibilité générale* and felt that, through the persistence of this property, an animal deprived of its lobes is still capable of perceiving the sensation of pain. Thus, an animal deprived of its cerebral lobes could still to some extent perceive sensory impressions, that is, it could still manifest sensation. He acknowledged that the sensations, lacking what he called 'l'élaboration intellectuelle'[202] for which the lobes were necessary, would be modified compared to the intact animal.

Claude Bernard, who as a young man had observed Longet and Magendie carrying out their experiments, was in his 'Leçons sur la Physiologie et la Pathologie du Système Nerveux', unequivocal in his assertion that 'la sensibilité', the property associated with the posterior spinal roots, could take two forms. As he expressed it, 'La sensibilité, qui des racines postérieures se transmet par la moëlle, peut revêtir deux formes: elle peut être sans conscience ou avec perception.'[203] Bernard then went on to distinguish between these two forms of 'sensibilité'. One was described as *sensibilité perçue* associated with voluntary action; to the other form he gave Magendie's term 'sensibilité sans conscience' and added 'ce qu'on a appelé depuis mouvement réflexe'.[204]

For his part Lewes defined his terms and then constructed his theory around them. In his view, and echoing Hall, while no one denied that most reflex actions often have conscious sensations preceding and accompanying them, the general view was that these sensations were not essential to the performance of the action, on the grounds that these conscious sensations may be absent and yet the actions still take place. This might be true, said Lewes, if the term consciousness is limited to one of the special modes of sensibility, or if the term sensation is restricted to this limited meaning of consciousness. But if the terms consciousness and sensation are redefined then the former situation no longer obtains. Lewes proposed the more general definitions of, '... *consciousness* as the total of all combined sensibilities, [and] *sensation* as the reaction of the sensory mechanism'.[205] He acknowledged that whether or not the reaction of a sensory mechanism should be called a sensation was a question of terminology. In his opinion the term must be used 'if anything like coherence is to be preserved in physiological investigations'.

Lewes rejected the model of central connections postulated by Hall and others, on the grounds of inadequate anatomical evidence,

> We have already seen that what Anatomy positively teaches is totally unlike the reflex mechanism popularly imagined. The sensory nerve is not seen to enter the spinal cord at one point and pass over to a corresponding point of exit; it is seen to enter the grey substance, which is continuous throughout the spinal cord; it is lost to view, its course being untraceable.[206]

Hall's excito-motory system was referred to as 'Imaginary Anatomy'. Indicating that while this aspect of Hall's theory was not supported, other equally imaginary aspects had survived, Lewes said,

> It is needless nowadays to point out that the existence of a distinct system of excito-motor nerves belongs to Imaginary Anatomy; but it is not needless to point out that the Imaginary Physiology founded on it still survives.[207]

According to Lewes this 'hypothetical' physiology was no less at variance with observation and induction than the hypothetical structure invented for its basis.[208] He was in fact nearer the contemporary position with his opinion that all processes are reflex processes, some of them having a voluntary and others an involuntary character. Physiologists were criticized and in fact his criticism was then aimed directly at Marshall Hall as the originator, who distinguished involuntary actions as reflex and invented for this class a special mechanism.[209]

In spite of the opposition from Hall and his supporters it has to be admitted that the behaviour of the spinal animal has about it an air of purposiveness. It is hardly surprising, with so many individuals attempting to interpret this behaviour that a controversy arose concerning the psychical powers of the spinal cord. The dispute is most commonly associated with the names of two German physiologists Eduard Pflüger[210] and Rudolph Heinrich Lotze, a contemporary of Bernard and Lewes. Their publications relevant to this issue began with an attack on Hall by Pflüger in 1853, the year of Hall's retirement. Thus, by the time this argument, sometimes referred to as the 'spinal soul (Rückenmarkseele) controversy' had really begun, Hall had withdrawn from the scene and did not join in the debate.

Pflüger had studied in Berlin and in 1859 became Professor of Physiology and Director of the Physiological Institute of Bonn. He had launched his initial attack in a monograph claiming that what was true in Hall's theory was already known and what was new in it was not true.[211] The central issue as Pflüger saw it was whether or not the brain was, as Hall believed, the exclusive organ of consciousness. Clearly there was associated with this, the question of whether or not in spinal, that is decerebrate, animals the spinal cord was capable of consciousness.

Young Pflüger's own belief, presented in vigorous language, was reminiscent of the older views of Whytt. He believed consciousness to be coextensive with the nervous system and to be divisible. Thus, not only did consciousness persist in a decapitated animal but also in isolated parts, provided they contained segments of the nervous system. Pflüger enjoyed Lewes' support on this.

A more moderate tone marked the utterances of his opponent Lotze who had studied at Leipzig and gained the chair of philosophy at Göttingen. Lotze held that the apparently purposive behaviour exhibited by spinal animals was not due to intelligence and sensation but to the persistence in the nerve tissue of physiological traces established within the nervous system when habitual responses were performed by the intact animal. These effects on the nervous system, although enabling the animal to respond to familiar stimulation in an apparently purposeful manner, are to little avail in protecting the animal when it is presented with more complex situations involving unfamiliar stimuli. Lotze presented his view in a paper on 'Instinct' in Wagner's *Handwörterbuch.*[212] The debate persisted throughout the second half of the nineteenth century and at least it provoked renewed experimentation and useful discussions on the nervous system.

In 1930 Fearing drew attention to Dewey, who in 1896 had pointed out the distortions which the classification of animal responses had introduced. Fearing believed that the division of animal responses into arbitrary classes such as habitual, reflex, volitional and automatic acts, a classification which began with the Greek physiologists distinguishing between voluntary and involuntary action, now tended to result in an assumption that these classes were independent functional entities. Such a classification, to which Hall had contributed by emphasizing the separateness of reflex actions presented, said Fearing, a wholly artificial picture of the adjusting organism.[213]

The Prochaska Controversy and the Charge of Plagiarism

The earlier two years 1837–8 had been among the most eventful in Hall's life. In March and April 1837 his rejected *Memoir* (1837) had been read before the Royal Society and in August of the same year he performed a series of experiments on a turtle at La Pitié Hospital in Paris when M. Serres, a member of the Institute of France, was present.[214] Furthermore, in addition to maintaining his practice, Hall was presenting his lectures on the Theory and Practice of Medicine at the Webb Street School, Southwark and on Medicine at the Sydenham School. It was during these lectures that Hall had precipitated the controversy in which George Newport was accused of plagiarism from

Professor Grant concerning the nervous system of the *Articulata.* During the same session Hall became involved in his second more protracted and at times extremely bitter controversy. This time, Hall himself was the target of an implied charge of plagiarism, the alleged victim being the physiologist George Prochaska (1749–1820).[215]

Hall was frequently accused of ignoring the contributions of others to the development of the concept of reflex action. Jules Cayrade felt that although Johannes Müller had arrived at certain similar deductions concerning the characteristics of reflex action he was not given due recognition by Hall. Combining an unflattering comment on the British in general with one about Hall in particular, Cayrade declared,

> ... faisant table rase du passé, sans aucun respect pour ceux qui l'avaient précédé dans cette étude, Marshall-Hall s'attribue, avec un orgueil tout britannique, la gloire de cette importante découverte, n'admettant même pas l'illustre Jean Müller à y participer.[216]

This second controversy, also conducted in the medical press, was provoked by an article from John Durance George, published in two parts, 7 and 14 April 1838 in the *London Medical Gazette.*[217] George was President of the student Medical Society of University College London to which he had recently read the paper. Apparently his paper was initiated by a remark from Professor Sharpey 'in lecturing upon the reflex or excito-motory function'. Sharpey had used Hall's terminology of 'excito-motory function' and had said, 'that, of all previous physiologists, Prochaska had approached most nearly in his views to those at present entertained'.[218] This remark does not necessarily suggest that Prochaska would be a useful target for plagiarism by a contemporary physiologist. George was surprised that Marshall Hall had not mentioned Prochaska in his publications up to 1837; he was obliged, therefore, to study Prochaska's own writings. How many typical students would have displayed such intellectual curiosity and initiative? In pursuing his investigations, George had found that 'Dr Hall's account was very imperfect'.[219] Since the whole issue was currently exciting much interest, the paper had been presented in the hope of making good some of the deficiencies rather than in the hope of rendering the history complete. George professed to hope that his paper would be of help to Hall in his preparation of the *History of the Nervous System,* to which Hall referred on more than one occasion but which he did not publish.

The paper was based on a study of Prochaska's original dissertation on the functions of the nervous system, first published in Latin in

1784[220] and subsequently, in Part Two of Prochaska's *Operum Minorum* published in 1800.[221] It is this work which is central to the 1838 controversy and which later stimulated the Sydenham Society to commission a translation by Thomas Laycock. This was published with a translation of Unzer's *Principles of Physiology* in 1851.[222] The interval of some 11 years between the dates of the controversy and publication of the translation is probably explained by Laycock's comment that he, 'laboured, it is true, with the usual drawbacks of an active professional life; ....'[223] Laycock was in fact an extremely busy individual.

George implied that Hall had gained his ideas on reflex function, which he paraded as uniquely his own discoveries, from Prochaska. This charge was echoed even more explicitly by the editorial comments accompanying George's paper in the *London Medical Gazette* and also in the *British and Foreign Medical Review.* Neither journal was ever as supportive of Hall as was the *Lancet,* through Wakley its editor. Under the heading 'The Reflex Function an Old Discovery', the editorial in the *London Medical Gazette* drew attention to George's paper and added, 'From this, it is clear that Dr Marshall Hall was completely anticipated by Prochaska in his views of the reflex function.'[224] Prochaska's absence from those sections of Hall's work devoted to the 'Opinions of Former Physiologists' was also noted. The journal then drew what it clearly felt to be its ace against Hall, namely that there existed a copy of Prochaska's work in the library of the Medical and Chirurgical Society, of which Hall was a member. Though this work was little known to others, 'one member, at least, has shown that he duly appreciated its value, by the fact of his having repeatedly taken it out: that member, need we add, is – Dr Marshall Hall!'[225] This charge was referred to by Hall in his Webb Street lectures. Thus, the controversy was also aired in the pages of the *Lancet.*

An anonymous correspondent to the next issue of the *London Medical Gazette* wrote in defence of Marshall Hall pointing out that Prochaska had in fact been cited by Hall.[226] This was true. The reference was made in a lecture on 10 February 1838 which was brief and not without its irony. Hall was attempting to overthrow an opinion of Sir Charles Bell and to establish precedence for Prochaska over Monro, as the first observer of the relationship between the ganglia and the *portio major* branch of the Vth cranial and posterior spinal nerves.[227] The editor seized the opportunity to add a note emphasizing that Hall was clearly familiar, therefore, with Prochaska's dissertation.[228]

Hall's reply to this charge of familiarity with this particular copy of Prochaska's work appeared in the *London Medical Gazette*[229] on 21 April and a week later in the *Lancet*, along with his final Webb Street lecture on the nervous system. Wakley's introductory note to the lecture referred to Hall's 'envious and interested detractors'.[230] Correspondence between Hall and the Assistant Librarian of the Medical and Chirurgical Society, concerning the dates on which he had borrowed Prochaska's work, had been submitted to Wakley. These were 30 December 1835, 30 January 1837 and 20 January 1838. Readers were reminded that Hall's paper to the Zoological Society had been read in November 1832 and his first memoir to the Royal Society in June 1833. Hall thus highlighted the intervals of three and two and half years between these events and his borrowings of Prochaska's work.[231] The somewhat hostile editorial then implied that Hall might, at an earlier date, have seen other copies of Prochaska's work. This suggestion was vehemently refuted. There could be few copies of Prochaska in London and Hall defied friend or foe to prove that he had ever seen another. In any case, he had consulted Prochaska's work to read the passages quoted by Sir Charles Bell.[232]

What were the claims made by George on behalf of Prochaska and how did he, apparently so persuasively, present them? Furthermore, to what extent did George's exposition really represent Prochaska's conception of reflex action, and to what extent had Prochaska in fact anticipated Hall? Prochaska's explanation of the physiology of the nervous system and in particular of involuntary, reflex actions, contained two major concepts. Neither was new. The first concept was that of the *vis nervosa*, an invisible entity, a latent power present in the pulp of the nerves and spinal cord and dependent on the effect of a stimulus as an excitory agent; the second was that of the *sensorium commune* already closely associated with Robert Whytt.[233] George had great admiration for Whytt to whom Prochaska paid almost less attention than did Hall. Of the role of the *sensorium commune* Prochaska had observed,

> This part, in which, as in a centre, the sensorial nerves, as well as the motor nerves, meet and communicate, and in which the impressions made on the sensorial nerves are reflected on the motor nerves, is designated by a term, now adopted by most physiologists the *sensorium commune*.[234]

In so far as an immaterial agent was being considered, this was very much in line with Whytt's notion of a sentient principle. But Prochaska's concept of the *sensorium commune* was as much a location as an immaterial influence. He discussed possible sites and came to the conclusion that the *sensorium commune* extended throughout the medulla oblongata, the crura of the cerebrum and cerebellum, part of the thalami optici and the whole of the medulla spinalis. Thus, for Prochaska it was coextensive with the origin of the nerves and as such corresponded to Hall's 'true spinal marrow'. This was probably the most significant passage associated with the attack on Hall. But Hall himself placed greater emphasis on the mechanistic aspect of his concept of reflex function for which there was no precise analogue in Prochaska's theory. He had dismissed Whytt's notion of a sentient principle with which he would have grouped Prochaska's *sensorium commune.*

It must be acknowledged, though, that Prochaska's concept was not entirely unmechanistic. He spoke of the reflexion of sensorial into motor impressions taking place 'according to certain laws; writ as it were by nature on the medullary pulp of the sensorium'.[235] Although he clearly recognized the dominance of the spinal cord as the coordinating centre in automatic motions which were not under the control of the will, Prochaska weakened his theory by also entertaining the possibility of peripheral anastomoses, providing connections between sensory and motor nerves and hence acting as an alternative site of reflexion. Whytt had already at an earlier date questioned the notion of peripheral anastomoses. Prochaska's *vis nervosa*, which he himself described as 'mysterious and unknown'[236] and which was not supplied by the brain, differed from that of Haller in being divisible and in persisting in segments of neuro-muscular system long after the death of the animal. Through this divisibility, Prochaska explained how isolated parts of the body, provided they were linked to a region of the nervous system, could still function.

The George paper opened with a fairly lengthy historical survey of neuro-muscular phenomena as recorded in the literature, from classical to contemporary times. It contained explanations accounting for these phenomena and the generalizations derived from them. Considerable attention was paid to Whytt's ideas but since the real aim in the main part of the paper was to demonstrate the correspondence between the eighteenth-century views of Prochaska and those published by Hall almost half a century later, translated passages of Prochaska's work were then presented, 'to show how just were the views of Prochaska, and how nearly allied to those at present entertained'.[237]

The paper was completed with a presentation side by side of corresponding passages from both Prochaska and Hall with George's own evaluations. Readers were left to deduce for themselves Hall's act of plagiarism since there was added somewhat maliciously that the paper would,

> ... serve to exhibit at the same time a proof of how two minds may, by pursuing the same subject, arrive at similar conclusions, independently(?) of each other.[238]

George clearly intended to give the impression that his piece as a whole was based mainly on the framework suggested by Prochaska's dissertation and that the issues selected by George were systematically and fully detailed by Prochaska. Indeed, in the second part of his paper George insisted that Prochaska was worthy of the highest praise for his scientific generalizations and explanation of the phenomena dealt with. It was claimed that, 'They are not alluded to incidentally in a part of the work devoted to another subject, but treated specifically and distinctly.'[239] Surprise was expressed when, as a result of tracing his route through the writings of both men, George ended up with almost identical statements, claiming that 'The reflex function appears, in a word, to be the *complement* of the functions of the nervous system hitherto known.'[240]

In reality though, in his presentation George had adopted as his framework the explicitly stated principles enunciated by Hall in the construction of his theory of reflex action. These principles included the divisions of the nervous system, the identification of a principle of action through which involuntary actions were brought about, the nature and location of this principle and its role in the animal economy. As if to prepare for any criticism that Prochaska's views were scant in detail, which they were, George explained that they were in fact concisely written.

Prochaska was credited with doing what Hall had not, that is, fully respect the views of his predecessors, for instance as to the seat of the common sensorium. In fact, in the original, this so called full account amounted to a list of names together with the location associated with each individual,

> Non unus locus est, in quo sedem sensorii communis constituerunt Clariss. Viri. *Bontekoe, Lancisius, De la Peyronie* in corpus callosum posuerunt sensorium commune; *Willisius* perceptionem sensuum et motus scaturiginem a corporibus striatis repetiit; *Cartesius* Glandulae pineali munus sensorii cummunis tribuit; Vieussensius centro ovali; *Boerhaave* illud aggregatum punctorum pro sensorio communi statuit, in quod omnes nervi sensorii terminantur ....[241]

In no sense should this have been described as a full account – and Whytt was not even mentioned.

A specific issue on which George accused Hall of obtaining from Prochaska an idea which he paraded as his own, was that of the concept of self-preservation as a function of reflex action. A translation of the following passage from Prochaska was included,

> Generalis tamen lex, qua commune sensorium impressiones sensorias in motorias reflectit, est nostri conservatio: ita ut impressiones externas corpori nostro nocituras sequantur certae impressiones motoriae, motus producturae eo collimantes, ut nocumentum a corpore nostro arceatur, amoveaturque; ....[242]

George's translation reads,

> Yet the general law by which the Common Sensorium reflects sensory impressions into motor, is self-preservation (*nostri conservatio*) so that certain motor impressions follow external hurtful impressions upon our body; the motions to be produced aiming at this, that the noxious thing shall be repelled and driven away: ....[243]

The passage does indeed refer to the general laws by which the *sensorium commune* reflects sensorial into motor impressions and is associated with the preservation of the individual by provoking motions which could for example deflect noxious impressions and is away from harming the body. But this general statement is Prochaska's total contribution on the issue. In any case Hall never claimed the enunciation of this concept as part of his own discoveries concerning reflex action. Furthermore, it is interesting that George did not cite Whytt as an even earlier and more eloquent source of awareness among physiologists of reflex actions being associated with preservation of the individual. Whytt made several references including,

> ... if we suppose our minds so formed and connected with our bodies, as that, in consequence of a stimulus affecting any organ, or of an uneasy perception in it, they shall immediately excite such motions in this or that organ or part of the body as may be most proper to remove the irritating cause; and this without any previous conviction of such motions being necessary or conducive to this end. Hence men do not eat, drink or propagate their kind, from deliberate views of preserving themselves or their species but merely in consequence of the sensations of hunger thirst, etc. ... the general and wise intention of all the involuntary motions, is the removal of everything that irritates, disturbs, or hurts the body; ....[244]

Prochaska and Hall were compared on the issue of experimental work. And here George really was out of his depth. Hall was widely recognized and acknowledged – and pilloried – as a most ingenious experimental physiologist. George referred to Prochaska carrying out on a frog what was portrayed as a crucial experiment and hinted at experimental work being a significant component in Prochaska's writing.[245]

In fact, in marked contrast to all Hall's publications, nowhere in his dissertation, apart from the reference to the effects in the lower limbs of cutting the spinal cord of a frog,[246] did Prochaska present the details of any of his own experiments. Curiously, immediately following the frog reference, Laycock's translation contains the comment, 'I am compelled to defer a more detailed and accurate account of these and similar experiments to another opportunity as this is not the proper place.'[247] But this statement is absent from Prochaska's original text.

The excuses for what were clearly, and in some cases understandable, short-comings in Prochaska's theorizing went on. Since Prochaska was writing well before Bell's work on the functions of the double roots of the nerves, it was considered not surprising that Prochaska failed to identify the particular nerves by which impressions were conveyed centrally and those into which they were reflected to bring about motions. Other omissions, according to George, such as those of 'the action of the sphincters, muscle tone and other *details,* impair his merit as the propounder of the *system* so little, that they need scarcely to be mentioned'.[248] George even went so far as to make the exaggerated claim that,

> ... the great principle, the theory, is laid down; and the chief improvement left to his successors are those of perfecting the details and extending the system.[249]

In his reply to George, Hall addressed himself to the fact that Prochaska's observations and views had been so long overlooked. Hall rightly felt that the material's being in Latin, when already predecessors such as Whytt had used the vernacular, limited its dissemination. Furthermore, while Hall was competent in reading Latin he claimed that even if he had glanced at the relevant sections of Prochaska, its vocabulary with its reference to the *sensorium commune* would have caused him to dismiss Prochaska, along with Whytt, despite Prochaska's use of the term *reflectuation.*[250]

But for Hall the real reason for Prochaska's hitherto low profile was that until the development of that 'harmonious combination' of existing ideas and doctrines on involuntary actions 'into a uniform system, and the application of it to the explanation of many phenomena which were not formerly regarded as explicable on such principles ... [no one] had *eyes* to *see* the full value of these things'.[251] Hall justifiably felt that he had made a significant contribution to this 'harmonious combination'. And, as he remarked, between 1832 when he first proposed his ideas until the present, his view had been totally neglected, 'No one thought them worth claiming for himself or for others.'[252] Also, before 1837 Prochaska scarcely appeared in histories of the nervous system. After 1837 he was never absent from them.

Hall felt that his experience was but another instance of a general phenomenon in the field of discovery and in support of his own assertion regarding Prochaska's seemingly unjust emergence from obscurity into a blaze of glory, Hall took comfort from Paris, who observed that,

> ... evidence may be often strained from the writings of philosophers in support of prior claims to late discoveries; but upon a candid review, these loose statements, or obscure hints, will generally be found wholly destitute of the pretensions which an unfair spirit of rivalry has too often laboured to support. Many of such hints indeed, so far from advancing the progress of truth, *had never even attracted notice, until AFTER the discoveries to which they have been supposed to relate*(!)[253]

The gist of Paris' statement applied to the Prochaska controversy: Hall did have a tendency to underestimate the contribution of some predecessors but his contemporaries were, with their retrospective judgements, claiming far too much for Prochaska.

Hall did refer to the work of predecessors but, as his critic Canguilhem has noted, '(il) ne leur avait attribué aucune importance'.[254] While Prochaska's work does contain reference to what in his own day were novel aspects of an explanation for motions already described by Whytt as involuntary motions, these references were generally tentative and sometimes presented alongside older, already rejected principles. For example, his retention of the notion of peripheral connections between sensory and motor nerve fibres. Furthermore, the work of Bell and Magendie in distinguishing sensory and motor nerves, inevitably afforded Hall the opportunity of including in his theory of reflex action reference to afferent and efferent nerves which was not available to Prochaska. Hall's theorizing was also well founded on first hand observations of his own experimental investigations. Not so Prochaska who, despite retrospective support from Hall's detractors, did not present a coherent and confident generalization about reflex action. Hall did, and was able to broaden his generalization embracing a range of hitherto unclassified motions and to apply it to the realm of disease explanation and therapy.

There remains though, one point of special interest about the Prochaska controversy. This concerns the status and background of the writer of the George article. John Durance George was educated at University College and became a Member of the Royal College of Surgeons in May 1838 the year in which he had been, like Hall at an earlicr period in Edinburgh, President of the student Medical Society at University College. It was in April that same year that George's article was published. George, who died in November 1851 at the age of 36, was still a student and only 22 years old when the article was published. The obituary in the *London and Provincial Medical Directory* (1852) described him as having possessed in those early days 'extraordinary quickness and readiness, considerable wit and fancy and a natural eloquence which training and cultivation might have developed into oratory of a high order'.[255] His fellow students were said to have predicted a brilliant career for him in medicine. This was not to be; he joined an uncle's dental practice and rapidly acquired a large fortune. But he retained some academic connection when in 1843 he became Lecturer in Dental Surgery at University College, where he held a class at nine o'clock on Saturday mornings – presumably to fit in with his practice work. The Prochaska article was his sole publication. But this article was based on considerable reading, analysis and criticism of a wide range of sources, skill in Latin and, with its somewhat snide remarks at Hall, an insider's knowledge of current jockeying for scientific recognition. Despite his potential brilliance, did George unaided really write the article or was he at least helped by someone else only too pleased to seize an opportunity to discredit Hall?

Notes

1. Glisson, F., *Tractatus de Ventriculo et Intestinis,* 1677.
2. Whytt, R., *Vital and Other Involuntary Motions,* 1751, pp.344–92. (In *The Works of Robert Whytt,* 1768, published by his son, the section is numbered 14.)
3. *Ibid.*, p.356.
4. Hall, M., *Diseases and Derangements of the Nervous System,* 1841, p.vii.
5. Hall, M., 'On the Reflex Function of the Medulla Oblongata and Medulla Spinalis, or the Principle of Tone in the Muscular System', *Phil. Trans.*, 123, part ii, 1833, pp.635–65. It was published with Hall's rejected 1837 Memoir as *Memoirs on the Nervous System,* 1837. Hereafter the 1833 article will be referred to both in the text and the notes as *Memoir* (1833) and the pagination will be that in *Memoirs* (1837). The 1837 Memoir will be referred to as *Memoir* (1837) in the text and *Memoirs* (1837) in the notes.
6. Hall, *Memoir* (1833), p.14.
7. Thomas Willis, anatomist and physician, was Sedleian Professor of Natural Philosophy at Oxford. His work, *Cerebri Anatome,* 1664, which was illustrated by Sir Christopher Wren, was regarded as the most complete account of the nervous system then available.
8. Huxley's admiration for Descartes is expressed in his essay 'On

Descartes' "Discourse Touching the Method of Using One's Reason Rightly and of Seeking Scientific Truth" ', in Huxley, T. H., *Lay Sermons,* 1870, pp.351–78, and in his lecture to the British Association at its meeting in Belfast in 1874.

9. According to Lhermitte 'La conception de l'arc réflexe n'appartient ni à Prochaska, ni à Marshall Hall, elle est née du genie de Descartes.' Lhermitte, J., *Les Fondements Biologiques de la Psychologie,* 1925, p.19.
10. Fearing, F., *Reflex Action,* 1930, p.28.
11. Richet, C., *Physiologie des Muscles et des Nerfs,* 1882, p.494.
12. *Ibid.,* p.659.
13. Review in *Brit. and For. Med. Rev.,* 9, 1840, p.98.
14. Richet, *Physiologie des Muscles et des Nerfs,* p.661.
15. Beaunis, H., *Nouveaux Elements de Physiologie Humaine,* 2 vols, 2nd edn, 1881, 1, p.558.
16. Eckhard, C., 'Geschichte der Entwickelung der Lehre von den Reflexerscheinungen', *Beit. zur Anat. und Physiol.,* 9, 1881, pp.31–192.
17. Whytt, *Vital and Other Involuntary Motions,* p.384.
18. Whytt, R., *Observations on the Sensibility and Irritability of the Parts of Men and Other Animals,* in [Whytt], *Works of Robert Whytt,* p.284.
19. *Ibid.,* p.285.
20. Grainger, R .D., *Structure and Functions of The Spinal Cord,* 1837.
21. See Hall, *Memoirs* (1837).
22. Legallois, J. J. C., *Expériences sur le Principe de la Vie,* 1812.
23. *Ibid.,* p.i.
24. Legallois speculated that the weaker movements might be due to the pathological condition of the spinal medulla following decapitation and hence seems to have envisaged the condition of surgical shock. *Ibid.,* p.x.
25. *Ibid.,* pp.iv–v.
26. *Ibid.,* p.viii.
27. *Ibid.,* p.xix.
28. *Ibid.,* p.34.
29. *Ibid.,* p.xii.
30. Legallois included the Report of the Institute, with his *Expériences sur le Principe de la Vie.* See pp.252–326.
31. Flourens, P., *Recherches Expérimentales,* 1824.
32. See review of Bell, C., 'On the Nerves of the Face', (published in *Phil. Trans.,* 119, 1829, pp.317–30), in *Lond. Med. and Phys. Journ.,* n.s. 7, 1829, pp.525–37.
33. *Brit. and For. Med. Rev.,* 9, 1840, pp.96–7.
34. *Ibid.*
35. Paul F. Cranefield, in his major work, *The Way In and the Way Out,* 1974, p.xiv, attributes the discovery to Magendie.
36. Schafer, E. A., *Textbook of Physiology,* 2 vols, 1898–1900, 2, p.786.
37. Eldridge-Green, F. W., 'The Claim of Sir Charles Bell to the Discovery of Motor and Sensory Nerve Channels', *Lancet,* vol.ii,

1912, pp.253–4; Waller, A. D., 'On the Claim of Sir Charles Bell to the Discovery of Motor and Sensory Nerve Channels', *ibid.*, pp.900–5; 'The Bell-Magendie Controversy', Letter, p.968; 'The Bell-Magendie Controversy', Letter, *ibid.*, pp.1035–6; 'The Bell-Magendie Question', Letter, *ibid.*, p.1395.

38. See Olmsted, J. M. D., *François Magendie,* 1944, pp.93–122, for a discussion of the issue. According to Olmsted, the term 'Bell-Magendie rule', 'remains to represent a compromise between the points of view of the older type of anatomist who arrived at function by way of inference and the physiologically-minded investigator who insists upon experimental verification'. After 1856, Claude Bernard 'gave Magendie full credit for the discovery of the difference in function between the dorsal and ventral roots of the spinal nerves'. Prior to this he had thought that Magendie had been the first to provide experimental proof of a distinction already understood by Charles Bell. Bernard expressed his change of mind in Bernard, C., *Rapport sur les Progrès et la Marche de la Physiologie Générale en France,* Paris, 1867, pp.10–14. See also Olmsted J. M. D. and Olmsted E. H., *Claude Bernard,* 1952, pp.155–6.
39. *Edin. Med. and Surg. Journ.*, 21, 1824, p.156. This comment appeared in a detailed Review Article on current work on the 'Physiology of the Nervous System' by Flourens, Rolando, Coster, Fodera and Magendie.
40. *Ibid.*, pp.142–3.
41. Müller, J., 'Nouvelles Expériences sur L'Éffet que Produit l'Irritation Mécanique et Galvanique sur les Racines des Nerfs Spinaux', *Annal. des Sci. Nat.*, 23, 1831, pp.95–112. Müller's appointment to the Berlin Chair of Physiology in 1833 was the first of its kind in Europe.
42. Müller, J., *Handbuch der Physiologie,* 2 vols, 1833–40. This work went through further editions and was translated into French and English. William Baly translated and annotated an early edition, Müller, J., *Elements of Physiology,* 2 vols, 1837–42.
43. *Ibid.*, I, pp.688–701.
44. 'On Professor Müller's Account of the Reflex Function of the Spinal Marrow.' Communicated by Marshall Hall, *Lond. and Edin. Phil. Mag. and Journ. of Sci.*, 10, 1837, pp.51–7; 124–9; 187–93.
45. Hall, M., 'Über die Reflectirende Function (Reflex Function) des Verlängerten und Rückenmarks', *Archiv für Anat. Physiol.*, 1, 1834, pp.374–84.
46. 'Brief über das Nervensystem an Professor J. Müller. Von Marshall Hall.' 'Erster Brief. Über die Vis Nervosa Haller's', *Ibid.*, 14, 1840.
47. Hall, 'Postscript', *Memoirs* (1837), p.xiii.
48. Hall, 'On Muller's Account', *Lond. and Edin. Phil. Mag.*, 10, p.52.
49. *Ibid.*, p.56.
50. *Ibid.*, pp.52–3.
51. Fearing, *Reflex Action,* p.142.

52. Hall, *Memoirs* (1837), p.113.
53. *Ibid.*
54. Hall, M., 27 November 1832, 'A Brief Account of a Particular Function of the Nervous System', *Proc. Comm. Sci. and Corresp. Zoo. Soc. Lond.*, part ii, 1832, pp.190–2. Hereafter this will be referred to in text and notes as 'Account Zoological Society'.
55. *Ibid.*, p.190.
56. Hall, *Memoir* (1833), p.5.
57. Hall, *Memoirs* (1837), p.12.
58. *Ibid.*, p.13.
59. *Ibid.*
60. See Hall, *Diseases and Derangements of the Nervous System.*
61. This claim was made by Hall at the beginning of his Croonian Lecture, delivered at the Royal College of Physicians in 1850. Hall, M., *Synopsis of the Diastaltic Nervous System,* 2nd edn, [1850], p.1. Hall's wife felt that if his additional seven years of research were added then a further ten thousand hours should be added and if his leisure time recently devoted *in practice* to his research were considered then the original figure should be doubled. Hall, C., *Memoirs*, p.87.
62. Canguilhem, G., *La Formation du Concept de Réflexe,* 1955, p.172. Canguilhem had therefore reduced Hall's claim by 1,000 hours.
63. Hall, *Memoir* (1833), pp.5–6.
64. *Ibid.*, pp.30–1; 38–9.
65. Hall, *Memoir* (1833), p.1.
66. Legallois, *Expériences sur le Principe de la Vie,* p.xi.
67. Hall, 'Account Zoological Society', p.190.
68. Hall, M., *New Memoir on the Nervous System,* 1843, p.36. Hereafter this memoir both in the text and in the notes will be referred to as *New Memoir* (1843).
69. Hall, M., *Lectures on the Nervous System and its Diseases,* 1836, p.12.
70. *Ibid.*, p.11.
71. Comment by Fearing, *Reflex Action,* p.140 on Sherrington and taken from Stirling, W., *Some Apostles of Physiology,* 1902, p.86. According to Sherrington, Hall considered this notion of the separateness of these two sub-systems to be his chief contribution. But this is not quite accurate.
72. Hall, *Memoirs* (1837), pp.51–2.
73. *Lancet,* vol.i, 1837–8, pp.127–8.
74. Hall, *Memoirs* (1837), pp.49–50.
75. Bell, C., *New Anatomy of the Brain,* [1811], p.5.
76. Bell, C., *An Exposition,* 1824, p.8.
77. Bell, *New Anatomy of the Brain,* p.4.
78. *Ibid.*, p.25.
79. *Ibid.*, p.4.
80. Bell, *An Exposition,* p.2.
81. Bell, C., *Nervous System of the Human Body,* 3rd edn, 1836, p.iv.

82. Bell, C., 'On the Nerves: Giving an Account of Some Experiments on their Structure and Functions, which Lead to a New Arrangement of the System', *Phil. Trans.*, 110, 1821, pp.398–424.
83. *Ibid.*, pp.416–18.
84. Bell, *An Exposition,* pp.2–3.
85. *Ibid.*, pp.vi–vii.
86. Hall, *Lectures on the Nervous System and its Diseases,* p.vii.
87. *Ibid.*
88. Hall, *Synopsis of the Diastaltic Nervous System,* p.5.
89. *Ibid.*, p.vii.
90. Hall, *Memoirs* (1837), p.51.
91. *Ibid.*, p.50.
92. *Ibid.*, p.51.
93. Hall, *New Memoir* (1843), p.37.
94. *Ibid.*
95. Hall referred to an experiment carried out by himself and Messrs Broughton and Field, supported by a small grant from the British Association in Edinburgh, and which had led to this conclusion about the pneumogastric nerve. Hall, M., Broughton, S. D., Field, [G]., 'Report of Progress made in an Experimental Inquiry regarding the Sensibilities of Cerebral Nerves, recommended at the Last Meeting of the Association', *Report of the Fourth Meeting of the British Association Held at Edinburgh,* 1835, pp.676–80.
96. Hall, *Memoir* (1833), p.37.
97. *Ibid.*
98. *Brit. and For. Med. Rev.,* 5, 1838, p.538.
99. Dutrochet, R. H. J., *Récherches Anatomiques et Physiologiques,* 1824, pp.164–70.
100. Virchow, R., *Cellular Pathology,* 1860, pp.243–82.
101. *Ibid.*, p.268.
102. Hall, *New Memoir* (1843), p.5.
103. See 'The Reflex Theory', in Lewes, G. H., *The Physical Basis of Mind,* 1877, pp.413–93.
104. *Ibid.*, p.413.
105. Haller, A., *Dissertation on the Sensible and Irritable Parts of Animals,* 1755. See the 1936 edition of same work with an Introduction by Owsei Temkin.
106. Clark, F. Le Gros, *Practical Anatomy and Elementary Physiology,* 1836, p.72.
107. *Ibid.*, pp.72–3.
108. Hall, *Memoir* (1833), p.38.
109. Hall, *Memoirs* (1837), p.112.
110. Blane, Sir Gilbert, *Select Dissertations,* 1822. Blane had already hinted at this exclusion of sensation in his Croonian Lecture on Muscular Motion read at the Royal College of Physicians in 1788.
111. Van Deen's ideas are discussed in Stilling, B., *Untersuchungen über die*

Functionen des Rückenmarks und der Nerven, 1842.

112. Hall, *Memoir* (1833), pp.8–9.
113. *Ibid.*, p.9.
114. Volkmann, A. W., 'Über Reflexbewegungen', *Archiv für Anat. Physiol.*, 1838, pp.15–43. This was translated and summarized: Volkmann, A. W., 'On Reflex Motions', *Brit. and For. Med. Rev.*, 6, 1838, pp.211–18.
115. Hall, *New Memoir* (1843), p.34.
116. [Whytt], *Works of Robert Whytt*, pp.205–6.
117. *Ibid.*, p.206.
118. *Ibid.*
119. *Ibid.*, p.206 in a footnote.
120. *Ibid.*
121. *Ibid.*, p.207.
122. Hall, *New Memoir* (1843), p.35.
123. Flourens, *Recherches Expérimentales*, p.vii.
124. Calmeil, L. F., 'Sur la Structure, les Fonctions et le Ramollissement de la Moëlle Épinière', *Journ. des Progrès*, 11, 1828, pp.77–124.
125. Mayo, H., *Outlines of Human Physiology*, 4th edn, 1837, reverse of final, unnumbered page. In his subsequent reference to the page containing this note Hall called it p.534. The pagination of the volume actually ceases at 532 and, most probably to deflect inquiry from his motive in carrying out this unusual procedure of adding an isolated piece on Marshall Hall at the end of a completed work, Mayo added, on what would be p.533, an extract from Müller's *Archiv für Anatomie, Physiologie und Wissenschaftliche Medicin* of the previous year and relating to a chapter on generation in birds.
126. *Ibid.*, The quotation comes originally from Mayo, H., *Anatomical and Physiological Commentaries*, parts i and ii, 1822 and 1823, part ii, pp.135, not p.138 as cited by Mayo and by [Alexander Walker] in *Documents and Dates of Modern Discoveries in the Nervous System*, 1839, p.132.
127. Mayo, *Outlines of Human Physiology*, p.534.
128. Hall, *Memoirs* (1837), p.56.
129. *Ibid.*, pp.56–7. The passage comes from Mayo, *Outlines of Human Physiology*, p.212.
130. [Walker], *Documents and Dates of Modern Discoveries in the Nervous System*, 1839, p.133.
131. See Chapter 4 *supra* on 'Marshall Hall and the Royal Society – 1830–50'.
132. Grainger, *Structure and Functions of The Spinal Cord*, p.6.
133. *Brit. and For. Med. Rev.*, 5, 1838, pp.486–540.
134. *Ibid.*, p.490.
135. *Ibid.*, p.515.
136. *Ibid.*, pp.514–15.
137. Müller, *Elements of Physiology*, 1, p.721.

138. Hall, *New Memoir* (1843), p.32.
139. Hall, *Synopsis of the Diastaltic Nervous System*, p.47.
140. *Ibid.*
141. Paton, G., 'On the Perceptive Power of the Spinal Cord as Manifested by Cold-Blooded Animals', *Edin. Med. and Surg. Journ.*, 65, 1846, p.260.
142. Mayo, *Anatomical and Physiological Commentaries,* part i, p.7.
143. Grainger, *Structure and Functions of The Spinal Cord,* pp.54–5. Grainger cites Mayo's finding which he is directly disputing.
144. Hall, *New Memoir* (1843), p.31.
145. *Ibid.*
146. Clark, *Practical Anatomy and Elementary Physiology,* p.70 in a note.
147. Review, 'Article XIII', *Brit. and For. Med. Rev.*, 5, 1838, p.515.
148. *Ibid.*, p.514.
149. Hall, M., *Essay on the Circulation of the Blood,* 1831, pp.2–3.
150. *Brit. and For. Med. Rev.*, 5, p.515 which gives the earlier citation as *ibid.*, 3, 1837, p.38.
151. *Edin. Med. and Surg. Journ.*, 48, 1837, p.33.
152. Hall, *Memoirs* (1837), p.101.
153. Hall, *Diseases and Derangements of the Nervous System,* p.235.
154. *Ibid.*, p.234. According to Brodie 'Priapism occurs even where the sensibility is entirely destroyed, and may be induced by the mechanical irritation caused by the introduction of the catheter, where the patient is entirely unconscious of the operation.' *Med.-Chir. Trans.*, 20, 1836, p.140.
155. In *hemiplegia* there is complete loss of sensation and motion in the upper and lower extremities of one side. In *paraplegia* the affliction affects both sides.
156. Hall, *Diseases and Derangements of the Nervous System,* p.232. Hall's source was Baly's translation of Müller's, *Elements of Physiology,* 1, p.721.
157. Hall, *Memoirs* (1837), p.101.
158. Hall, *New Memoir,* [1843], p.35. Hall used this case to exemplify his own assertion that 'The soul is one, and cannot be divided; it cannot exist or manifest itself in two portions of the same animal.' *Ibid.*
159. Preston, C., 'An account of a Child Born Alive Without a Brain, and the Observables in it on Dissection', *Phil. Trans.*, 9, 1697, pp.457–67.
160. Ollivier, C. P., 'Déformation de l'éxtrémité supérieure de la moëlle dans les cas d'ancephalie et acephalie', *De la Moëlle Épinière et de ses Maladies,* 1824, pp.96–105.
161. Lawrence, W., 'Account of a Child Born Without a Brain, Which Lived Four Days', *Med.-Chir. Trans.*, 5, 1814, pp.165–224.
162. Hall, *Lectures on the Nervous System and its Diseases,* p.47.
163. Lawrence, 'Account of a Child', p.221.
164. MS *Researches into the Effects of Certain Physical and Chemical Agents on the Nervous System;* by Marshall Hall MD, F.R.S. etc., *Archives 1847, AP. 29 5–6,* Royal Society, London. The first part of the paper

was received 28 January and the second 14 April 1847.

165. Clarke, E. and Jacyna, L. S., *Nineteenth-Century Origins of Neuroscientific Concepts,* 1987, pp.163–73.
166. Clarke, E., and O'Malley, C. D., *The Human Brain and Spinal Cord,* 1968, p.177.
167. 'Introductory Observations', MS *Researches into the Effects of Certain Physical and Chemical Agents on the Nervous System, Archives 1847, AP.29.5–6,* Royal Society, London.
168. Clarke & Jacyna, *Nineteenth-Century Origins of Neuroscientific Concepts,* p.195.
169. Leopoldo Nobili, an Italian physicist, improved the galvanometer which had been developed by Ørsted and Schweigger to detect and measure animal electricity. With his astatic galvanometer Matteucci detected the 'corrente della rana' – the frog current – for the first time in 1825. Carlo Matteucci another Italian physicist extended Nobili's work and demonstrated that the frog current was what we recognize today as the 'current of injury'. Clarke & O'Malley, *The Human Brain and Spinal Cord,* pp.185–6.
170. *Ibid.*, p.178. Clarke & Jacyna, *Nineteenth-Century Origins of Neuroscientific Concepts,* p.210.
171. Ritter's work on electricity lasted for just over a decade,1792–1810, and his publications included, Ritter, J. W., *Das electrische System der Korper,* 1805. Ritter later became increasingly involved with the occult and believed that the interdependence of inorganic nature and human physiology was determined by a subterranean electrical force. Clarke & Jacyna, *Nineteenth-Century Origins of Neuroscientific Concepts,* p.177.
172. Clarke & Jacyna, *Nineteenth-Century Origins of Neuroscientific Concepts,* p.179.
173. Matteucci, C., 'Electrophysiological Researches – First Memoir. The Muscular Current', *Phil. Trans.*, 1845, Pt. II, pp.283–95; 'Second Memoir. On the proper Current of the Frog', *ibid.*, pp.297–301; 'Third Memoir. On Induced Contractions', *ibid.*, pp.303–17. Matteucci, C., 'Electrophysiological Researches. – Fifth Series. Part I, Upon Induced Contractions', *ibid.*, 1847, Part II, pp.231–4, 'Part II, Upon the Phenomena elicited by the passage of the current through the Nerves of a living Animal or an Animal recently killed, according to the direction of the Current', *ibid.*, pp.234–7; 'Sixth Series. Laws of the Electric Discharge of the Torpedo and other Electric Fishes – Theory of the production of Electricity in these animals', *ibid.*, pp.239–42; 'Seventh and last Series. Upon the relation between the intensity of the Electric Current, and that of the corresponding physiological effect', *ibid.*, pp.243–8. Matteucci completed this paper with a postscript beginning 'As I do not propose to resume my electro-physiological researches, at least for some time to come ....' Matteucci, C., 'Electro-physiological Researches. – Eighth Series.' *ibid.*, 1850, Pt. I, pp.287–96; 'On Induced

Contractions – Ninth Series', *ibid.*, Pt II, pp.645–9. Matteucci, who explained that he had been rushed into publishing this paper as a result of the announcement to the Académie des Sciences by Du Bois-Reymond of his work on the Law of Muscular Current, finished the paper with the words '... we wait the light of new experiments to proceed safely further in this difficult field of science'. Du Bois-Reymond, E., 'On the Discovery of the Law of the Muscular Current', in *On Animal Electricity: Being an Abstract of the Discoveries of Emil Du Bois-Reymond,* London, 1852, pp.127–60. See also Du Bois-Reymond, E., 'On the Time Required for the Transmission of Volition and Sensation through the Nerves. A Lecture given at the Royal Institution', Appendix I in Jones, H. B., *Croonian Lectures on Matter and Force,* 1868, pp.97–132.

174. The term 'electrogenic' which, Hall claimed, described 'at once the origin and the independence of this condition', was another example of Hall's short-lived new terminology introduced into his work on the nervous system. He sent an explanatory note about its derivation from the Greek, after the paper had been submitted to the Royal Society. Hall explained, 'Dr Marshall Hall compliments and begs the favour of Mr Weld to have the following note inserted in his paper at the word electrogenic.' MS Note, Hall to Weld, 12 May 1847, *AP. 29.5–6,* Royal Society, London.
175. *Ibid.*
176. Clarke & Jacyna, *Nineteenth-Century Origins of Neuroscientific Concepts,* p.204.
177. MS *AP.29.5–6,* Royal Society, London. Hall repeated this claim when he later published the paper elsewhere, *Edin. New Phil. Journ.,* 45, 1847, p.253.
178. *Ibid.*
179. *Ibid.*
180. *Ibid.*
181. *Ibid.*
182. Clarke & Jacyna, *Nineteenth-Century Origins of Neuroscientific Concepts,* pp.209–10.
183. Cranefield, P. F. 'The Organic Physics of 1847 and the Biophysics of Today', *Journ. Hist. Med.,* 12, 1957, pp.407–23. Mendelsohn, E., 'Physical Models and Physiological Concepts: Explanation in Nineteenth-Century Biology', *Brit. Journ. Hist. Sci.,* 2, part 3, 1965, pp.201–19.
184. Clarke & Jacyna, *Nineteenth-Century Origins of Neuroscientific Concepts,* p.210.
185. This 'school' owed much to Michael Foster, (1836–1907), later knighted, who was born into a nonconformist family and who studied medicine at University College London where William Sharpey invited him to become a teacher in practical physiology. He gave the first course in the subject in England. He succeeded Huxley as Fullerian Professor of

Physiology at the Royal Institution and then, with Huxley's support, moved to Cambridge. Foster founded (1878) the *Journal of Physiology*, and was influential in establishing the Physiological Society – such as Hall had called for over half a century earlier – in 1875. Through his students, his teaching and his wide influence rather than through his original research, Foster influenced the development of physiology both at home and in America. See Geison, G. L., *Michael Foster and the Cambridge School of Physiology*, 1978.

186. 'Vivisection, by Dr M. Hall', *Charleston Med. Journ. and Rev.*, 9, 1853, pp.127–8.
187. Cartwright, S. A., 'Motive Power of the Blood Proved by Experiments on Four Crocodiles – one Brought to Life', *Boston Med. and Surg. Jour.*, 48, 1853, p.433. Cartwright also published 'Decisive Experiment', *ibid.*, pp.474–9. The same journal also published a reprint from the *New Orleans Delta* of Cartwright and Dowler's experimental work, 'More Alligator Experiments', *Boston Med. and Surg. Journ.*, *ibid.*, p.423.
188. Cartwright, 'Decisive Experiment', *ibid.*, p.479.
189. Cartwright, 'Motive Power', *ibid.*, p.435.
190. *Ibid.*
191. *Ibid.*, 49; pp.330–1.
192. 'Dr Marshall Hall on the Nervous System', *ibid.*, p.366.
193. Cartwright, 'Motive Power', *ibid.*, pp.436–7.
194. Cartwright, 'Decisive Experiment', p.477.
195. Hall, C., *Memoirs*, pp.348–9.
196. Longet, F. A., *Traité de Physiologie*, 2 vols, 2nd edn, 1860, 2, p.210.
197. *Ibid.*, p.211.
198. *Ibid.*
199. *Ibid.*, p.213.
200. *Ibid.*
201. *Ibid.*
202. *Ibid.*
203. Bernard, C., *Leçons sur la Physiologie et la Pathologie du Système Nerveux*, 2 vols, 1858, 1, p.339.
204. *Ibid.*, pp.339–40.
205. Lewes, *Physical Basis of Mind*, p.431.
206. *Ibid.*, p.425.
207. *Ibid.*
208. *Ibid.*
209. *Ibid.*, p.424.
210. In 1868 Pflüger founded the *Archiv für die gesamte Physiologie des Menshen und der Tiere* in which were published many significant contributions to physiology in the latter part of the nineteenth century.
211. Pflüger, E. F. W., *Die Sensorischen Functionen des Rückenmarks der Wirbelthiere*, 1853. Hall has a very high profile throughout the first half of the monograph.
212. Lotze, R. H. 'Instinct', in Wagner, R. *Handwörterbuch der Physiologie*,

4 vols in 5, 1844, 2, pp.191–209.

213. Fearing, *Reflex Action,* p.314.
214. Wakley reported this event and said the experiments were made at the request of a member of the French Institute. He contrasted this treatment of Hall with that of the Council of the Royal Society who refused to witness similar experiments. *Lancet,* vol.i, 1837–8, p.648.
215. In his sketch of Marshall Hall, Edwin Clarke pointed out that this episode has never been fully explained. Gillispie, C. C. (ed.), *Dictionary of Scientific Biography,* 14 vols, 1970–6, 6, pp.58–61.
216. Cayrade, J., *Recherches sur les Mouvements Réflexes,* 1864, 2, pp.1–179. *Quotation,* pp.30–1. Cayrade always wrote 'Marshall-Hall'.
217. George, J. D., 'Contribution to the History of the Nervous System', *Lond. Med. Gaz.,* vol.ii, 1837–8, pp.40–7; pp.93–6.
218. *Ibid.,* p.40.
219. *Ibid.,*
220. Prochaska, G., *Adnotationum Academicarum,* 1784. This contained Prochaska's dissertation on the nervous system together with contributions on pathology and pathological anatomy.
221. Prochaska, G., 'Commentatio de Functionibus Systematis Nervosi', *Operum Minorum,* Pars II, 1800. This is the edition referred to in the present book.
222. Unzer, J. A., *The Principles of Physiology,* and Prochaska, G., *A Dissertation on the Functions of the Nervous System,* 1851. Laycock based his translation on the first edition. In a footnote Laycock stated that there were few alterations in the 1800 edition. The present writer confirms this claim. (Hereafter Laycock's translation cited as Prochaska, *Dissertation.*)
223. *Ibid.,* Preface p.XII.
224. *Lond. Med. Gaz.,* vol.ii, 1837–8, p.73.
225. *Ibid.*
226. [Justitia], 'Dr M. Hall and Prochaska', *ibid.,* p.128.
227. Hall, M., 'The Ganglionic System', *Lancet,* vol.i, 1837–8, p.682.
228. Editorial note appended to 'Justitia's' letter, *Lond. Med. Gaz.,* vol.ii, 1837–8, p.128.
229. 'Dr M. Hall and Prochaska', *ibid.,* p.160.
230. *Lancet,* vol.ii, 1837–8, pp.129–36.
231. *Lond. Med. Gaz.,* vol.ii, 1837–8, pp.248–9.
232. Hall had consulted Prochaska's work for the preparation of the first edition, in April 1836, of the lectures he was currently giving. *Lancet,* vol.ii, 1837–8, p.134.
233. Prochaska, *Dissertation,* p.390.
234. *Ibid.,* p.429.
235. *Ibid.,* p.430.
236. *Ibid.*
237. George, 'History of the Nervous System', p.43. George explained that he had made his translation as literal as possible in order to ensure accuracy.

238. *Ibid.*, p.46.
239. *Ibid.*, p.93.
240. *Ibid.*, p.95.
241. Prochaska, *Operum Minorum,* pp.151–2.
242. *Ibid.*, p.154.
243. George, J. D., 'Contribution to the History of the Nervous System', *Lond. Med. Gaz.*, vol.ii, 1837–8, pp.44–5.
244. [Whytt], *Works of Robert Whytt,* pp.151–2.
245. George, 'History of the Nervous System', p.44.
246. Prochaska, *Operum Minorum,* p.84.
247. Prochaska, *Dissertation,* p.398.
248. George, 'History of the Nervous System', p.93.
249. *Ibid.*
250. Hall, 'Lectures', *Lancet,* vol.ii, 1837–8, pp.130–6. Hall presented this response to his audience at the termination of his series of lectures on the Theory and Practice of Medicine which he had been delivering at the Webb Street Theatre of Anatomy and Medicine in Southwark.
251. *Ibid.*, pp.131–2.
252. *Ibid.*, p.132.
253. *Ibid.* The original quotation is from Paris, J. A., *Sir Humphry Davy,* 1831, pp.163–4. The italics, capitals and exclamation mark are Hall's additions.
254. Canguilhem, *La Formation du Concept,* p.133.
255. 'John Durance George, Esq.', *Lond. Prov. Med. Dir.,* 1852, p.659.

6

Hall's Final Years

Retirement and Tour of America

Around 1850 Hall's throat trouble of the 1830s worsened and in 1852 he decided to leave London and let the house at 38 Grosvenor Street, the best house in the best street, as he described it and to which he had moved from Manchester Square only in 1850. Hall had originally intended to spend three months each year in London as a way of keeping in touch with his patients and probably also as a way of dying in harness as was often the case with medical men. In the event, Dr J. Russell Reynolds (1828–96), who was interested in Hall's work was persuaded to leave Leeds and return to London where he was recommended to many of Hall's patients. He took on the lease of Hall's house and there was speculation that a fortune had been made out of the deal. In reality Reynolds paid one thousand pounds for furniture and fittings and one hundred pounds for a grand piano.[1] Grandson of a Physician in Ordinary to George III, Reynolds was a graduate of University College London. In London he gained many of the positions coveted by Hall and quickly succeeded to his leading place as a physician specially interested in nervous diseases. He became Holme Professor of Clinical Medicine at University College in 1862 and five years later succeeded Sir William Jenner as Professor of the Principles and Practice of Medicine. He was also appointed Physician in Ordinary to the Queen's household in 1878 and was created a baronet a year before his death.

Having disposed of house and practice, Hall and his wife prepared to travel to the United States where their son was travelling at the time. On 12 February 1853, four days before Hall's 63rd birthday, the couple embarked at Liverpool on board the Cunard Royal Mail steamship 'Arabia' and set sail for New York, which they reached on 23

February.[2] During the voyage the boat ran into a heavy storm which provoked considerable seasickness on board. 'It is impossible to convey any idea of the complete prostration that the patient experiences during an attack of seasickness, both morally and physically.' So ran an extract in the *Charleston Medical Journal*[3] based on Hall's original paper published in the *Virginia Medical and Surgical Journal.* The extract had begun with an autobiographical note, 'I have recently made a voyage from Liverpool to the United States, and seized upon the occasion to study the physiology of sea-sickness.'[4] Hall also published a note in French on the subject.[5]

Having arrived in New York, Hall and his family set out on a tour of the United States, Canada and Cuba. The *Boston Medical and Surgical Journal* gave details of parts of Hall's itinerary and published abstracts of the lectures, sometimes in French and sometimes accompanied by experiments, which Hall delivered *en route.*[6] During their 15 months' peregrination they had, according to Hall's wife, 'crossed the United States, between the Atlantic and the Mississippi, four times and twice traversed its length, besides visiting Canada and Cuba.'[7] They travelled unceasingly, by railroad, by steamer up and down the Mississippi and across the Great Lakes and by coach through swamps and forests, sightseeing at natural phenomena such as the Mammoth Cave in Minnesota and the Niagara Falls, catching a glimpse of an expedition just about to set off across the Rocky Mountains prospecting for the new railroad to California and visiting communities of America's captive social groups, its Indians and Negro slaves. Thus he witnessed a Council meeting of the chiefs of the Sioux, Chippewas and Winnebagoes tribes. And, more importantly for Hall who had done some preparatory reading on slavery before travelling to the United States, he visited slave plantations during his passage through Charleston and New Orleans. It was also at this stage of the journey that he carried out the experiments with Cartwright and Dowler on local alligators in the Southern States.

Their punishing journey combined with Hall's impaired health and the fact that he had contracted malaria in Havana, for which he was taking large doses of quinine, certainly took its toll and seems to have aged him. One reporter in Boston observed, 'Dr Hall is a man probably between sixty and seventy years of age; his hair is nearly white with age, and his eye possibly a little dimmed .... He spoke in a low voice ....'[8] The throat ailment was definitely causing trouble during the journey because in Buffalo *en route* to the Falls, Hall asked for the planned dinner to be replaced by a more casual soirée at which he would have found it easy to eat little or nothing. (The post-mortem account of Hall's condition indicated that during swallowing in his final years 'much regurgitation could be heard by those sitting near him.') Towards the end of their journey, in January 1854, another reporter regretted that he could not even give an outline of Hall's remarks because 'the distance from him, at which we sat and the low pitch of his voice ... prevented us from hearing distinctly'.[9] Hall even had to cancel some of his lectures. Despite the fact that he hoped to return to the United States he and his wife must have been relieved to go aboard the Collins' steamer, appropriately named 'Atlantic', for their return crossing to Liverpool, setting sail on 15 April 1854.

On Slavery

Introduction

Prior to their departure for the United States Hall had indeed, according to his wife, spent some time concentrating on the subject of slavery and the condition of Negroes in the United States. He had determined to study the subject at first hand during their North American journey but it is not suggested that this interest in slavery was the prime motive for their visit.

Despite the fact that Hall had frequently shown an enthusiasm for making his presence felt, on occasions for self-promotion, there is little reason to doubt the sincerity of his abhorrence of slavery. He had already given up his professional life, he was turned 60 and he was a sick man.

Having been brought up in an evangelical home and having moved in their circles in London, Hall would have been well aware of the involvement of the politically more influential wing of evangelicals who gained the support of public opinion in persuading the Government to end the British slave trade in 1807. The abolition of slavery in British colonies was not accomplished until 26 years later. According to Bradley, the abolition of slavery had been first and foremost a missionary effort and the abolition movement was almost entirely carried out under evangelical leadership and inspiration.[10] Hall's attitude towards slavery could have been influenced by his direct association over the years with Quakers such as Thomas Hodgkin. In America, too, the Quakers had played a leading role in efforts to abolish slavery through developing anti-slavery attitudes. They quickly recognized, as Hall was to observe, that white Americans were deeply prejudiced against Negroes.[11]

In America the two decades prior to Hall's visit and before the Civil War, usually portrayed as a battle to end slavery, were seen as the age of reform. The period was certainly marked by an energetic pursuit of the dollar but it also saw the beginnings and growth of an array of reform movements. These included an outburst of evangelicalism which promoted a range of causes such as pacifism, temperance, prison reform and above all anti-slavery or abolition. But the reform movements did not attract really mass followings.

Not surprisingly for a Francophile, Hall had turned to French literature as a source of information on anti-slavery. Throughout the eighteenth and nineteenth centuries some of France's leading figures had considered the status of the Negro. The Abbé Prevost, Montesquieu and Voltaire had all severely criticized slavery and the slave trade and in the growing liberalism of the eighteenth century, the subject of abolition had occupied an important place in national affairs and in the writing of the period.

In 1835 there were published in France what Tinnin described as two striking and prophetic studies of American government and society. These were Alexis de Tocqueville's *De la Democratie en Amérique*[12] and Gustave de Beaumont's *Marie, ou l'Ésclavage aux États-Unis,*[13] a work which was half novel, half sociological treatise having for its theme slavery and racial injustice in the United States. Hall was familiar with both works.

Beaumont and Tocqueville, two young French aristocrats and magistrates under Charles X and later Louis Philippe, had left France in October 1830 following the July Revolution of that year, for a tour of the United States ostensibly to study the penitentiary system with a view to applying it in France. As in England at the time social reform activity included efforts directed at changing the penal code. In fact, as clandestine moderate liberals, their real interest had been in studying democracy in action. Tocqueville, like many others, believed America to be more egalitarian than it actually was. He failed to see the extent to which money, family and status exerted power in the New World just as in the Old. Nevertheless, before publishing their individual contributions they produced a joint work on the American penitentiary system, based on observations made during their nine months' tour. This journey has been detailed by Pierson in *Tocqueville and Beaumont in America.*[14]

Hall's Observations on Slavery

The Southern States deeply resented the philosophy and activities of the Abolitionists in all their different degrees of abolitionism. Nevertheless, according to his wife Hall found that, in general, Louisiana, with its 'code noir' of social rules for Negroes, displayed a more humane attitude towards slaves than did the State of Mississippi[15], although in both, slaves were regarded as the property of their owners. In New Orleans, Hall had witnessed the severe flogging of a slave (slaves were rarely flogged to death as this would be destruction of property) and he remembered the experience long after. In a letter dated 3 April 1854 to Dr G. Webster of Dulwich he wrote,

> I have not *ceased* to study the fate of the poor African in the United States for a single day, and my mind is *now* quite decided on the subject of self-emancipation, and *against* abolition with discipline, and colonisation .... Lectures and public dinners have occupied me in every large city, but chiefly, as I have said, the afflictive case of the poor Negro.[16]

During his tour, Hall had submitted letters on slavery to a number of American journals. On his return to England these were re-published, with additional material, as a small volume *The Two-Fold Slavery of the United States.*[17] The 'two-fold' slavery embodied in the title was that of primary and acknowledged slavery together with the more subtle form actually suffered in many states by so-called free blacks. They were often not allowed to remain in the state of their former slavedom and found it difficult to obtain employment. Some were even returned to slavedom. Setting forth his perception of the position of the Negro in America and commenting on a number of schemes for abolishing slavery, Hall presented his own proposal for achieving this end, through a process of self-emancipation.

The problem of slavery in the United States Hall saw as essentially a religious issue and one for the conscience of the people concerning their attitude towards more than three and a half million fellow human beings. He was extremely troubled by the prohibition, in some states, of marriage between slaves. Emphasizing the degrading aspects of slavedom, Hall placed great stress on the shame of its being,

> ... a system of unholy and sinful adultery – of unions of the sexes without marriage .... As slavery is assuredly the dark spot on the United States, the absence of marriage – such marriage as is holy and indissoluble – is the dark spot on slavery.[18]

The enforced separation of parents and children was deplored because, as well as depriving slaves of their rights as parents and as citizens, it encouraged second unions or partnerships. He found almost blasphemous the doctrine of the Savannah River Baptist Association in reply to the query of whether or not married parties subsequently separated could marry again. The reply had been, 'That such separation is, civilly, a separation by death, and they believe that, *in the sight of God,* it would be so viewed.'[19] It is not surprising, given

Hall's essentially Christian code of personal ethics that, in proposing a scheme of education as part of the programme of self-emancipation, his main objective was to provide 'free and open access to the sources of instruction in holy religion ....'[20] Education, therefore, was to include proselytizing.

Hall compared the existing, widely advocated unconditional emancipation of slaves with his own proposal for self-emancipation. Unconditional emancipation he saw as setting free many who would not possess the skills and behaviour to take care of themselves. Suffering hunger and deprivation they could fall into intemperance and perhaps even crime. Worse, the infirm would be left completely unsupported. Such issues had already divided the ranks of the Abolitionists who, while united in their desire to end slavery, differed on the means by which they believed this objective should be achieved.

Hall's concern matched that of an increasing number of his contemporaries on both sides of the Atlantic; they saw slavery as a moral contradiction and as incompatible with Christianity. According to Davis the growth of such anti-slavery opinion could itself be seen as evidence of moral progress.[21] But whereas many nineteenth-century writers tended to exaggerate the tension between slavery and Christianity and to disregard the crucial economic function of slavery in American development, Marshall Hall both recognized this function and sought to perpetuate it in a more humane context.

Hall's Proposal for Self-Emancipation

On account of the climate, the white man needed Negro labour to grow cotton, sugar and rice. Hall questioned whether the white man could even cope with the conditions required to produce these goods. Since Negroes could clearly withstand them it was considered imprudent to try deporting free Africans to Liberia, as advocated by the American Colonization Society. They should be given, nevertheless, the opportunity to emigrate if they wished. Hall was aware that during the previous 30 years, while ten thousand free Africans had emigrated from the United States they had left behind three and a half million slaves still in bondage. The repatriation policy of the Colonization Society was harshly criticized[22] for what Hall saw as unjust, forced, deportation even though in Baltimore he had spoken with Mr Latrobe described as 'the talented Secretary of the Liberian Colonization Society'.

Hall believed that his proposal for the abolition of slavery was based on justice and self-interest which would free the slave from bondage to the slave owner and at the same time liberate the slave owner from his guilt of sin against God and against his fellow human. Because of the subsequent systematic oppression still suffered by free American Negroes, the condition of the free African was not so desirable as to make the acquisition of freedom of real value to the majority of slaves. This oppression was more marked in the so-called Free States of the North than in the Slave States which, when he arrived in America, had about 250,000 free blacks. In some States this racial discrimination by white Anglo-Saxons prohibited, under pain of imprisonment, free Negroes from residing and working in them. In other states black people on board ship were, on entering harbour, imprisoned in the local gaols until the ship set sail again. In Philadelphia no African could drive a public vehicle and in New York to be seen arm in arm with a black man could provoke mobbing. These constraints were for Hall part of the second, more subtle dimension of the two-fold slavery suffered by supposedly free people.

Hall recommended that slaves should be given the opportunity of being educated in order to acquire characteristics such as thrift, and be enabled to reach positions whereby they could be paid a fair wage and encouraged to save money towards buying their freedom. Thus, the evolution of a peasantry of hired labour such as was common in Europe was envisaged. Those individuals who did not or could not avail themselves of these self-help measures should remain slaves. This continuing condition of slavery for some was not to be seen as punitive but as being in their best interest.

Hall's plan for self-emancipation with its call for training and discipline, together with special provision for the infirm, was perceived as obviating the hazards and disadvantage of unconditional emancipation. But cooperation between Americans and Africans was essential. In addition, and Hall was certainly concerned for the interests of the slave owner, the value of the slave would be paid to the owner and the freed slave would have the opportunity of continuing to serve his former master as a paid employee. Financial help would be needed for the execution of this proposal, and Federal and State aid was called for, in addition to the assistance of philanthropists. To coordinate the activities involved, Hall proposed the formation of a 'Society in aid of Self-Emancipation'.[23]

Thomas Hodgkin, an ardent activist in the anti-slavery movement in London, was also a gradualist in terms of seeing the need for adequate government provision for slaves to be in place well before freedom was granted. Hall's scheme did not ignore the principle of human equality as enunciated in the opening passage of the Declaration of American Independence drafted by Thomas Jefferson and amended by Benjamin Franklin and John Adams. But it displayed ambivalence on the issue; Hall's views on Negroes and slavery may, to some extent, have coincided with those of Thomas Jefferson. He combined a heartfelt hostility to slavery with a deep conviction that Negroes were inferior. Although in the founding of America Jefferson gave pride of place to the idea of equality he did not mean that he believed that all people have the same talent or sensibility or virtue. Neither was Hall advocating real equality for the Negro. J. R. Pole has examined the question of whether the idea of equality has in fact played an important role in American history.[24]

In spite of drawing an analogy with the Jews who lived 'in the midst of another race of men yet separate', Hall's aspirations for the Negroes definitely envisaged them as being subservient to whites. A more elevated status than that of plantation worker would be rare, even though Hall did once rather ambiguously say of the slave, 'Let him be prepared and qualified to carry on a little occupation, shop or farm, if he should ever attain to one. Let the way to greater elevation be open to him.'[25]

Hall's scheme was based on a concept of mutual self-interest. It could have been acquired from Hodgkin who opposed the unconditional Abolitionists and maintained that regard was due to,

> ... the legally-secured property of those who, on the faith of our colonial laws, have invested their property in the West Indies, many of whom I cannot help regarding as innocent and therefore unfortunate participants in the iniquitous system of slavery.[26]

In Hodgkin's view 'These rights it is our bounden duty not to trample upon, whilst laudably engaged in endeavouring to redress the wrongs of the Negroes.'[27] Hall would also have been aware of Hodgkin's positive, but to Hall unacceptable suggestions based on national economic interests. Hodgkin supported colonization of the coast of West Africa by Britain, seeing it as a means of inhibiting slave trading between Africa and the rest of the world. Furthermore, it would curb what he claimed to be the more serious problem, that of slave trading within Africa itself.

Hodgkin believed that such colonization would open up markets for Britain to the mutual benefit of the British and Africans. The American Colonization Society, which was already making territorial gains in Africa, would exploit the West African potential if Britain did not. America, referred to as 'Britain's daughter', was clearly seen as a rival in spite of Hodgkin's assurance that,

> It is not to induce our country to undermine, or compete with this influence which America is obtaining, but to lead her to participate in it as the reward of her exertions for the good of Africa, that the British Colonization Society is established.[28]

In any case Hodgkin was honest enough to spell out again the importance of the economic gains for Britain, faced with losing some of its overseas markets,

> ... it is not only for the interest of our national credit, as successful philanthropists, but of our national prosperity, as manufacturers and merchants, to promote the establishment of new markets, which may receive our commodities when the older markets shall close against them.[29]

Hall's small work caused little stir either in America or in Britain. Certainly it was not followed by the furore in America which had followed the publication almost 20 years earlier of Harriet Martineau's *Society in America*,[30] based on her two-year stay in the United States from September 1834 to August 1836. During this period she had covered much the same route as Marshall Hall and in her wider treatment of American institutions had included observations and personal comment on the subject of slavery, which as a Unitarian she could not support. Following her return to England she had retained her involvement in the issue and was described by John Collins, an active American Abolitionist, as 'one of the most devoted among the Abolitionists in Great Britain'.[31]

For the public at home, Hall's publication provided a useful reminder of an issue still relevant in spite of legislation half a century earlier to abolish slavery. In preparing an extract of this work for inclusion in his mother's *Memoirs,* Marshall Hall's son appealed to 'our Transatlantic Brethren' to weigh his father's words and apply his scheme unless a better one should be found. His son expressed the hope that,

> ... the present dangerous crisis in the history of the United States may be passed in safety, and its citizens be led to legislate conscientiously for the future of both races; ....[32]

Although the work of slaves could have been performed by a wage earning, free labour force such as Hall envisaged, few people confronted the issue. This was partly for economic reasons but more importantly, as Pole has indicated, this resistance was largely due to the fact that even those who deplored slavery, and told one another it would one day disappear, were unable to separate it from the problem of race. His 'second slavery', oppression due to prejudice and persecution, Hall continued mistakenly to see as a phenomenon of the Northern States believing it to be absent from the South, 'where the domestic relation between the African and the European are so much more intimate'.[33] Hall had clearly failed to see that this intimacy too was based on a master-servant conception of inequality between European and African – a conception which Hall, although unaware of it, had shared.

Return to England

On their return to England, Marshall Hall and his wife stayed with relatives in Nottingham where a medical soirée was arranged and Hall lectured on the nervous system. A threatened outbreak of cholera in Nottingham sent Hall back to London with the intention of spending the latter part of the summer on the Continent had not further outbreaks of cholera postponed this visit. Delayed thus in London, Hall attended the opening of the Crystal Palace and accepted an invitation to travel to Manchester to deliver three lectures at the Chatham Street School on the Spinal System. These he illustrated with demonstrations on live material.[34] An attack of laryngitis somewhat marred the third lecture, the last he ever delivered in England.

A vote of thanks was proposed by Mr Wilson and seconded by Dr Noble who told of having worked on Hall's discoveries for 20 years, as well as having witnessed some of his experiments. Noble had himself been the target of attack from critics including Carpenter even though he questioned the role of vivisection in making discoveries in physiology.[35]

Late in 1854 Hall and his wife travelled to Europe visiting Louis on their way to the South of France and thence to Rome, where Hall studied Hebrew and lectured to Italian physicians and members of the British Colony.[36] In Naples the newly active Vesuvius impressed Hall almost more than had Niagara Falls. Returning to Paris, Hall spent three months completing his small work, *Aperçu du Système Spinal*,[37] written in French and dedicated to Flourens. This work summarized Hall's investigations on the nervous system.

On Resuscitation

Although he seemed to have benefited from the period abroad, Hall's health began again to deteriorate about mid-December 1855 and thereafter continued to decline. By June 1855 his throat condition had steadily worsened, for he had replied to a dinner invitation from Sir Joseph and Lady Oliffe begging to be allowed to join the party *after* dinner.[38] Nevertheless, Hall retained his curiosity and enthusiasm to embark on new ventures.

As a committee member of the Royal Humane Society, which had been established in 1774 for the purpose of restoring those apparently dead from drowning, Hall became active in the Society during his retirement. Through the Society's Annual Report for 1855 Hall's attention was drawn to the problems associated with resuscitating the apparently drowned.[39] Cases of drowning through accidents on inland waters and at sea, information on the design of lifeboats and on the location of receiving houses where those drowned in the Thames could be treated, were contained in the Report. It also provided details of the Society's recommended method of treating the drowned. Essentially, this consisted of moving the body carefully and speedily to the nearest house, cleaning the mouth and nostrils and attempting, through the application of heat to restore the natural warmth of the body. This was to be achieved with a warm bath at blood heat or if not available, with warming pans, hot water bottles or hot flannels and rubbing of the body.

Through his earlier research on the physiology of respiration and on the reflex control of breathing, Hall was struck by what he saw as the inadequacy of the 'Rules' of the Society which had, he felt, the wrong objective in view. What the apparently drowned person needed was immediate restoration of lung action, that is inspiration and expiration of air, in order firstly to rid the blood of the accumulated poisonous carbon dioxide which was causing the condition of asphyxia and secondly to restore oxygen to the blood. By recommending submersion in heated water to warm it up and restore life, the Society was tackling first the problem of coldness of the body. But the cold and insensible conditions of the body was a secondary effect which could be eliminated if breathing could be restored first.

The founders of the Society had, in 1774, advertised their original recommended method for attempting resuscitation.[40] Like their successors three-quarters of a century later they insisted on careful handling of the body and condemned certain current practices, such as hanging drowned persons up by the feet or placing them over a barrel, warning that such treatment could be injurious. However, more significantly, while they recommended speedy transfer of the body to a warm place and offered financial inducement of one guinea to publicans, bakers and the like if they would receive drowned bodies into their warm premises, it was made clear that this was preliminary to the ensuing stage in which 'various stimulating methods should next be employed'.[41] The Society gave as the most efficacious method, what is an excellent description of mouth to mouth resuscitation,

> ... to blow with force into the lungs, by applying the mouth to that part of the patient, closing his nostrils with one hand, and gently expelling the air again by pressing the chest with the other imitating the strong breathing of a healthy person.–The medium of a handkerchief or cloth may be used to render the operation less indelicate.[42]

Thus in 1774 active attempts at artificial respiration were an important part of the method recommended by the Humane Society. Over the years the bellows had come into use as a means of attempting to induce respiration. But by about 1800[43] emphasis was placed on recommending the application of external heat or friction to restore the circulation. Mouth to mouth resuscitation was to be attempted if no bellows were at hand 'but whenever a bellows can be procured, it is to be preferred, as air forced in by this means, will be much more serviceable than air which has already been breathed'.[44]

However, in the 1832 Report the use of the bellows in the hands of inexperienced persons was shown not to be without its dangers.[45] An alternative method involving compression and relaxation of the thorax with a bandage-like structure, quickly made on the spot by partially tearing a sheet or tablecloth wrapped around the thorax was introduced. This method, although no longer part of the main recommendations was still included among a series of additional notes in the 1853 Report.[46]

Hall looked for a scientific rationale to support his attempts to re-emphasize the primary importance of artificial respiration over warming. He initiated a number of experiments which were carried out at St George's Hospital where he was helped by three students, Charles Hunter, Edward L. Fox and Robert L. Bowles.[47] According to a note from Hunter to Charlotte Hall for her *Memoirs,* Hall himself performed many of the experiments, including some which involved the inhalation of various gases.[48] The experiments were repeated by two other individuals, Dr Ogle and Mr Bullock.

In January 1856 a pamphlet was published giving details of these experiments along with Hall's proposed procedure for handling drowned subjects. A copy was presented to the Royal Humane Society and referred to in the 1856 Report,

> Dr Marshall Hall, a member of the Committee, has just presented to the Society a Manuscript Copy of a Treatise on 'Asphyxia; being an abstract of an investigation of its nature, carbonic acid blood poisoning, and its remedy, prone and postural respiration,' He is also

> having it printed, at his own expense, with a view of freely supplying the Committee with a sufficient number of copies for the purpose of being sent to all the Medical Assistants of the Society, for which he has received the thanks of the Committee, as also for the lively interest he has thereby evinced in the objects of the Society.[49]

Despite Hall's efforts, the method for resuscitation recommended by the Society for that year was identical with that published in the previous year; it still recommended the use of the warm bath which Hall had condemned.[50]

Hall's method which came to be known by his name, and which survived in domestic medical books until well into the present century, involved first clearing the mouth and windpipe. The operator was then to watch for spontaneous breathing to restart or to intervene by deliberately stimulating respiration. This needed a careful rhythmic combination of placing the patient in the prone position on his chest, to promote expiration and prevent blocking by the tongue and then turning him onto his side to promote inspiration, **Plate VIII.** In the *Lancet,* Hall explained that,

> These movements being made alternatively, gently and equably, *sixteen* times a minute, and no more, respiration is efficiently performed, and this without bellows, syringe, or any apparatus, all of which are not only unnecessary, but for reasons which cannot now be detailed, dangerous.[51]

A full set of *New Rules for the Treatment of Asphyxia* was promised for a later date. The next contribution 'Asphyxia, its rationale and its remedy', appeared in the *Lancet* in April 1856, setting forth Hall's plan with its different priorities compared to those of the Royal Humane Society, whose procedure was openly criticized,

> To elevate the temperature, or to accelerate the circulation, *without* having *first* secured the return of respiration, is therefore *not to save,* but in reality to *destroy life*![52]

Hall's plan was in effect an application of the work of Milne Edwards, always admired by Hall. Milne Edwards had advanced the principle that,

> 1. The *higher* the temperature, within physiological limits, the sooner the animal *perishes* if deprived of air.
> 2. The *lower* the temperature, also within physiological limits, the *longer* the animal *survives* the privation of air.[53]

This being so, Hall had reasoned, in the patient afflicted with apnoea and not breathing, every means of raising the temperature, such as a long spell in a warm bath and several other means as recommended by the Royal Humane Society are actually deleterious. These illustrations were used to criticize medical colleagues for their lack of awareness of the significance for clinical practice of research results such as those of Edwards,

> The truth is, the propriety of the warm bath in apnoea is a vulgar error; an error which physiology in the volume of Edwards ought long ago to have corrected. Who will pretend, after this, that medicine is not physiological, ...?[54]

Plate VIII: Book plate

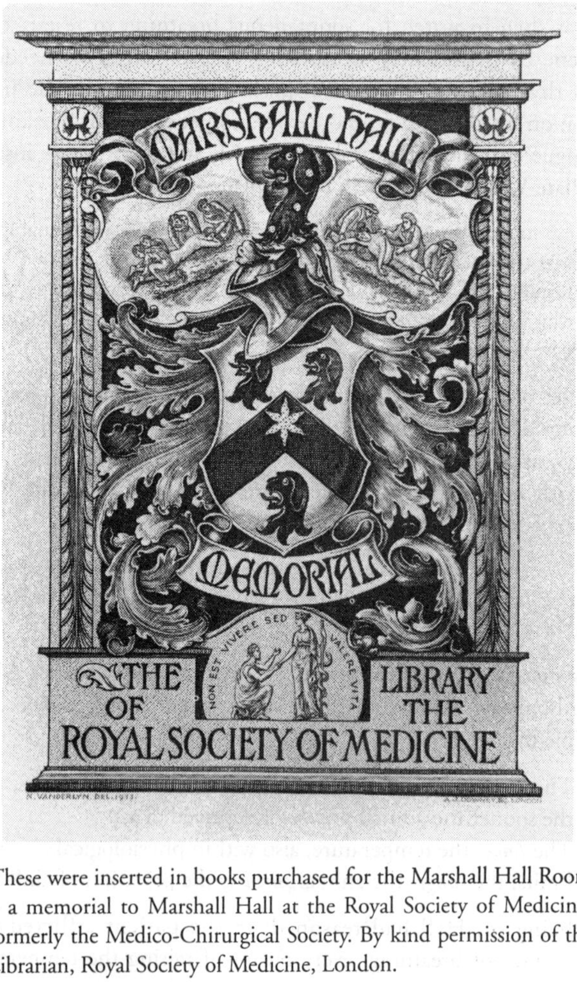

These were inserted in books purchased for the Marshall Hall Room – a memorial to Marshall Hall at the Royal Society of Medicine, formerly the Medico-Chirurgical Society. By kind permission of the Librarian, Royal Society of Medicine, London.

And so, contrary to the Royal Humane Society recommendations, a cold environment was advocated for carrying out the procedures. This was partly to obviate the danger of delay, but also for the more scientific reason that cold air was recognized to be an undoubted stimulator of respiration. Actually, a sudden short but immediate plunge into a bath of hot water at 100° to 104°F would, if available, prove an even more powerful excitor of respiration but is rarely if ever available. This point had already been made in the Society's 1774 pamphlet.

Although Hall's so called 'Ready Method' was first envisaged as a means of treating the apparently drowned, it came to have more extensive application to three main groups of mishap where respiration was arrested. Firstly were cases of impending death from accidental causes including poisoning, or the maladministration of chloroform during surgical operations. Secondly, were other cases of impending death from disease conditions such as convulsion, epilepsy, choking and serious head injury. Finally, Hall's 'Ready Method' could be a suitable remedy for the still-born infant. Of these Hunter later said,

> Whatever may be the number of lives added to the population by the Marshall Hall method, in the two classes just described, they fall very short of that of the infants apparently dead that have been brought to life by the long and persevering efforts of a vast number of the professional men of England.[55]

During Hall's lifetime the *Lancet* published letters and accounts citing instances of lives saved through his method. On 18 July 1857, Frederick James Reilly of Mile End in East London outlined a difficult labour where he had resorted to Hall's method in an attempt to revive the still-born infant which was eventually expelled. Reilly wrote '... I cannot help sensibly feeling that my little patient is truly indebted for its life to Dr Marshall Hall.'[56]

In the meantime, Hall had returned to consider the physiological issues involved in apnoea and asphyxia, topics related to his early Royal Society paper on respiration and hibernation. He now carried out a series of experiments over a six-month period. Looking ahead he proposed carrying out 'on the first opportunity' an investigation involving the introduction, into his experimental animal, of two specially-made thermometers, one into the jugular vein on the right and the other into the carotid artery on the left to detect what he speculated would be a lower temperature on the left.

It was following Hall's death that his original pamphlet together with articles and correspondence on Hall's Ready Method from the *Lancet* were assembled by his son[57] together with other views on respiration and apnoea, especially those of Milne Edwards. According to Charles Hunter, Hall's proposal appealed on account of its simple, ready and unencumbered procedures which required 'nothing save the observing eye and the active hands of the bystander, who would thus proceed to employ the method'.[58]

On Strychnine Detection

Just as Hall's contribution on artificial respiration was a beneficial application of his research into the physiology of respiration, so his long-standing interest in the effect on the nervous system of drugs such as strychnine, used as a stimulant and tonic, came to have practical application. As early as 1837 in his *Memoir* (1837), under 'The therapeutics of the excito-motory system' Hall had referred to the use of strychnine as medication in the treatment.[59] Furthermore his conclusion contained suggestions for new inquiries on the nervous system and he had already submitted papers to the *Comptes Rendus* in 1847 and 1853 on the effect of strychnine on the frog.[60]

In 1855 and 1856 a number of criminal poisonings in which the use of strychnine was suspected were reported. One of the most celebrated murder trials concerned William Palmer, accused in November 1855 of murdering John Parsons Cook with strychnine. Hall became involved in the lively debate on the Palmer-Cook trial which considered the issue of strychnine and its detection. The case was published in the *Lancet.* It led to Hall proposing a physiological test as a more sensitive alternative to the usual colour test for strychnine.[61] According to Hale-White everyone in the country was excited by Palmer's trial since it was the first time strychnine had been used for murder.[62] Hall became interested in at least the science of the case and Hale-White claimed that he was among the many doctors briefed for Palmer's defence but was not called 'for he would have had to say that the symptoms were those of strychnine poisoning'.[63]

For his strychnine detection test, Hall worked on animals other than frogs but claimed that frog nerve-muscle tissues in which strychnine produces the violent tetanic convulsions typical of strychnine poisoning in humans, were more sensitive. On account of this susceptibility and their use in strychnine detection, Hall described the frog as strychnoscopic. Experimental procedure was refined to such an extent that while initially Hall could detect 1/50 of a grain of acetate in strychnine he came to claim that he could detect 1/16,000 part of a grain of the salt.[64] Both Hall's method and the colour test had their supporters;[65] one anonymous contributor to the *Lancet* claimed to have detected 1/50,000 part of a grain with the colour test.

At the Palmer-Cook inquest there had been a difference of opinion between the analytical chemists on whether or not strychnine is indestructible within the body, or undergoes positive change following its absorption into the blood stream. The press seized on this conflict with suitably eye-catching headlines. One discussant in the debate even blamed this publicized difference of opinion for the death of a woman found soon afterwards to have been murdered with strychnine.[66]

Further evidence revealed that the substance is not immediately destroyed in the body. The professor of Chemistry at the Royal Institution in Manchester revealed that he had once investigated the deaths of several dogs from a Cheshire pack of hounds which had been wilfully poisoned in 1849. He had suspected strychnine poisoning and his subsequent investigations included work on the disinterred bodies of some of the animals. He succeeded in preparing for exhibition in court, a sample of the hydrochlorate of strychnine, thus demonstrating that strychnine will resist decomposition in the stomach for nearly a month.[67]

There were other legal cases involving suicide both intentional and accidental with *nux vomica,* which contains strychnine and which was more readily available from druggists than arsenic, following the 'Sale of Arsenic Prevention Act'. They provoked J. Nichols to question the suitability of strychnine as a medicine. Nichols held that potentially fatal effects should not exclude from the pharmacopoeia any drug with otherwise highly beneficial uses. But he felt that strychnine was not in this category since its initial, anticipated usefulness in the treatment of paralytic affections had not been fulfilled and other drugs could perform its beneficial functions. If it were found impossible to remove it as a medicine, Nichols appealed for strychnine to be kept in such a state of solution that while a physician could use it as a remedy it would be 'unfit for the deadly purpose of crime'.[68]

Nichols' plea was taken up by Hanotte Vernon from St Mary's Hospital who claimed that legislation had always failed to prevent crime. He deplored the suggestion that any poison should be withheld, although he agreed that the use of powerful agents such as strychnine should be controlled.[69] In his lectures George Harley, who taught practical physiology at University College, considered the mode of action of strychnine.[70] Having given Brown-Sequard's opinion that 'the action of strychnine' consists in an increase of nutrition of the nervous centres which leads to augmentation of the reflex faculty, Harley explained that he himself was,

> ... inclined to believe that strychnine acts by destroying the power of the tissues and fluids of the body to absorb oxygen and exhale carbonic acid; in a word, from thus arresting nutrition by preventing the interchange of the gases in the animal economy.[71]

Harley was on the right lines.

In July 1856, Glover, a former lecturer in Toxicology, provocatively posed the question 'Was Cook killed by strychnia?' In Glover's view it seemed that while circumstantial evidence strongly implicated Palmer as the murderer, the agent could have been another poison such as a compound of cyanogen, which produces symptoms similar to those exhibited by the victim Cook and to those of strychnine poisoning. He felt ironically that Palmer, in protesting that he was unjustly condemned since he definitely did not kill Cook with strychnine, could have been innocent of precisely that crime.[72] Hall's work had been the earliest physiological test for strychnine detection. It was still recommended in textbooks on toxicology during the second half of the nineteenth century.

On Sewage Disposal

In 1852, in an essay, *Suggested Work on the Thames,* one of his earliest retirement publications, Hall set forth a building programme addressing and linking two pressing transport problems in London.[73] First, was the problem of sewage collection and its subsequent handling and profitable dispersal as manure. Second, was that of providing enhanced means of transport for London's increasing population moving around and within the City area. Hall's programme involved using the River Thames and its banks to solve the two problems which, if implemented as he proposed, could on grounds of cost saving, share certain facilities. Hall returned to this issue again in 1856[74] when little had been done to ease the transport problem and sewage was proving to be an increasingly serious health hazard in Plumstead, South West London.

Changes in Britain's economy and the growth in its population from the mid-eighteenth century on, were putting enormous strains on the resources and institutions of all its cities, especially London whose population more than doubled over the first 40 years of the nineteenth century from 958,000 to 1,948,000. The response of local government to the demands on it and the uncoordinated, and frequently corrupt behaviour of the bodies responsible for housing, roads and sewage, was frequently inefficient. Housing speculation had led to the erection of high-density cheap urban housing with inadequate sanitary arrangements and associated roads which lacked drainage. Back-to-back housing with one privy to 30 buildings and water supplied for one hour a day from a public stand-pipe was a common feature.

Accumulating deposits of human excrement and other domestic waste combined with industrial refuse and smoke from mills and furnaces led to a marked deterioration in urban living conditions. In London there was a marked contrast between the mortality rates of low lying areas such as Southwark, Bermondsey and Rotherhithe, which lacked even sub-surface drainage, and more elevated districts such as Hampstead.[75] These developments reversed an earlier trend whereby the mortality rate had been falling during the 1820s. It began to rise and in the new manufacturing towns at this time more than one in every two children died before it reached its fifth birthday.

Already in 1827, in response to the mounting public criticism of the foul water supply in London, Roget had been ordered by George IV to set up a commission to investigate the water supply of parts of West London. Roget and his commission met with stiff resistance from the private water companies. And the government itself limited Roget's brief to a consideration only of the quantity and quality of the water. Roget produced a most competent if politically embarrassing document, the first of its kind.[76] The report was ignored and disappointingly, but hardly surprisingly, Hall made no reference to it.

Two individuals who became deeply involved in public health issues, including the sewage problem, were Edwin Chadwick[77] and John Simon.[78] Chadwick, who had trained as a barrister, became England's first paid Commissioner for Public Health and Secretary of the Poor Law Office. He and Simon became antagonists over the respective roles and importance of engineers and the medical profession on issues of public health. In 1848 Simon, a medical man, was elected Medical Officer of Health for the City of London; he represented the claims of preventive medicine and science in sanitary affairs. Chadwick had dismissed the ideas of medical men in favour of an engineering solution, for coping with the City's refuse and sewage. He pressed the need for a constant supply of water to implement his plan, but acknowledged the problems, including medical ones, of pollution created by discharging sewers into the Thames.

The medical profession itself was divided and Bancroft of St George's Hospital led a powerful pressure group which declared that putrefying animal matter was innocent of the charge of producing contagious fever.[79] This was certainly contrary to Hall's views and did nothing to enhance the perception of medical men as reliable specialists on the issue of sewage disposal.

A report by Chadwick, produced after two years' intensive work and eventually published under his own name in 1842,[80] voiced merciless strictures on the London Water companies. It also criticized local administrations and displayed contempt for the medical profession. But it contained some excellent suggestions for house and street drainage. Like Roget's report it was ignored.

Both Chadwick and Simon ran into political difficulties with government and other interest groups despite, in the case of Chadwick, having Charles Dickens as an outspoken champion on the Health of Towns Association.[81] Each retired a disappointed man before either the transport or sewerage problems had been satisfactorily advanced. Hall's contribution to the whole debate was but one of many which continued and indeed still continue to appear on the issue of London's sewage and transport problems.

At the time of the appearance of Hall's first building proposal there was already under consideration a new project for an additional large central station in the City connecting all London's existing railway stations.[82] This development was to be combined with a new street in the City and another bridge across the Thames. Compulsory purchase orders had already been served on owners of houses and other property which would have to be demolished for the scheme to go ahead. The idea of a new central station and associated road was dismissed by Hall on the grounds, quite rightly, that this would only increase the existing unbearable congestion on already overcrowded streets in the area.

Hall's own project, which would, he claimed, bring commercial prosperity to the City so that investors would soon reap rewards, would not involve the demolition of any existing property. It was aimed to utilize the transport facilities already present and to simplify existing sanitary, agricultural and commercial arrangements. There was proposed a less expensive circuit railway, analogous to the present-day Circle Line on the underground system, connecting all the main London stations and to which could easily be linked the existing outer railway circuit system on the outskirts of London. Hall's plan, which even Brunel initially misunderstood to involve the demolition of properties, wharves and warehouses along the banks of the Thames, was not envisaged as an overground construction but one involving the construction of tracks within the banks of the river above high water mark. Hall seems to have imagined that the track could remain open like an attractive colonnade rather than running in closed tunnels like the present underground system. These tracks in the river banks were to be linked above high water mark by connections across the river, running along the sides of London and Westminster bridges. The system as a whole would enable people and goods to be moved more rapidly within the City and would bring it increased prosperity.

The other part of Hall's project, the handling of London's sewage, involved the construction, again within the river bank from Chelsea to Limehouse but below low water mark, of cloacae or main sewers into which would run all the subsidiary sewers of the metropolis. Following the suggestion of his son, Hall advocated the separation of solid faeces from other domestic waste water. The cloacae would need to be set in strong spacious mason work and this, it was suggested, could provide the foundation for the Thames railway circuit. Added to the provision for separating household waste, were plans for the proper drainage of marshy and other undrained areas in South East London and for the erection of a Thames flood barrier, which he called a Tide-excluding wall. Hall would have been familiar with the London flooding in 1841. The barrier, built at Woolwich and opened in 1982 did not, therefore, become a reality for more than 130 years after its appearance in Hall's proposal.[83]

Familiar with Liebig's *Chemistry in its Application to Agriculture and Physiology*,[84] Hall was struck by the cost of imported guano and other fertilizers for agricultural purposes. He considered that, apart from the dangerous polluting of rivers and fouling of the atmosphere, the country was losing a valuable resource in not retaining and using its own sewage. By the middle of the nineteenth century the impoverished state of the soil was already causing concern. As Liebig had pointed out, hundredweights of phosphates were flowing annually into the sea from the Thames and other rivers. He had speculated that if Britain harnessed these materials it could soon dispense with its importation of foreign corn. In effect, the cost of importing guano, which in 1849 was half a million pounds and rose to over a million in 1851, amounted to a duty on corn. Hall computed that 'The quantity of cattle, of corn, of green vegetables, etc. introduced into London, must approximately represent the quantity of sewage.'[85] It would indeed be worth recycling the sewage.

To establish an effective and efficient sewage system, it was necessary to construct a double circulatory plumbing system of pipes or 'arteries' carrying fresh water to dwellings and a venous system for carrying used water and excreta from them. The venous system of pipes would need to be wider to accommodate the rain. Returning to this whole project in 1856 in a debate carried out mainly in the *Lancet*,[86] Hall again emphasized his son's suggestion of separating the closet excreta, of about 5 million gallons a day from other domestic waste water of about 45 million gallons. The excreta were to be channelled into the cloacae while the rest was to be allowed to flow separately into the river. By 1856 these volumes had considerably increased.

Using, if possible, the same engines as those which would move the trains, the Thames cloacae were to be emptied nightly of their solid contents, which would tend to sink. The solid material was then to be transported each night by rail between the first and last mail train, in closed waggons all over the country, from 50 to 500 miles away and perhaps even in some places, to the very fields on which the material was to be spread as manure. Emphasizing the potential commercial and agricultural gain which could flow from his proposal, Hall claimed that,

> The loss sustained by our fields will be restored, *pari passu;* the importation of bones and of guano may cease; the pollution of the atmosphere and of the river will be prevented; the threatened nuisance and infection at Plumstead will be avoided.[87]

Hall was concerned about the fate of the sewage water following the removal of solids. It could itself provide valuable fertilizer material but in the absence of resources to handle it, Hall saw no alternative but to allow it to flow back into the river. But neither had Robert Stephenson nor others addressing this problem, made any alternative suggestion for using this water. Hall's plan was criticized on the grounds of cost and also on the inadequate arrangements for flushing out the sewers themselves.

Four other new sewage disposal plans had recently been considered by the Metropolitan Board of Works which still lacked a medical voice and which the *Lancet* accused of spending more time on frivolous matters like discussing the heraldry for its seal than on the importance of sewage disposal.[88] The plan selected as most suitable involved the disposal of sewage in the Erith Marshes. The *Lancet,* which had championed Hall's plan, cautioned the Board to seek medical opinions of at least a wider group of engineers, a chemist and a medical professional.[89]

Thus Hall's proposals went the way of those of Roget and Chadwick and little more happened. He had completed his own contributions to the debate with a call for the setting up of a 'Society of State Medicine'.[90] This could investigate the relationship between the problems of sewage disposal and the health of the nation's cities, rural districts and armed forces. State medicine did develop in England but not until the decade following Hall's death. There is current concern about city sewers and sewage disposal, about water recycling, the safety of drinking water and of the beaches and sea water around the country. Of the sewers, one Manchester spokesman in the 1970s predicted, in a radio programme, that for several cities coping with systems a century and a half old, if action is not taken soon 'They will fall into their own sewers.'

Hall's Terminal Illness

Following a final visit to the Continent in 1855, Marshall Hall and his wife returned to London in September that year. His grandson was born soon afterwards. Although Benjamin Brodie had, a decade earlier, detected no obstruction when he passed a bougie into Hall's pharynx, his throat trouble deteriorated still further before the end of the year and his expectoration became tinged with blood. But Hall continued to work and to publish on all his areas of interest. A few visits were made to the south coast and Boulogne although a hoped-for visit to Jerusalem and Egypt was not to be. In the meantime his pharyngitis and its attendant dysphagia increased and he lost weight. Hall's medical advisers, including Dr Williams and Mr Caesar Hawkins, suspected ulceration of the larynx or pharynx and Hall wrote of this diagnosis to a number of individuals including Louis to whom he sent a detailed account of his whole case.[91] Paget, who examined him in July 1857 had received a similar account. Hall clearly felt that his end was approaching although, according to his wife, he was not over-preoccupied with his condition. Indeed he kept up an active correspondence including communication with members of the 'Society of Friends' on the abolition of slavery and with the *Lancet* on a variety of subjects. Nevertheless, by March 1857 Hall had already let it be known to a nephew that he wished to be interred at Nottingham in the General Cemetery and in a spot looking towards the location of his former home and his father's works.[92]

Throughout 1857 his worsening condition was exacerbated by recurrences of the malarial fever first experienced during his American tour. Swallowing became so difficult that he himself thought he would die of inanition. Describing at length Hall's final months, his wife spoke of the distress caused by his hunger during much of the time. Hall was active to the end having dictated, six days before his death, a letter for Dr Hunter giving his opinion on a respiratory condition which they had both been discussing.[93] One of the Brighton doctors in attendance during this period was Dr Alfred Hall who asked permission to name his next son after Hall. In this way Sir Edward Marshall Hall (1858–1927) the renowned criminal lawyer became Hall's namesake.[94]

Plate IX: Hall's Memorial and Tomb – Nottingham

By kind permission of Dr Richard Jones, Newcastle, who took this photograph for the author.

Hall died 11 August 1857. The post-mortem examination and report, made by Dr Ranson of Nottingham, revealed a stricture together with ulceration of the oesophagus. Histological examination of one of the patches or deposits found in the mucous membrane of the trachea revealed cell appearances which together with the rest of the report led D'Arcy Power, writing in 1891, to conclude that Hall died 'from cancer of the upper part of the oesophagus'.[95] An account of the history of Hall's case together with the post-mortem appearances was published in the *Lancet*[96] and repeated in the *British Medical Journal.*[97] Hall was buried 19 August 1857 in Nottingham according to his wishes. His wife, who died 14 January 1885, aged 83, was buried in the same vault and a large monument was later dedicated to them both, **Plate IX.**

Hall's death was widely reported in journals and newspapers in England and abroad and a number of them also presented obituaries of him.[98] Having already been the subject of biographical articles in his own day, he continued to be the subject of others for some time after his death. During the century and a quarter since his death, references to Hall have continued in the form of chapters within larger works and within theses, some of which have been devoted to him. He has continued to figure consistently in encyclopaedias and dictionaries, especially for his work on the nervous system.

Notes

1. According to the *Dictionary of National Biography,* Hall was exposed to censure because he had entered into an agreement, contrary to accepted practice among physicians, to transfer his patients to Reynolds. Hall had put straight the record of the financial transaction in a dictated note just before his death. Hall, C., *Memoirs,* p.280.
2. The arrival was reported as 'Arrival of the Arabia .... The Cunard steamship *Arabia,* Captain Yudkins, arrived at this port at 2 o'clock yesterday afternoon. She left Liverpool at 12 o'clock on Saturday, the 12th. inst. She has thus made the passage in eleven days and two hours.' *The New York Herald,* 24 February 1853, p.2.
3. Hall, M., 'The Physiology of Sea-Sickness', Charleston *Med. Journ. and Rev.,* 8, 1853, pp.682–3.
4. *Ibid.*
5. Hall, M., 'Note sur le Mal de Mer', *Comptes Rendus,* 35, 1853, pp.600–1.
6. See *Boston Med. and Surg. Journ.*, 48, 1853, p.146; *ibid.*, 49, 1854, p.171. The latter note, 21 September 1853, informed readers that 'this distinguished medical writer ... arrived safely in Boston on Saturday last, from an extensive tour of the West'.
7. Hall, C., *Memoirs,*, pp.320–1.
8. *Boston Med. and Surg. Journ.*, 49, 1854, p.532.

9. 'Complimentary dinner to Dr Marshall Hall', *ibid.*
10. Bradley, I., *Call to Seriousness,* 1976, p.80. Bradley focuses on the Anglican Evangelicals who were not subjected to the same discrimination as the Methodist Evangelicals to which Hall's family belonged.
11. Jordan, W. D., *White over Black,* 1968, p.xii. A recent New York report recommending changes in the way social studies are taught suggests that 'race' should now be seen as a cultural not a physical description; this, it is claimed, would lead to a 'better understanding of the stunting effect of racism on US society and its people'. *Time,* New York, 8 July 1991, p.25. Jordan published a drastic abridgement with minor modifications of his large earlier work, six years later. *Idem, The White Man's Burden,* 1974.
12. [De Tocqueville, J. A.], *De la Démocratie en Amérique,* 2 vols, 2nd edn, 1835. Hall probably read the original work although it was quickly translated and became a transatlantic masterpiece. It has subsequently been translated many times e.g. De Tocqueville, J. A., *Democracy in America,* 1969.
13. De Beaumont, G., *Marie ou l'Ésclavage aux États-Unis,* 2 vols, 1835. This work went through five editions but was not translated until 1958. De Beaumont, G., *Marie or Slavery in the United States,* 1958.
14. Pierson, G. W., *Tocqueville and Beaumont in America,* 1938.
15. Hall, C., *Memoirs,* pp.315–16.
16. *Ibid.*, p.321.
17. Hall, M., *Two-Fold Slavery of the United States,* 1854.
18. *Ibid.*, pp.65–6.
19. *Ibid.*, p.68.
20. *Ibid.*, p.69.
21. Davis, D. B., *The Problem of Slavery in Western Culture,* 1970, p.40. Professor Davis' Inaugural Lecture at Oxford was provocatively entitled: *Was Thomas Jefferson an Authentic Enemy of Slavery?,* 1970.
22. Hall, *Two-Fold Slavery of the United States,* pp.54–8. Hall believed that 'the Colonization Society is the calumniator of the African race!' *Ibid.*, p.55.
23. The full title proposed by Hall was 'Society in Aid of Self-Emancipation from Slavery in the United States.' *Ibid.*, p.73.
24. Pole, J. R., *The Pursuit of Equality,* 1978. According to Pole it has not been any particular conception of equality but the abstract and philosophical question of the nature of true equality, which has dominated American history. See also, Pole, J. R. (ed.), *Slavery, Race and Civil War in America,* 1974. Pole, J. R., 'Equality', in Green, J. P. and Pole J. R., *The Blackwell Encyclopaedia of the American Revolution,* 1991, pp.616–19.
25. Hall, *Two-Fold Slavery of the United States,* p.60.
26. [Hodgkin, T.], *On Negro Emancipation,* 1832, p.19. The main portion of this pamphlet consists of a letter dated 27 July 1832 and addressed 'To a member of the Anti-Slavery Society.'

27. *Ibid.*
28. Hodgkin, T., *On the British Colonization Society,* 1834, p.2.
29. *Ibid.* Hodgkin had already published in the previous year, in defence of the Society, *An Inquiry into the Merits of the American Colonization Society,* 1833.
30. Martineau, H., *Society in America,* 3 vols, 1837.
31. Collins, J. A., *Right and Wrong among the Abolitionists of the United States,* 1841, p.3.
32. Hall, C., *Memoirs,* p.336–7.
33. Hall, *Two-Fold Slavery of the United States,* p.17.
34. The lectures were published in the *Lancet* by Mr Southam who expressed his opinion that Hall's views on the spinal system had the same relative importance to medical science, as the atomic theory of their late illustrious townsman, Dalton, had to the science of chemistry. Hall, M. 'Three Lectures on the Spinal System and its Diseases', Delivered at the Chatham Street School of Medicine, Manchester, *Lancet,* vol.ii, 1854, pp.391–3; pp.411–12; pp.451–3. Hall had told his audience that the lectures could be regarded as a specimen of work in preparation for students, with the title *Physiology in the Laboratory and the Clinical Ward.* The work did not appear.
35. Noble, D., *The Brain and its Physiology,* 1846, pp.30–5.
36. Hall, C., *Memoirs,* pp.347–9.
37. Hall, M., *Aperçu du Système Spinal,* 1855.
38. MS Letter, M. Hall to Sir Joseph and Lady Oliffe, dated 22 June 1855, London, Wellcome Institute for the History of Medicine.
39. *Eighty First Annual Report of the Royal Humane Society,* [1855].
40. *Plan and Reports of the Society .... For the Recovery of Persons Apparently Drowned,* 1775.
41. *Ibid.,* p.8.
42. *Ibid.*
43. Pamphlet of the Royal Humane Society, *Directions for Recovering Persons Who are Supposed to be Dead, from Drowning,* [1800].
44. *Ibid.,* p.2.
45. *Fifty Eighth Annual Report of The Royal Humane Society,* 1832, pp.125–6.
46. 'Artificial Respiration', *Seventy Ninth Annual Report of The Royal Humane Society,* 1853, pp.84–5.
47. Hall's wife names the three students in a footnote to an Appendix by Charles Hunter, which she includes in *Memoirs,* p.502. Hall's posthumously published work on his Ready Method is dedicated to these three students. See note 55 *Infra.*
48. Hall, C., *Memoirs,* p.502. Hall would have found common cause with J. B. S. Haldane who more than half a century later, during physiological research associated with the First World War, on the effects of inhalation of gas, carried out some of the crucial experiments on himself.

49. *Eighty Second Annual Report of the Royal Humane Society,* 1856, p.24.
50. *Ibid.*, p.91.
51. Hall, M., 'On a New Mode of Effecting Artificial Respiration', *Lancet,* vol.i, 1856, p.229.
52. Hall, M., 'Asphyxia, its Rationale and its Remedy', *ibid.*, pp.393–4.
53. Hall, M., (ed.), *Prone and Postural Respiration in Drowning,* 1857, p.177. The work was dedicated to George Webster and 'to those students of St George's Hospital who, with such zeal, ability and perseverance, worked out the problem of postural respiration'.
54. *Ibid.*, p.178.
55. Hunter, C., 'On the Marshall Hall Method in Drowning and Other Forms of Apnoea or Suspended Respiration, Appendix IV' in Hall, C., *Memoirs,* p.511.
56. 'The Marshall Hall Method of Treatment in Asphyxia', Letter dated June 1857, *Lancet,* vol.ii, 1857, p.73.
57. See note 53 *Supra.*
58. It was Hunter who had suggested to Hall the name 'Ready Method'. Hunter, 'Appendix IV', in Hall, C. *Memoirs,* pp.507–8.
59. Hall, *Memoirs* (1837), p.103.
60. Hall, M., 'Comparaison Entre les Éffets Tétanoides des États Électrogéniques, et ceux de la Strychnine', *Comptes Rendus,* 24, 1847, p.1054; 'Des Éffets de l'Acetate de Strychnine', *ibid.*, 37, 1853, pp.289–91.
61. According to Hall, 'As the nerve and muscles of the frog's leg, properly prepared, have been very aptly designated as galvanoscopic, so the whole frog, properly employed, becomes strychnoscopic.' Hall, M., 'Note on the Detection of Strychnine', *Lancet,* vol.i, 1856, p.36.
62. Hale-White, Sir W., *Great Doctors of the Nineteenth Century,* 1935, p.96.
63. *Ibid.*, Hall's experimental work on strychnine was referred to in an article 'Strychnine' in Charles Dickens's *Household Words,* 12, 1856, p.424.
64. See two letters from Marshall Hall to the Editor, 'Note on the Detection of Strychnia'; 'The Physiological Test for Strychnia', *Lancet,* vol.i, 1856, p.36; p.623.
65. See range of letters to the Editor, *Lancet,* vol.i, 1856, pp.670–1.
66. 'Detection of Strychnine', Letter, to Editor from H. G. Whight, MD, *Lancet,* vol.i, 1856, p.718.
67. Calvert, F. C., 'On the Detection of Strychnine', *Lancet,* vol.i, 1856, p.511.
68. Nichols, J., 'Should Strychnia be Employed as a Medicine?', Letter, dated May 1856, *ibid.*, p.623.
69. Vernon, H., 'Should Strychnia be Employed as a Medicine?' Letter, dated 2 June 1856, *ibid.*, pp.670–1.
70. 'Notes of Three Lectures on the Physiological Action of Strychnia. Delivered by George Harley', *ibid.*, pp.619–20; pp.647–9.

71. *Ibid.*, p.649.
72. Glover, R. M., 'Was Cook Killed by Strychnia?' Letter, dated July 1856, *ibid.*, vol.ii, 1856, p.56.
73. Hall, M., *Suggested Works on the Thames,* 1852. For further details and comment on London's sewage and attendant problems at the time, see: Chapman, C. B., 'The Year of the Great Stink', *Pharos,* July, 1972, pp.90–105. Sheppard, F., *The Infernal Wen,* 1971. Pelling, M., *Cholera, Fever and English Medicine,* 1968.
74. Hall, M., 'On the Principle of the Sewerage in London and Other Large Cities', Letter, dated September 1856, *Lancet,* vol.ii, 1856, p.363.
75. Hall, *Suggested Works on the Thames,* pp.11–12. Hall provided a Table giving data on elevation, cholera mortality and general mortality rates in the London regions referred to.
76. [Roget, P. M.], 'Water Supply to the Metropolis: Copy of the Commission Issued for Inquiring into the State of the Supply of Water to the Metropolis', *Parliamentary Papers,* 9, 1828, pp.47–53.
77. Brundage, A., *Edwin Chadwick,* 1988.
78. Lambert, R., *Sir John Simon,* 1963.
79. Finer, S. E., *Life and Times of Sir Edwin Chadwick,* 1980, p.217.
80. Chadwick, E., *Sanitary Condition of the Labouring Population,* 1842. This report was known as the Sanitary Report. Brundage, *Edwin Chadwick,* p.83.
81. Finer, *Life and Times of Sir Edwin Chadwick,* p.239. Dickens was initially mistrustful of Chadwick's part in Poor Law administration but once won over, in 1847, he was Chadwick's most outspoken public champion.
82. Hall, *Suggested Works on the Thames,* p.3.
83. Herbert Spencer had also, in the 1850s, advocated the building of a Thames Barrier at Woolwich. The danger of flooding in London from oceanic surges into the Thames is constantly increasing due to the melting of the polar ice caps, the tilting of Britain towards the South East at about 1 foot each century and the slow sinking of London on its bed of clay. Tides are thus rising at the rate over 2 feet per century. The Flood Barrier, whose gates can be closed in 30 minutes cost about £500 million. Weinreb, B., and Hibbert, C. (eds), *The London Encyclopaedia,* 1990, pp.286–7.
84. Liebig, J., *Organic Chemistry,* 1840.
85. Hall, *Suggested Works on the Thames,* p.13.
86. Hall, M., 'On the Principle of the Sewerage in London and other Large Cities', *Lancet,* vol. ii, p.363; 'The Five Points of the Sewerage', *ibid.*, p.617. (This contribution contained his son's suggestion.) The sewerage issue was raised in at least four editorials during the period, *ibid.*, pp.516–17; p.604; p.628; p.682.
87. *Ibid.*, p.363.
88. Editorial, *ibid.*, p.516.
89. 'Medical Annotations', *ibid.*, p.684.

90. Hall, *Suggested Works on the Thames,* p.22.
91. MS Letter, M. Hall to P. C. A. Louis, dated 17 Avril 1857, Archives, Paris, Académie des Sciences, Institut de France. In this letter Hall offers to publish extracts of Louis' work in the *Lancet.* He also tells Louis of the success of his 'Ready Method' in having 'sauvé la vie à cinquante êtres humaines!' Of these, the majority were still-born infants but four had been drowned and three had been poisoned with chloroform.
92. Hall, C., *Memoirs,* p.366; pp.371–2.
93. *Ibid.,* p.417.
94. Marjoribanks, E., *Life of Sir Edward Marshall Hall,* 1929, pp.16–17. The present writer has also corresponded with another Marshall Hall, whose parents named him after the lawyer. He lives in the North East and has written on artists from the region. Hall, Marshall, *Artists of Northumbria,* 1973.
95. Power, D'Arcy, 'Dr Marshall Hall and the Decay of Bloodletting', *The Practitioner,* 82, Jan.–June, 1909, pp.320–31.
96. 'Dr Marshall Hall: The History of his Case and the Post-Mortem Appearances', *Lancet,* vol.ii, 1857, pp.253–6. This report referred to 'bodies ... resembling retrograde cancer-cells'.
97. 'Dr Marshall Hall; The History of his Case and the Post-Mortem Appearances', *Brit. Med. Journ.,* 1857, pp.760–3.
98. 'Death of Marshall Hall', *Lancet,* 15 August 1857, vol.ii, 1857, pp.172–5. 'The Late Dr Marshall Hall', *Med. Times and Gaz.,* n.s., 15, 1857, pp.225–8. Most of this obituary was repeated in *Edin. Med. Journ.,* vol. ii, 1857–8, pp.377–80. *Gentleman's Magazine,* n.s., 3, 1857, p.350. Flourens announced Hall's death to the French Academy of Science, *Archiv. Gén. de Méd.,* Ve série, 10, 1857, p.495. *Comptes Rendus,* 45, 1857, p.38; p.281. 'Marshall Hall sa Vie et Ses Écrits', *Gaz. Méd. d'Orient,* no. 9, 1858, pp.153–9. *Edin. New Phil. Journ.* n.s. 7, 1858, pp.347–9. *Annual Register 1857,* 1858, pp.329–30. *Proc. Roy. Soc. Lond.,* 9, 1859, pp.52–6. 'Marshall Hall, MD, F.R.S.L. and E.', in Pettigrew, T. J., *Medical Portrait Gallery,* 4 vols, [1840], 4, pp.1–14. 'Biographical Sketch of Marshall Hall, M.D., F.R.S.', *Lancet,* vol.ii, 1850, pp.120–8. This article included a daguerreotype of Hall taken by Mr Mayall of the Strand. 'Marshall Hall, and the Discovery of Reflex Action' in Bettany, G. T., *Eminent Doctors,* 2 vols, 1885, 1, pp.264–85.

Bibliography

Books

A Lover of Science, *The Knighthood of Sir J. Smith – a Deathblow to Science,* London, R. Crozier, 1830.

A Statement of Circumstances Connected with the Late Election for the Presidency of the Royal Society, London, Richard Taylor, 1831.

Abernethian Code: *The Abernethian Code of Health and Longevity, or Everyone's Health in his own Keeping, by the Proper Regulation of the Stomach and Bowels, in order to the Attaining and Securing of those Invaluable Blessings, founded on the Principles and Practice of John Abernethy, Esq., FRS ... Concluding with some most remarkable instances of Longevity, and its Comparative Estimate in Different Classes of Society,* London, J. Williams, 1829.

[Abernethy, J.], *The Surgical Works of John Abernethy,* 2 vols, London, Longman, Hurst, Rees, Orme & Brown, 1811.

Ackerknecht, E. H., *A Short History of Medicine,* revised edn, Baltimore, The Johns Hopkins University Press, 1982.

Acton, W., *A Complete Practical Treatise on Venereal Diseases, and their Immediate and Remote Consequences: Including Observations on Certain Affections of the Uterus attended with Discharges,* London, Henry Renshaw, 1841.

Acton, W., *A Practical Treatise on Diseases of the Urinary and Generative Organs (in Both Sexes.) Part I. – Non-Specific Diseases. Part II – Syphilis,* 2nd edn, London, John Churchill, 1851.

Acton, W., *Prostitution, Considered in its Moral, Social and Sanitary Aspects, in London and Other Large Cities. With Proposals for the Mitigation and Prevention of its Attendant Evils,* London, John Churchill, 1857.

Addresses Delivered at the Anniversary Meetings of the Royal Society, on Thursday November 30, 1831 and Friday November 30, 1832, by His Royal Highness The Duke of Sussex, K.G., etc, etc, etc. The President, London, Richard Taylor, 1833.

Address Delivered at the Anniversary Meeting of the Royal Society, on Wednesday, November 30, 1836, By His Royal Highness The Duke of Sussex, K.G., etc, etc, etc, etc. The President, London, Richard Taylor, 1836.

An Appeal to the Fellows of the Royal Society, Concerning the Measures taken by Sir Joseph Banks, their President, to Compel Dr Hutton to Resign the Office of Secretary to the Society for their Foreign Correspondence, By a Friend to Dr Hutton, London, J. Derbrett, 1784.

An Important Address to Wives and Mothers on the Dangers and Immorality of Man-Midwifery, [Anon.], Finsbury, Lewis & Company, 1830.

Andrade, E. N. da C., *A Brief History of the Royal Society,* London, The Royal Society, 1960.

Babbage, C., *Reflections on the Decline of Science in England, and on Some of its Causes,* London, B. Fellowes, 1830.

Babbage, C., *The Exposition of 1851: or the Views of the Industry, the Science, and the Government of England,* London, John Murray, 1851.

Barrow, Sir John, *Sketches of the Royal Society and Royal Society Club,* London, John Murray, 1949.

Bartlett, E., *An Essay on the Philosophy of Medical Science,* Philadelphia, Lea & Blanchard, 1844.

Beaunis, H., *Nouveaux Elements de Physiologie Humaine Comprenant les Principes de la Physiologie Comparée et de la Physiologie Générale,* 2 vols, 2nd edn, Paris, Librarie J. B. Baillière et Fils, 1881.

Bell, C., *An Exposition of the Natural System of the Nerves of the Human Body,* London, A. & R. Spottiswoode, 1824.

Bell, C., *Idea of a New Anatomy of the Brain, Submitted for the Observations of his Friends,* London, Strahan & Preston, [1811].

Bell, C., *The Nervous System of the Human Body: as Explained in a Series of Papers read before the Royal Society of London,* 3rd edn, Edinburgh, Adam & Charles Black, 1836.

Bellot, H. Hale, *University College London 1826–1926,* London, University of London Press, 1929.

Berman, M., *Social Change and Scientific Organisation, The Royal Institution, 1799–1844,* London, Heinemann Educational Books, 1978.

Bernard, C., *Leçons sur la Physiologie, et la Pathologie du Système Nerveux,* 2 vols, Paris, J. B. Baillière et Fils, 1858.

Bernard, C., *Rapport sur les Progrès et la Marche de la Physiologie Général en France,* in *Recuil de Rapports sur les Progrès des Lettres et des Sciences en France,* Paris, A L'Imprimerie Impériale, 1867.

Bettany, G. T., *Eminent Doctors: Their Lives and Their Work,* 2 vols, London, John Hogg, 1885.

Blane, Sir Gilbert, *Select Dissertations on Several Subjects of Medical Science,* London, Thomas & George Underwood, 1822.

Blumenbach, J. F., *The Institutions of Physiology,* London, Bensley & Son, 1817.

Blumenbach, J. F., *Institutiones Physiologicae,* Gottingae, Apud J. Christ. Dieterich, 1787.

Bradley, I., *The Call to Seriousness: The Evangelical Impact on the Victorians,* London, Jonathan Cape, 1976.

Briscoe, J. P. (ed.), *Nottinghamshire Facts and Fictions,* 2nd series, Nottingham, Shepherd Brothers, 1877.

British Association for the Advancement of Science. Eleventh Meeting. 1841. Journal of Sectional Proceedings, Plymouth, R. White Stevens, 1841.

British Association for the Advancement of Science. General Meeting, Cambridge, Hodson & Brown, 1833.

British Association for the Advancement of Science, List of Members, Journal, etc., 1832–54, (London, 1854).

Brundage, A., *England's 'Prussian Minister' Edwin Chadwick and the Politics of Government Growth, 1832–1854,* University Park, The Pennsylvanian State University Press, 1988.

Bulwer Lytton, Bart., Sir E., *The Caxtons. A Family Picture,* 3 vols, Edinburgh & London, William Blackwood & Sons, 1849.

Bynum W. F., *Science and the Practice of Medicine in the Nineteenth Century,* Cambridge, Cambridge University Press, 1994.

Bynum, W. F., Porter, R. and Shepherd, M. (eds), *The Anatomy of Madness. Essays in the History of Psychiatry,* 2 vols, London, Tavistock Publications, 1985.

Bynum, W. F. and Porter, R. (eds), *Medical Fringe and Medical Orthodoxy 1750–1850,* London, Croom Helm, 1987.

Canguilhem, G., *La Formation du Concept de Réflexe aux XVIIe et XVIIIe Siècles,* Paris, Presses Universitaires de France, 1955.

Cannon, S. F., *Science in Culture: The Early Victorian Period,* New York, Dawson & Science History Publications, 1978.

Catalogue de la Bibliothèque de l'Académie Nationale de Médecine, Paris, Doin, 1971.

Catalogue of the Library of the Royal Medical Society, Edinburgh, The Darien Press, 1896.

Cayrade, J., *Recherches Critiques et Expérimentales sur les Mouvements Réflexes,* in *Collection des Thèses Soutenues à la Faculté de Médecine de Paris,* Paris, A. Parent, Imprimeur de la Faculté de Médecine, 1864.

Chadwick, E., *Report on the Sanitary Condition of the Labouring Population,* London, HMSO, 1842.

Charlton, H. B., *Portrait of a University 1851–1951. To Commemorate the Centenary of Manchester University,* Manchester, Manchester University Press, 1951.

[Christison, R.], *The Life of Sir Robert Christison, Bart.,* Edited by his Sons, 2 vols, Edinburgh, William Blackwood & Sons, 1885–6.

Clark, F. le Gros, *The Practical Anatomy and Elementary Physiology of the Nervous System. Designed for the Use of Students in the Dissecting-Room,* London, Rees, Orme, Brown, Green & Longman, 1836.

Clark, Sir James, *A Memoir of John Conolly, M.D., D.C.L., Comprising a Sketch on the Treatment of the Insane in Europe and America,* London, John Murray, 1869.

Clarke, C. M., *Observations on Some Diseases of Females which are attended by Discharges. Illustrated by Copper-Plates of the Diseases, etc.,* 2 vols, London, Longman, Hurst, Rees, Orme, & Brown, 1814.

Clarke, E. and Jacyna, L. S., *Nineteenth-Century Origins of Neuroscientific Concepts,* Berkeley, Los Angeles, London, University of California Press, 1987.

Clarke, E. & O'Malley, C. D., *The Human Brain and Spinal Cord. A Historical Study Illustrated by Writings from Antiquity to the Twentieth Century,* Berkeley & Los Angeles, University of California Press, 1968.

Collins, J. A., *Right and Wrong among the Abolitionists of the United States: or the Objects, Principles and Measures of the Original American Anti-Slavery Society, Unchanged:... With an Introductory Letter by Miss Martineau and an Appendix,* Glasgow, Geo. Gallie, 1841.

Common Sense; or, the Abernethian Code of Health and Longevity; founded on the Principles of John Abernethy, Esq., F.R.S. Senior Surgeon to St. Bartholomew's Hospital. With extracts from Hunter, Conaro, etc., etc., Likewise an Introductory View of the Living Functions of the Animal Economy, London, T. R. Drury, 1830.

Conolly, J., *A Remonstrance with the Lord Chief Baron Touching the Case Nottidge versus Ripley,* 3rd edn, London, John Churchill, 1849.

Conolly, J., *A Study of Hamlet,* London, Edward Moxon, 1863.

Conolly, J., *An Inquiry Concerning the Indications of Insanity with Suggestions for the Better Protection and Care of the Insane,* London, John Taylor, 1830.

Conolly, J., *An Introductory Lecture Delivered in the University of London, on Thursday, October 2,* 1828, London, John Taylor, 1828.

Conolly, J., *Four Lectures on the Study and Practice of Medicine; Delivered on Different Occasions in the University of London,* London, Sherwood, Gilbert & Piper, 1832.

Conolly, J., Morgan, Augustus de, Lardner, Dionysius, Long, G. and M'Culloch, J. R., *A Letter to the Shareholders and Council of the University of London, on the present State of that Institution,* London, Thomas Davison, 1830.

Cope, Z., *Some Famous General Practitioners,* London, Pitman Medical Publishing, 1961.

Cope, Z., *The History of St. Mary's Hospital Medical School,* London, William Heinemann Ltd, 1954.

Corry, J., *The Detector of Quackery; or, Analyser of Medical, Philosophical, Political, Dramatic and Literary Imposture,* 1st and 2nd edns, London, T. Davison, both 1802.

Cranefield, P. F., *The Way In and the Way Out. François Magendie, Charles Bell and the Roots of the Spinal Nerves,* New York, Futura Publishing Company, 1974.

[Cullen, W. (ed.)], *Synopsis Nosologiae Methodicae,* Edinburgh, 1769. [No publisher given.]

[Cullen, W.], *Nosology: or, a Systematic Arrangement of Diseases, by Classes, Orders, Genera, and Species; with the Distinguishing Characters of Each, and Outlines of the Systems of Sauvages, Linnaeus, Vogel, Sagar, and Macbride. Translated from the Latin of William Cullen, M.D. Late Professor of the Practice of Physic in the University of Edinburgh,* Edinburgh, W. Creech, 1800.

Davis, D. B., *The Problem of Slavery in Western Culture,* Middlesex, Pelican Books, 1970.

Davis, D. B., *Was Thomas Jefferson an Authentic Enemy of Slavery?,* Oxford, Clarendon Press, 1970.

De Beaumont, G., *Marie or Slavery in the United States. A Novel of Jacksonian America.* Translated from the French by Barbara Chapman with an Introduction by Alvis L. Tinnin, Stanford, Stanford University Press, 1958.

De Beaumont, G., *Marie ou l'Ésclavage aux États-Unis, Tableau de Moeurs Américaines,* 2 vols, Bruxelles, Louis Hauman et Comp., 1835.

[De Tocqueville, J. A.], *De la Démocratie en Amérique,* par Alexis de Tocqueville, Avocat à la Cour Royale de Paris, L'un des auteurs du livre intitulé: Du Système Pénitentiaire aux., États-Unis. Orné d'une Carte, D'Amérique, 2 vols, 2nd edn, Paris, Libraire de Charles Gosselin, 1835.

De Tocqueville, J. A., *Democracy in America,* A new translation by George Lawrence, edited by J. P. Mayer, Garden City, New York, Doubleday & Co., 1969.

Desmond, A., *Archetypes and Ancestors. Palaeontology in Victorian London 1850–1875,* London, Blond & Briggs, 1982.

Dictionary of National Biography ... from the earliest times to 1900, 66 vols, London, Oxford University Press, first published 1885–1901, reprinted [1949–50].

Dictionary of Scientific Biography, 18 vols, New York, Charles Scribner's Sons, 1970–90.

Directions for Recovering Persons Who are Supposed to be Dead, from Drowning, London, Royal Humane Society, [1800].

Dissertations by Eminent Members of the Royal Medical Society, Edinburgh, David Douglas, 1892.

Dr Abernethy's Code of Health and Long Life, with the Cause and Cure of Indigestion; Every One His Own Physician. With Copious Dietic Tables Showing the Comparative Digestibility of Various Kinds of Food; Approved Formulae, etc. A new edition with Medical Mottos, Maxims, and Proverbs Printed round Every Page, London, J. Cornish, [1857].

Du Bois-Reymond, E., *On Animal Electricity: Being an Abstract of the Discoveries of Emil Du Bois-Reymond,* London, John Churchill, 1852.

Dutrochet, R. H. J., *Recherches Anatomiques et Physiologiques sur la Structure Intime des Animaux et des Végétaux, et sur Leur Motilité,* Paris, Chez J. B. Baillière, 1824.

Edwards, W. F., *De l'Influence des Agens Physiques sur la Vie, Paris,* Chez Crochard, 1824.

[Edwards, W. F.], *On the Influence of Physical Agents on Life, By W. F. Edwards.* Translated from the French by Dr Hodgkin and Dr Fisher, London, S. Highley, 1832.

Eliot, G., *Middlemarch A Study of Provincial Life,* 4 vols, Edinburgh and London, William Blackwood & Sons, 1871–2.

Emblen, D. L., *Peter Mark Roget. The Word and the Man,* New York, Thomas Y. Crowell Company, 1970.

Every One his Own Physician or the Pocket Aesculapius, Containing the Abernethian Code of Health and Longevity, with Notes and Illustrations, to which is added Copious Dietetic Tables, Exhibiting the Comparative Digestibility of Various Kinds of Food, Approved Formulae, etc., London, S. Cornish and J. Chidley, 1837.

Everyone his own Physician or the Abernethian Code of Health and Longevity, with Observations on Diet and Regimen and Copious Dietic Tables, Exhibiting the Comparative Digestibility of Various Kinds of Food, with Approved Formulae from Several Eminent Physicians, London, John Kettle, 1836.

Fearing, F., *Reflex Action. A Study in the History of Physiological Psychology,* London, Baillière, Tindall & Cox, 1930.

Felkin, W., *A History of the Machine-Wrought Hosiery and Lace Manufactures,* Cambridge, W. Metcalfe, 1867.

[Felkin, W.], *Felkin's History of the Machine-Wrought Hosiery and Lace Manufacturers.* Centenary Edition, Newton Abbot, David & Charles, 1967.

Finer, S. E., *The Life and Times of Sir Edwin Chadwick,* London, Methuen & Co., Ltd, 1980. (First published 1952.)

Flaubert, G., *Madame Bovary Mœurs de Province,* Paris, Michel Lévy Frères, 1857.

Flourens, P., *Recherches Expérimentales sur les Propriétes et les Fonctions du Système Nerveux dans les Animaux Vertébrés,* Paris, Chez Crevot, Libraire-Editeur, 1824.

Foster, J., *Men-at-the-Bar; a Biographical Hand-List of the Members of the Various Inns of Court, Including Her Majesty's Judges, etc.,* London, Reeves & Turner, 1885.

Geikie, A., *The Scenery of Scotland Viewed in Connection with its Physical Geology,* London, Macmillan & Co., 1865. A second edition was published in 1887 and a third in 1901.

Geison, G. L., *Michael Foster and the Cambridge School of Physiology. The Scientific Enterprise in Late Victorian Society,* Princeton, Princeton University Press, 1978.

General List of the Members of the Medical Society of Edinburgh, Edinburgh, Hay, Gill & Co., 1823.

Gillispie, C. C. (ed.), *Dictionary of Scientific Biography,* 18 vols, New York, Charles Scribner's Sons, 1970–90.

Glisson, F., *Tractatus de Ventriculo et Intestinis cui Praemittitur alius, de Partibus Continentibus in Genere; et in Specie, de iis Abdominis,* London, Henry Brome, 1677.

Glover's Nottingham Directory, for 1825, Being a General List of the Merchants, Tradesmen and Principal Inhabitants of Nottingham, New Radford and New Snenton, Nottingham, S. Glover, 1825.

Grainger, R. D., *Observations on the Structure and Functions of The Spinal Cord,* London, S. Highley, 1837.

Granville, A. B., *The Royal Society in the XIXth Century; Being a Statistical Summary of its Labours During the Last Thirty-Five Years With Many Original Tables and Official Documents...and a Plan for its Reform,* London, Mr Churchill, 1836.

[Granville, A. B.], *Science Without a Head; or the Royal Society Dissected. By One of the 687 FRS,—sss.,* London, T. Ridgway, 1830.

Gray, J., *History of the Royal Medical Society, 1737–1937,* Douglas Guthrie (ed.), Edinburgh, University Press, 1952.

Green, J. P. and Pole J. R., *The Blackwell Encyclopaedia of the American Revolution,* Cambridge Massachusetts, Basil Blackwell, 1991.

Hale-White, Sir William, *Great Doctors of the Nineteenth Century,* London, Edward Arnold, 1935.

Hall, C., *Memoirs of Marshall Hall, by his Widow,* London, Richard Bentley, 1861.

Hall, M. (ed.), *Prone and Postural Respiration in Drowning and Other Forms of Apnoea or Suspended Respiration; by Marshall Hall,* London, John Churchill, 1857.

Hall, M., *A Critical and Experimental Essay on the Circulation of the Blood; Especially as Observed in the Minute and Capillary Vessels of the Batrachia and of Fishes,* London, R. B. Seeley and W. Burnside, 1831.

Hall, M., *A Descriptive, Diagnostic and Practical Essay on Disorders of the Digestive Organs and General Health, and Particularly on their Numerous Forms and Complications, Contrasted with some Acute and Insidious Diseases; Being an Attempt to Prosecute the Views of Dr Hamilton and Mr Abernethy, and a Second Edition of the Essay on the Mimoses, with Additions,* London, Longman, Hurst, Rees, Orme, & Brown, 1820.

Hall, M., *A Letter, Addressed to the Earl of Rosse, President-Elect of the Royal Society,* dated 15 November 1848, 2nd edn, London, J. Mallet, November 1848.

Hall, M., *A Letter Respectfully Addressed to His Royal Highness Prince Albert,* London, J. Mallet, 1851.

Hall, M., *A Note Addressed to the Members of the Council of the Royal Society,* London, J. Mallet, 7 June 1837.

Hall, M., *Aperçu du Système Spinal ou de la Série des Actions Réflexes dans Leurs Applications à la Physiologie, à la Pathologie et Spécialement à l'Épilepsie,* Paris, Libraire de Victor Masson, 1855.

Hall, M., *Cases of a Serious Morbid Affection, Chiefly Occurring after Delivery, Miscarriage, etc. From Various Causes of Irritation and Exhaustion; and of a Similar Affection, Unconnected with the Puerperal State,* London, Longman, Hurst, Rees, Orme & Brown, 1820.

Hall, M., *Commentaries on Some of the More Important of the Diseases of Females. In Three Parts,* London, Longman, Rees, Orme, Brown & Green, 1827.

Hall, M., *Commentaries Principally on those Diseases of Females which are Constitutional,* 2nd edn, London, Sherwood, Gilbert & Piper, 1830.

Hall, M., *Disputatio Inauguralis De Febribus Inordinatis; quam Annuente Summo Numine, ex Auctoritate Reverendi Admodum Viri,* Edinburgh, D. Georgii Baird, SS. T.P., 1812.

Hall, M., *Introductory Lecture to a Course of Lectures on the Practice of Physic; to be Delivered at the Medical School in Aldersgate Street,* London, [1830].

Hall, M., *Lectures on the Nervous System and its Diseases,* London, Sherwood, Gilbert & Piper, 1836.

Hall, M., *Medical Essays,* London, Longman, Hurst, Rees, Orme & Brown, 1825.

Hall, M., *Medicine, its Divisions, its Rewards, and its Reforms: Being the Annual Oration Delivered at the British Medical Association on the 8th of October, 1840,* 2nd edn, London, J. Mallett, 1840.

Hall, M., *Memoirs on the Nervous System,* London, Sherwood, Gilbert & Piper, 1837.

Hall, M., *New Memoir on the Nervous System,* London, Hippolyte Baillière, 1843.

Hall, M., *Observations on Bloodletting Founded Upon Research on the Morbid and Curative Effects of Loss of Blood,* London, Sherwood, Gilbert & Piper, 1836.

Hall, M., *Observations on Medical Reform, Addressed to the Provincial Medical Association, Assembled at York on August the 4th 1841,* London, 1841.

Hall, M., *On a Morbid Affection of Infancy, arising from Circumstances of Exhaustion, but resembling Hydrencephalus,* London, R. B. Seeley and W. Burnside, 1829.

Hall, M., *On Diagnosis. In Four Parts. Part I. The Phaenomena of Health and Disease. Part II. The Diagnosis of the Diseases of Adults. Part III. The Diagnosis of Local Diseases. Part IV. The Diagnosis of Diseases of Children,* London, Longman, Hurst, Rees, Orme & Brown, 1817.

Hall, M., *On Medical Reform. A Letter ... Addressed to the Right Hon. Sir Robert Peel, Bart., 16 February, 1841,* London, J. Mallet, 1841.

Hall, M., *On the Diseases and Derangements of the Nervous System, in their Primary Forms and in their Modifications by Age, Sex, Constitution, Hereditary Predisposition, Excesses, General Disorder and Organic Disease,* London, H. Baillière, 1841.

Hall, M., *On the Mimoses; or a Descriptive, Diagnostic and Practical Essay, on the Affections usually Denominated Dyspeptic, Hypochondriac, Bilious, Nervous, Chlorotic, Hysteric, Spasmodic etc.,* London, Longman, Hurst, Rees, Orme & Brown, 1818.

Hall, M., *On the Mutual Relations Between Anatomy, Physiology, Pathology, and Therapeutics, and the Practice of Medicine. Being the Gulstonian Lectures for MDCCCXLII,* London, H. Baillière, 1842.

Hall, M., *Practical Observations and Suggestions in Medicine,* London, John Churchill, 1845.

Hall, M., *Practical Observations and Suggestions in Medicine, Second Series,* London, John Churchill, 1846.

Hall, M., *Principles of the Theory and Practice of Medicine including a Third Edition of the Author's work upon Diagnosis,* London, Sherwood, Gilbert & Piper, 1837.

Hall, M., *Prone and Postural Respiration in Drowning and Other Forms of Apnoea or Suspended Respiration,* (Edited by his Son, Marshall Hall, Esq.), London, John Churchill, 1857.

Hall, M., *Researches Principally Relative to the Morbid and Curative Effects of Loss of Blood,* London, R. B. Seeley and W. Burnside, 1830.

Hall, M., *Suggested Works on the Thames,* London, Longman & Co., 1852.

Hall, M. Synopsis of Apoplexy and Epilepsy; with Observations on Trachelismus, Laryngismus, and Tracheotomy; and the Proposal for a Hospital for Epileptics, 2nd edn, London, Joseph Mallet, [1852].

Hall, M., *Synopsis of Cerebral and Spinal Seizures of Inorganic Origin and of Paroxysmal Form as a Class; and of their Pathology as Involved in the Structures and Actions of the Neck,* 2nd edn, London, Joseph Mallet, [1851].

Hall, M., *Synopsis of the Diastaltic Nervous System; or the System of the Spinal Marrow, and its Reflex Arcs; as the Nervous Agent in all the Functions of*

Ingestion and of Egestion in the Animal Economy, 2nd edn, London, Joseph Mallett, [1850].

Hall, M., *The Principles of Diagnosis*, 2nd edn, London, Sherwood, Gilbert, & Piper, 1834.

Hall, M., *The Two-Fold Slavery of the United States; with a Project of Self-Emancipation*, London, Adam Scott, 1854.

[Hall, M.], *Eupædia; or Letters to a Mother on the Watchful Care of her Infant, in Reference to Diet, Clothing, Air, Exercise, Medicine, etc. By a Physician*, London, Sherwood, Gilbert, & Piper, [1831].

Hall, Marshall, *The Artists of Northumbria. A Dictionary of Northumberland and Durham Painters, Draughtsmen and Engravers, born 1647–1900*, in Artists of the Regions Series, Newcastle upon Tyne, Marshall Hall Associates, 1973.

Hall, M. Boas, *All Scientists Now. The Royal Society in the Nineteenth Century*, Cambridge, Cambridge University Press, 1984.

Haller, A., *A Dissertation on the Sensible and Irritable Parts of Animals.* Translated from the Latin. With a Preface by M. Tissot, MD, London, J. Nourse, 1755.

Haller, A., *A Dissertation on the Sensible and Irritable Parts of Animals.* [London, J. Nourse, 1755], Introduction by Owsei Temkin, Baltimore, Johns Hopkins Press, 1936.

Hamilton, A., *A Treatise on the Management of Female Complaints, 7th edn revised ... and enlarged with hints for the treatment of the principal diseases of infants and children, by Dr James Hamilton, Junior, Professor of Midwifery in the University of Edinburgh*, Edinburgh, George Ramsay & Company, 1813.

Hamilton, J., *Observations on the Seats and Causes of Diseases: Illustrated by the Dissections of the late Professor Morgagni of Padua*, Edinburgh, P. Hill, 1795.

Hamilton, J., *Practical Observations on Various Subjects Relating to Midwifery*, Parts I and II, Edinburgh, Bell & Bradfute, 1836.

Hamilton, J., *Practical Observations on Various Subjects Relating to Midwifery*, 2nd edn, revised and enlarged, Edinburgh, Bell & Bradfute, 1840.

Hamilton, J., *Selected Cases in Midwifery; Extracted from the Records of the Edinburgh General Lying-in Hospital. With Remarks*, Edinburgh, P. Hill, 1795.

Henry, T., *Experiments and Observations on the Following Subjects; On the preparation, calcination and medical use of Magnesia Alba,* London, Joseph Johnson, 1773.

Henry, W., *A Tribute to the Memory of the Late President of the Literary and Philosophical Society,* Manchester, R. and W. Dean, 1819.

Herschel, J. F. W., *A Preliminary Discourse on the Study of Natural Philosophy,* in *The Cabinet Cyclopaedia. Conducted by the Rev. Dionysius Lardner,... Assisted by Eminent Literary and Scientific Men. Natural Philosophy,* London, Longman, Rees, Orme, Brown & Green, 1831.

Hodgkin, T., *An Inquiry into the Merits of the American Colonization Society. And a Reply to the Charges brought Against it. With an Account of the British African Colonization Society,* London. J. and A. Arch, 1833.

Hodgkin, T., *On the British Colonization Society. To which are Added, Some Particulars Respecting the American Colonization Society; and a Letter from Jeremiah Hubbard, addressed to a Friend in England, on the Same Subject,* London, Richard Watts, 1834.

[Hodgkin, T.], *On Negro Emancipation and American Colonization. By Dr Hodgkin,* London, Richard Watts, 1832.

Hogg, J., *A Series of Chemical and Medical Tables, forming a Synopsis of Chemistry, Materia Medica, Pharmacy, and Nosology,* London, John Taylor, 1830.

Hogg, J., *London As It Is; Being a Series of Observations on the Health, Habits, and Amusements of the People,* London, John Macrone, 1837.

Horner, L., *Letter to the Council of the University of London,* London, J. Moyes, 1830.

Huxley, T. H., *Lay Sermons, Addresses, and Reviews,* London, Macmillan & Co., 1870.

Inkster, I. and Morrell, J. (ed.), *Metropolis and Province. Science in British Culture, 1780–1850,* London, Hutchinson, 1983.

Jacob, F. H., *A History of the General Hospital Near Nottingham, Open to the Sick and Lame Poor of any County,* Bristol, John Wright & Sons Ltd, 1951.

Jones, H. B., *Croonian Lectures on Matter and Force. Given at the Royal College of Physicians in 1868,* London, John Churchill & Sons, 1868.

Jordan, W. D., *The White Man's Burden. Historical Origins of Racism in the United States,* New York, Oxford University Press, 1974.

Jordan, W. D., *White over Black. American Attitudes Towards the Negro, 1550–1812,* Chapel Hill, University of North Carolina, 1968.

Keith, A., *Menders of the Maimed. The Anatomical and Physiological Principles Underlying the Treatment of Injuries to Muscles, Nerves, Bones and Joints,* London, Henry Frowde, Hodder & Stoughton, Oxford University Press, 1919.

Lambert, R., *Sir John Simon 1816–1904 and Social Administration,* London, Macgibbon & Kee, 1963.

Lawrence, W., *An Introduction to Comparative Anatomy and Physiology; Being the Two Introductory Lectures Delivered at the Royal College of Surgeons, on the 21st and 25th of March 1816,* London, J. Callow, 1816.

Lawrence, W., *Lectures on Physiology, Zoology, and the Natural History of Man, Delivered at the Royal College of Surgeons,* London, J. Callow, 1819.

Laws of the Medical Society of Edinburgh, Edinburgh, Hay, Gall & Co., 1823.

Lee, R., *Memoirs of the Ganglia and Nerves of the Uterus,* London, John Churchill, 1849.

Legallois, J. J. C., *Expériences sur le Principe de la Vie, Notamment sur celui des Mouvemens du Coeur, et sur le Siège de ce Principe; Suivies du Rapport Fait à la Première Classe de l'Institut sur Celles Relatives aux Mouvemens du Coeur,* Paris, Chez D'Hautel, 1812.

Lewes, G. H., *The Physical Basis of Mind, Being the Second Series of Problems of Life and Mind, with Illustrations,* London, Trubner & Co., 1877.

Lhermitte, J., *Les Fondements Biologiques de la Psychologie,* Paris, Gauthier-Villars et Cie., 1925.

Library of Useful Knowledge. Natural Philosophy, 4 vols, London, Baldwin & Cradock, 1829–38.

Liebig, J., *Organic Chemistry in its Application to Agriculture and Physiology,* Edited from the Manuscript of the Author by Lyon Playfair, London, Taylor and Walton, 1840.

Liebig, Justus von, *Familiar Letters on Chemistry, and its Relation to Commerce, Physiology, and Agriculture,* John Gardner, MD (ed.), London, Taylor & Walton, 1843.

List of Members, Laws, and Library – Catalogue of the Medical Society of Edinburgh, Edinburgh, William Aitken, 1820–27.

Lithographed Signatures of the Members of the British Association for the Advancement of Science, Who Met at Cambridge, June, MDCCCXXXIII with A Report of the Proceedings at the Public Meetings During the Week; and an Alphabetical List of the Members, Cambridge, Pitt Press, 1833.

Little, E. M., *History of the British Medical Association 1832–1932,* London, British Medical Association [1932].

Longet, F. A., *Traité de Physiologie,* 2 vols, 2nd edn, Paris, Germer Baillière, 1860.

Loudon, I., *Medical Care and the General Practitioner 1750–1850,* Oxford, Clarendon Press, 1986.

Louis, P. C. A., *Recherches sur les Éffets de la Saignée dans Quelques Maladies Inflammatoires, et sur l'Action de L'Émétique et des Vésicatoires dans la Pneumonie,* Paris, J. B. Baillière, 1835.

[Louis, P. C. A.], *Researches on the Effects of Bloodletting in some Inflammatory Diseases, and on the Influence of Tartarized Antimony and Vesication in Pneumonitis by P. Ch. A. Louis, Translated by C. G. Putnam, with Preface and Appendix By James Jackson,* Boston, Hilliard, Gray & Company, 1836.

Macleod, R. and Collins, P. (eds), *The Parliament of Science; The British Association for the Advancement of Science 1831–1981,* Middlesex, Science Reviews Ltd, 1981.

Marjoribanks, E., *The Life of Sir Edward Marshall Hall,* London, Victor Gollancz, 1929.

Martineau, H., *Society in America,* 3 vols, London, Saunders & Otley, 1837.

Maulitz, R. C., *Morbid Appearances. The Anatomy of Pathology in the Early Nineteenth Century,* Cambridge, Cambridge University Press, 1987.

Mayo, H., *Anatomical and Physiological Commentaries,* parts i and ii, London, Thomas & George Underwood, 1822 and 1823.

Mayo, H., *Outlines of Human Physiology,* 4th edn, London, Henry Renshaw & J. Churchill, 1837.

McMenemey, W. H., *A History of the Worcester Royal Infirmary*, London, Press Alliances, 1947.

McMenemey, W. H., *The Life and Times of Sir Charles Hastings, Founder of the British Medical Association*, Edinburgh and London, E. & S. Livingstone, 1959.

Meeting of the British Association at Newcastle. Report of the Local Secretaries, Newcastle, Charnley, Finlay & Charlton, Currie & Bowman, 1839.

Merrington, W. R., *University College Hospital and its Medical School: A History*, London, Heinemann, 1976.

Mollett, J. W., *Sir David Wilkie*, London, Sampson Low, Marston, Searle, Rivington, 1881.

Morrell, J. and Thackray, A., *Gentlemen of Science. Early Years of the British Association for the Advancement of Science*, Oxford, Clarendon Press, 1981.

Morris, R., *et al.* (eds), *The Edinburgh Medical and Physical Dictionary* ... as *Employed in the Present Improved State of Medical Science*, 2 vols, Edinburgh, Bell & Bradfute and Mundell, Doig & Stevenson, 1807.

Müller, J., *Elements of Physiology. Translated from the German, with Notes by William Baly, M.D.*, 2 vols, London, Taylor & Walton, 1837–42.

Müller, J., *Elements of Physiology. Translated from the German, by Wm. Baly, M.D. Arranged from the second London edition by John Bell, M.D.*, Philadelphia, Lea & Blanchard, 1843.

Müller, J., *Handbuch der Physiologie des Menschen für Vorlesgungen*, 2 vols, Coblenz, Verlag von J. Holscher, 1833–40.

Newman, C. E., *The Evolution of Medical Education in the Nineteenth Century*, London, Oxford University Press, 1957.

Noble, D., *The Brain and its Physiology; A Critical Disquisition on the Methods of Determining the Relations Subsisting between the Structure and the Functions of the Encephalon*, London, John Churchill, 1846.

Ollivier, C. P., *De la Moëlle Épinière et de ses Maladies*, Paris, Chez Crevot, 1824.

Olmsted, J. M. D. and Olmsted, E. H., *Claude Bernard and the Experimental Method in Medicine*, London, Abelard-Schuman, 1952.

Olmsted, J. M. D., *François Magendie,* New York, Schuman, 1944.
Paget, S. (ed.), *Memoirs and Letters of Sir James Paget, Edited by Stephen Paget one of his Sons,* London, Longman, Green & Co., 1901.

Paris, J. A., *The Life of Sir Humphry Davy, Bart. LL.D.,* London, Henry Colburn and Richard Bentley, 1831.
Parkes, Samuel, *Reply to Dr Henry's Letter respecting the Introduction of Bleaching by Oxymuriatic Acid,* London, C. Baldwin, January 1816.
Parliamentary Papers, London, HMSO.

Pelling, M., *Cholera, Fever and English Medicine, 1825–1865,* Oxford, Oxford University Press, 1968.

Percival, T., *Medical Ethics; or, a Code of Institutes and Precepts, Adapted to the Professional Conduct of Physicians and Surgeons; .... To which is added an Appendix; containing a Discourse on Hospital Duties; also Notes and Illustrations,* Manchester, J. Johnson, 1803.

Peterson, M. J., *The Medical Profession in Mid-Victorian London,* Berkeley, University of California Press, 1978.

Pettigrew, T. J., *Medical Portrait Gallery. Biographical Memoirs of the Most Celebrated Physicians, Surgeons etc. who have Contributed to the Advancement of Medical Science,* 4 vols, London, Fisher, Son & Co. and Whittaker & Co. [1840].

Pflüger, E. F. W., *Die Sensorischen Functionen des Rückenmarks der Wirbelthiere, Nebst Einer Neuen Lehre Über die Leitungsgesetze der Reflexionen,* Berlin, Verlag von August Hirschwald, 1853.

Pierson, G. W., *Tocqueville and Beaumont in America,* New York, Oxford University Press, 1938.

Plan and Reports of the Society Instituted at London in the Year M.DCC.LXXIV. For the Recovery of Persons Apparently Drowned, London, [The (Royal Humane) Society], 1775.

Pole, J. R. (ed.), *Slavery, Race and Civil War in America, American Historical Documents,* Vol. I, London, Harrap, 1974.

Pole, J. R., *The Pursuit of Equality in American History,* Berkeley, University of California Press, 1978.

Porter, D. and Porter, R., *Patient's Progress. Doctors and Doctoring in Eighteenth-Century England,* Oxford, Polity Press, 1989.

Porter, R. (ed.), *Patients and Practitioners. Lay Perceptions of Medicine in Pre-Industrial Society,* Cambridge, Cambridge University Press, 1985.

Porter, R., *Disease, Medicine and Society in England 1550-1860,* London, Macmillan Education, 1987.

Power, Sir D'Arcy (ed.), *British Medical Societies,* London, The Medical Press and Circular, 1939.

Poynter, F. N. L. (ed.), *The Evolution of Medical Education in Britain,* London, Pitman Medical Publishing Ltd, 1966.

Poynter, F. N. L. (ed.), *The Evolution of Hospitals in Britain,* London, Pitman Medical Publishing Ltd, 1964.

Priestley, J. B., *Victoria's Heyday,* London, Heinemann, 1972.

Prochaska, G., *A Dissertation on the Functions of the Nervous System,* and Unzer, J. A., *The Principles of Physiology,* translated and edited by Thomas Laycock, London, Printed for the Sydenham Society, 1851.

Prochaska, G., *Adnotationum Academicarum,* Fasciculus Tertius, Pragae, Apud Wolfgangum Gerle, 1784.

Prochaska, G., *Operum Minorum Anatomici Physiologici et Pathologici Argumenti,* Pars II, Viennae, Prostat Apud Wappler et Beck, 1800.

Rattray Taylor, G., *The Angel-Makers. A Study in the Psychological Origins of Historical Change 1750–1850,* London, Heinemann, 1958.

Rees, A., *Cyclopaedia: or an Universal Dictionary of Arts and Sciences. Containing an Explanation of the Terms and an Account of the Several Subjects, in the Liberal and Mechanical Arts, and the Sciences, Human and Divine. Intended as a Course of Ancient and Modern Learning. By E. Chambers, F.R.S. with the Supplement, and Modern Improvements, Incorporated in One Alphabet. By Abraham Rees, D.D.,* 4 vols, London, W. Strahan *et al.,* London 1778–83.

Reiser, S. J., *Medicine and the Reign of Technology,* Cambridge, Cambridge University Press, 1981.

Report of the First and Second Meetings of the British Association for the Advancement of Science; at York in 1831, and at Oxford in 1832: Including its Proceedings, Recommendations, and Transactions, London, John Murray, 1833.

Report of the Third Meeting of the British Association for the Advancement of Science; Held at Cambridge in 1833, London, John Murray, 1834.

Report of the Fourth Meeting of the British Association for the Advancement of Science; Held at Edinburgh in 1834, London, John Murray, 1835.

Report of the Fifth Meeting of the British Association for the Advancement of Science; Held at Dublin in 1835, London, John Murray, 1836.

Report of the Sixth Meeting of the British Association for the Advancement of Science; Held at Bristol in 1836, London, John Murray, 1837.

Report of the Eighth Meeting of the British Association for the Advancement of Science; Held at Newcastle in 1838, London, John Murray, 1839.

Report of the Seventeenth Meeting of the British Association for the Advancement of Science; Held at Oxford in 1847, London, John Murray, 1848.

Report of the Select Committee on Medical Registration, London, 1847.

Reznek, L., *The Nature of Disease,* London, Routledge & Kegan Paul, 1987.

Richardson, R., *Death, Dissection and the Destitute,* London, Penguin, 1989.

Richet, C., *Physiologie des Muscles et des Nerfs. Leçons Professées à la Faculté de Médecine en 1881,* Paris, Libraire Germer Baillière et Cie., 1882.

Risse, G. B., *Hospital Life in Enlightenment Scotland. Care and Teaching at the Royal Infirmary of Edinburgh,* Cambridge, Cambridge University Press, 1986.

Ritter, J. W., *Das electrische System der Korper, Leipzig, bey* C. H. Reclam, 1805.

Roget, P. M., *Animal and Vegetable Physiology Considered with Reference to Natural Theology, The Fifth Bridgewater Treatise on the Power, Wisdom and Goodness of God as Manifested in the Creation,* 2 vols, London, William Pickering, 1834.

Rolleston, Sir Humphry, *Centenary of the Nottingham Medico-Chirurgical Society On March 7th, 1828,* Nottingham, Cooke & Vowles, 1928.

Royal Society of London, *A Statement of Circumstances Connected with the late Election for the Presidency of the Royal Society,* London, 1831.

Royal Society of London, *Lists of Council, Fellows etc.* 30 November 1850, London, Richard and John Edward Taylor, 1851.

Royal Society of London, *Special General Meeting,* 25 January 1847.

Roszak, T., *The Making of a Counter Culture. Reflections on the Technocratic Society and its Youthful Opposition,* London, Faber and Faber, 1970.

Roszak, T., *Person/Planet. The Creative Disintegration of Industrial Society,* London, Victor Gollancz Ltd, 1979.

Rupke, N. A. (ed.), *Vivisection in Historical Perspective,* London, Croom Helm, 1987.

Russell, J., *A History of the Nottingham Subscription Library More Generally Known as Bromley House Library,* Nottingham, Derry & Sons Ltd, 1916.

Sadler, M. T., *The Law of Population: A Treatise in Six Books, in Disproof of the Superfecundity of Human Beings, and Developing the Real Principle of their Increase,* London, John Murray, 1830.

Sadler, T. (ed.), *Diary, Reminiscences, and Correspondence of Henry Crabb Robinson, Barrister at Law, F.S.A.,* 3 vols, London, Macmillan & Co., 1869.

Sauvages, F. B. de (ed.), *Nosologie Méthodique, ou Distribution des Maladies en Classes, en Genres et en Éspèces, Suivant l'Ésprit de Sydenham, [et] la Méthode des Botanistes. Traduite sur la dernière édition latine, par M. Gouvian, ... On à joint à cet Ouvrage celui du Chev. von Linné, intitulé Genera Morborum, avec la Traduction Françoise à coté,* Lyon, Jean-Marie Bruyset, 1772.

Schafer, E. A., *Textbook of Physiology,* 2 vols, Edinburgh and London, Young J. Pentland, 1898–1900.

Sheppard, F., *London 1808–1870: The Infernal Wen,* London, Secker & Warburg, 1971.

Showalter, E., *The Female Malady. Women, Madness and English Culture, 1830–1980,* London, Virago Press, 1985.

Smellie W., *A Treatise on the Theory and Practice of Midwifery,* 3 vols, London, D. Wilson, 1752–64.

South, Sir James, *Charges Against the President and Councils of the Royal Society,* London, Richard Clay, 1830.

South, Sir James, *Charges Against the President and Councils of the Royal Society*, 2nd edn, London, R. Fellowes, 1830.

Sprigge, S. Squire, *The Life and Times of Thomas Wakley*, London, Longman, Green & Co., 1897.

Statement by the Council of the University of London, Explanatory of the Nature and Objects of the Institution, London, Longman and John Murray, 1827.

Statement of Facts Concerning the Proposed Election of Officers and Council of the Royal Society for the Ensuing Year, [London, 1848].

Statutes and Directions for the Government of the General Hospital, Near Nottingham, 2nd edn, Nottingham, 1821.

Statutes and Rules, for the Government of the General Hospital, Near Nottingham: Open to the Sick and Lame Poor of Any County, Nottingham, S. Creswell, 1781.

Statutes of the Royal Society, London, Richard Taylor, 1831.

Stilling, B., *Untersuchungen über die Functionen des Rückenmarks und der Nerven. Mit specieller Beziehung auf die Abhandlungen J. van Deen's, zur Physiologie des Rückenmarks etc.*, Leipzig, Verlag von Otto Wigand, 1842.

Stirling, W., *Some Apostles of Physiology*, London, Waterlow & Sons Ltd, 1902.

Suttor, G., *Memoirs Historical and Scientific of the Right Honourable Sir Joseph Banks, Bart.*, Parramatta, E. Mason, 1855.

Taylor, G. Rattray, *The Angel-Makers. A Study in the Psychological Origins of Historical Change 1750–1850*, London, Heinemann, 1958.

The Fifty-Eighth Annual Report of The Royal Humane Society for 1832, London, The Royal Humane Society, 1832.

The Eighty First Annual Report of the Royal Humane Society, Instituted 1774, to Collect and Circulate the Most Approved and Effectual Methods for Recovering Persons Apparently Drowned or Dead; To Suggest and Provide Suitable Apparatus for, and Bestow Rewards on, those who assist in the Preservation and Restoration of Life. Supported by Voluntary Contributions ....For 1855, London, The Royal Humane Society, [1855].

The Eighty-Second Annual Report of the Royal Humane Society... for 1856, London, The Royal Humane Society, 1856.

The History and Statutes of the Royal Infirmary of Edinburgh, Edinburgh, Balfour & Smellie, 1778.

The Record of the Royal Society of London, London, Harrison & Sons, 1897.

The Record of the Royal Society of London, 2nd edn, London, Harrison & Sons, 1901.

Thomson, J., *An Account of the Life, Lectures and Writings, of William Cullen,* Edinburgh, William Blackwood, 1832.

Thomson, J., *An Account of the Life, Lectures and Writings, of William Cullen, M.D.,* 2 vols, Edinburgh, William Blackwood & Sons, 1859. Volume I states: First published in 1832 now re-issued along with the second volume and having prefixed to it a biographical notice of the Author. Volume II states: Commenced by Dr John Thompson and Dr William Thompson, and Concluded by David Craigie, M.D., F.R.C.P.E., F.R.S.E., etc., etc.

Thomson, J. (ed.), *The Works of William Cullen, Containing his Physiology, Nosology, and First Lines of the Practice of Physic: with Numerous Extracts from his Manuscript Papers, and from his Treatise of the Materia Medica,* 2 vols, Edinburgh, William Blackwood, 1827. In the endnotes this book is referred to as [Cullen], *Works.*

Tillotson, J., *Our Untitled Nobility,* London, Gall & Inglis, n.d.

Todhunter, I., *William Whewell, D.D., An Account of his Writings with Selections from his Literary and Scientific Correspondence,* 2 vols, London, Macmillan & Co., 1876.

Tracts On Delirium Tremens, on Peritonitis, and on some other Internal Inflammatory Affections, and on the Gout, London, Thomas Underwood, 1813.

[Underwood, M.], *A Treatise on the Diseases of Children with Directions for the Management of Infants; by the late Michael Underwood, Ninth Edition, with Notes, by Marshall Hall,* London, John Churchill, 1835.

University College Report for 1859, London, University College London, 1860.

Unzer, J. A., *The Principles of Physiology,* and Prochaska, G., *A Dissertation on the Functions of the Nervous System,* translated and edited by Thomas Laycock, London, Printed for the Sydenham Society, 1851.

Venn, J. A., *Alumni Cantabrigienses. A Biographical List of all Known Students, Graduates and Holders of Office at the University of Cambridge from the Earliest Times to 1900,* Cambridge, Cambridge University Press, 1947.

Virchow, R., *Cellular Pathology as Based Upon Physiological and Pathological Histology, Twenty Lectures Delivered in the Pathological Institute of Berlin during the Months of February, March and April, 1858.* Translated from the 2nd edn of the original by Frank Chance, London, John Churchill, 1860.

Vogel, M. J. and Rosenberg, C. E. (eds), *The Therapeutic Revolution. Essays in the Social History of American Medicine,* Pennsylvania, University of Pennsylvania Press, Inc., 1979.

Wagner, R., *Handworterbuch der Physiologie mit Rücksicht auf Physiologische Pathologie,* 4 vols in 5, Braunschweig, Druck und Berlag von F. Bieweg und Cohn, 1844.

[Walker, A.], *Documents and Dates of Modern Discoveries in the Nervous System,* London, John Churchill, 1839.

Watson, R., *Chemical Essays,* 5 vols, 7th edn, London, G. Woodfall, 1800.

Weinreb, B. and Hibbert, C. (eds), *The London Encyclopaedia,* London, Macmillan, 1990. (First published 1983.)

Whewell, W., *Address Delivered in the Senate House at Cambridge, June XXV, M.DCCC.XXXIII. On the Occasion of the Opening of the Third General Meeting of the British Association for the Advancement of Science,* Cambridge, John Smith, 1833.

Whytt, R., *An Essay on the Vital and Other Involuntary Motions of Animals,* Edinburgh, Hamilton, Balfour & Neill, 1751.

[Whytt, R.], *The Works of Robert Whytt,* published by his son, Edinburgh, *Becket,* De Hondt & Balfour, 1768.

Williams, C. J. B., *Memoirs of Life and Work,* London, Smith, Elder & Co., 1884.

Williams, C. J. B., *Principles of Medicine Comprising General Pathology and Therapeutics,* London, John Churchill, 1843.

Willis T., *Cerebri Anatome cui Accessit Nervorum Descriptio et Usus,* London, J. Martyn &J. Allestry, 1664.

Newspapers

New Orleans Delta, New Orleans.

New York Herald, New York.

Nottingham Journal, Nottingham.

Journals

Abs. Phil. Trans. – *Abstracts of the Papers Printed in the Philosophical Transactions of the Royal Society,* Royal Society, London.

Alp. Journ. – *The Alpine Journal,* London.

Annal. des Sci. Nat. – *Annales des Sciences Naturelles,* Paris.

Annals of Phil. – *The Annals of Philosophy; or, Magazine of Chemistry, Mineralogy, Mechanics, Natural History, Agriculture, and the Arts,* London.

Annual Register – *The Annual Register, or a View of the History and Politics of the Year 1857,* London, 1858.

Archiv für Anat. Physiol. – *Archiv für Anatomie, Physiologie und Wissenschaftliche Medicin,* Berlin.

Archiv für Ges. Physiol. – *Archiv für die gesamte Physiologie des Menshen und der Tiere,* Bonn.

Archiv. Gén. de Méd. – *Archives Générales de Médecine,* Paris.

Beit. zur Anat. und Physiol. – *Beitrage zur Anatomie und Physiologie, Geissen,* Verlag von Emil Roth.

Boston Med. and Surg. Journ. – *The Boston Medical and Surgical Journal,* Boston.

Brit. and For. Med. Rev. – *The British and Foreign Medical Review and Quarterly Journal of Practical Medicine and Surgery,* London.

Brit. Ann. Med. – *British Annals of Medicine, Pharmacy, Vital Statistics, and General Science,* London, Sherwood, Gilbert & Piper.

Brit. Journ. Hist. Sci. – *The British Journal for the History of Science,* London.

Brit. Med. Journ. – *British Medical Journal,* London.

Charleston Med. Journ. and Rev. – *The Charleston Medical Journal and Review,* Charleston, Steam Power Press of Walker and James.

Coffin's Bot. Journ. – *Coffin's Botanical Journal and Medical Reformer,* Manchester and London.

Comptes Rendus – *Comptes Rendus des Séances de l'Académie des Sciences,* Paris, and *Comptes Rendus, Hebdomadaires des Sciences de l'Académie,* Paris.

Edin. Med. and Surg. Journ. – *Edinburgh Medical and Surgical Journal,* Edinburgh.

Edin. Med. Journ. – *Edinburgh Medical Journal.* Combining the *Monthly Journal of Science* and *Edinburgh Medical and Surgical Journal.*

Edin. New Phil. Journ. – *The Edinburgh New Philosophical Journal,* Edinburgh.

Gaz. Méd. d'Orient – *Gazette Médicale d'Orient Publiée par la Societé Impériale de Médecine de Constantinople,* Constantinople.

Gent. Mag. – *The Gentleman's Magazine and Historical Review by Sylvanus Urban, Gent.* or *Monthly Intelligencer,* London.

Household Words – *Household Words,* a weekly journal conducted by C. Dickens, London.

Isis – *The Isis. A Quarterly Magazine,* London.

Journ. des Progrès – *Journal des Progrès des Sciences et Institutions Médicales en Europe en Amérique ....,* Paris.

Journ. Hist. Med. – *Journal of the History of Medicine and Allied Sciences,* New Haven.

Journ. Nat. Phil. Chem. and Arts – *Journal of Natural Philosophy, Chemistry and the Arts,* London. The journal was popularly known as *Nicholson's Journal* after its editor W. Nicholson.

Lancet – *The Lancet,* London.

Lond. and Edin. Mon. Journ. Med. Sci. – *London and Edinburgh Monthly Journal of Medical Science,* London.

Lond. and Edin. Phil. Mag. and Journ. of Sci. – *The London and Edinburgh Philosophical Magazine and Journal of Science,* London.

Lond. Med. and Phys. Journ. – *The London Medical and Physical Journal,* London. Formerly the Medical and Physical Journal.

Lond. Med. Gaz. – *The London Medical Gazette. Being a Weekly Journal of Medicine and the Colateral Sciences,* London.

Lond. Prov. Med. Dir. – *The London Provincial Medical Directory,* London.

Med. and Phys. Journ. – *The Medical and Physical Journal,* London. This continued as the *London Medical and Physical Journal* from 1815–26.

Med. Hist. – *Medical History,* London.

Med. Times and Gaz. – *The Medical Times and Gazette,* London.

Med.-Chir. Trans. – *The Medico-Chirurgical Transactions,* London.

Mem. Lit. and Phil. Soc. Manch. – *Memoirs of the Literary and Philosophical Society of Manchester,* London. Continued 1888–96 as *Memoirs and Proceedings of the Manchester Literary and Philosophical Society,* Manchester.

Pharos - *Pharos,* London.

Phil. Trans. – *Philosophical Transactions,* London.

Proc. Comm. Sci. and Corresp. Zoo. Soc. Lond. – *Proceedings of the Committee of Science and Correspondence of the Zoological Society of London,* London.

Proc. Manch. Lit. and Phil. Soc. – *Proceedings of the Manchester Literary and Philosophical Society,* Manchester. Later, 1888, joined with *Memoirs of the Literary and Philosophical Society of Manchester.*

Proc. Roy. Soc. – *Proceedings of the Royal Society of London.* Being a continuation of the Series entitled *Abstracts,* Royal Society, London.

Quart. Journ. – *Journal of Science and the Arts,* London. The title of this journal changed to *Quarterly Journal of Literature, Science and the Arts* in 1820. It was popularly known as *Quarterly Journal.*

The Practitioner – *The Practitioner. A Medical Journal,* London.

Time – *Time Magazine,* New York.

West. Rev. – *The Westminster Review,* London, Baldwin, Cradock & Joy.

Zeit. für klin. Med. – *Zeitschrift für klinische Medizin,* Herausgegeben von Dr Fr. Th. Frerichs, und Dr E. Leyden, Ester Band, Berlin.

Manuscripts

MS *Election Certificate 47, VIII,* 1830–40, Royal Society, London.

MS Grant, R.E., *Lectures on Comparative Anatomy. Notes for a Course of Lectures Delivered Oct.–Dec. 1830* (plus printed copy of examination paper on comparative anatomy, dated 8 Jan. 1831) and MS *Lectures on Comparative Anatomy Delivered in the University of London, session 1832–33 and 1833–34, MS.ADD.1,* University College London, London.

MS Hall, M., *A Note Addressed to the Members of the Council of the Royal Society; By Marshall Hall,* dated 7 June 1837, *MC.2.255,* Royal Society, London.

MS Hall, M., *On the Reflex Function of the Medulla Oblongata and Spinalis, or the Principle of Tone in the Muscular System, Archives 1832, AP.17.5,* Royal Society, London.

MS Hall, M., *On the True Spinal Marrow, and the Excito-Motory System of Nerves, Archives 1837,* Royal Society, London.

MS Hall, M., *On the Anatomy and Physiology of the Minute and Capillary Vessels. Archives 1831, AP.15,* 32, *6,* Royal Society, London.

MS Hall, M., *On the Effect of Water Raised to Temperatures Moderately Higher than that of the Atmosphere, upon Batrachian Reptiles, Archives 1831, AP.15,* 32, *6,* Royal Society, London.

MS Hall, M., *Researches into the Effects of Certain Physical and Chemical Agents on the Nervous System, Archives 1847, AP.29.5–6,* Royal Society, London.

MS Hall's Testimonials, *Archive 4564b,* University College London, London.

MS *Journal Book of Royal Society,* Royal Society, London.

MS Letter, George Newport to the President and Council of the Royal Society, 7 March 1838, *MC.2.279,* Royal Society, London.

MS Letter, H. Mayo to J. G. Children, 8 May 1837, *RR.1.86–88,* Royal Society, London.

MS Letter, Hall to Paget, n.d., *WMS/ALS,* Wellcome Institute for the History of Medicine, London.

MS Letter, J. D. Forbes to Marshall Hall, 9 November 1846. *Forbes Letter Box,* University Library, St Andrews, Scotland.

MS Letter, J. G. Children to J. Lubbock, 12 June 1833, *LUB. C. 177,* Royal Society, London.

MS Letter, Lee, R., 'To the President and Council of the Royal Society of London', *RR.1.150,* Royal Society, London.

MS Letter, M. Hall to Dr George Shattuck, *Shattuck Papers, 21a,* 1848–54, The Francis A. Countway Library of Medicine, Boston, Massachusetts.

MS Letter, M. Hall to J. Lubbock, 10 November 1831. *LUB. H. 17.* Royal Society, London.

MS Letter, M. Hall to C. R. Weld, 5 November 1847, *MC.4.189,* Royal Society, London.

MS Letter, M. Hall to P. C. A. Louis, dated 17 Avril 1857, *Archives,* Paris, Académie des Sciences, Institut de France.

MS Letter, M. Hall to Members of the Council, dated 8 June 1837, *MC.2.255,* Royal Society, London,

MS Letter, M. Hall to Sir Joseph and Lady Oliffe, dated 22 June 1855, London, Wellcome Institute for the History of Medicine, London.

MS Letter, M. Hall to the Council of the Royal Society, 1 February 1838, *MC.272,* Royal Society London. Hall included the reply which he was requesting Council to publish. *MC.273,* Royal Society London.

MS Letter, Marshall Hall to Sir Archibald Geikie, dated 3 May 1894, *MS General 524,* University of Edinburgh.

MS Letters, Marshall Hall to C. C. Atkinson, Esq., 1 May, 11 May, 22 July 1839, *Archive 4538a, 4538b, 4602,* University College London, London.

MS Letter, P. C. A. Louis to H. I. Bowditch, dated 1 October 1840. *B MS Misc.* 2 Folders, Folder 2. Francis A. Countway Library of the History of Medicine, Boston, Massachusetts.

MS Letter, R. S. Whitfield to Thomas Hodgkin, 20 July 1843, *Hodgkin Papers,* Ilmington, Warwickshire.

MS Letter, Thomas Hodgkin to [Dr Foville], 12 December 1842, *Hodgkin Papers,* Ilmington, Warwickshire.
MS Letter, W. Lawrence to H. Mayo, 1 May 1837, *RR.1.86–88,* Royal Society, London.

MS Letter, W. H. Walshe to the Council of University College, 6 January 1841. *College Correspondence Applications,* University College London, London.

MS Letters, 6 July, 18 July 1838, *Warren Papers, 19,* 1838–40, Massachusetts Historical Society, Boston.

MS Letters, J. F. W. Herschel to M. Hall, 16 October 1839, *HS.9.200–201,* Royal Society, London.

MS Letters, M. Hall to J. F. W. Herschel, 1 & 14 October, 21 & 27 November 1839, 3 April 1840, 23 November 1848, *HS.9.198/199/202–205,* Royal Society, London.

MS Letters, M. Hall to Dr Munk, 30 & 31 October 1850, Royal College of Physicians, London.

MS Letters, M. Hall to W. R. Grove, 13, 26 & 28 November 1848, Library of The Royal Institution, London.

MS Letters, Marshall Hall to J. D. Forbes, 6 & 14 November, 1846. *Forbes Letter Box,* University Library, St Andrews, Scotland.

MS Letters, Marshall Hall to Thomas Hodgkin, 22, 23, 24 & 28 July 1843, *Hodgkin Papers,* Ilmington, Warwickshire.

MS Letters, Marshall Hall to the Council of University College, 1 May, 23 July 1839, *Archive 4564a, 4604,* University College London, London.

MS Letters, Marshall Hall to Mr H. Lee, FGS, dated 14 & 15 April 1870, 23 May 1870, 14 June 1870, 30 July 1870, 27 August 1870, 9, 21, 24 & 26 September 1870, 24 October 1870 and 16 November 1870, *Documents 5382 – 1 to 12,* Wellcome Institute for the History of Medicine, London.

MS Letters, R. B. Todd to P. M. Roget, 28 June, 12 October 1843, *RR.1.265–6,* Royal Society, London.

MS Letters, Thomas Hodgkin to My Dear Friend (Marshall Hall), 22 July 1843, *Hodgkin Papers,* Ilmington, Warwickshire.

MS Letters, W. H. Walshe to Charles C. Atkinson, 8 April, 8 June 1841. *College Correspondence Applications,* University College London, London.

MS *List of Students Attending Lectures in the University of Edinburgh from 1790,* University of Edinburgh, Edinburgh.

MS *Lists of Students, General and Medical,* 1811 to 1813, *7–8,* University of Edinburgh, Edinburgh.

MS Managers' Extraordinary Meeting 11 August 1812 and Managers' Meetings 15 January 1811 and 7 February 1814, *Minute Book of Royal Infirmary Edinburgh,* 7–8, Royal Infirmary Edinburgh, 1813 and 1814.

MS *Medical Matriculation Lists, 1810–1811, 6,* University of Edinburgh, Edinburgh.

MS *Minute Book of the Faculty of Medicine of the University of London, 1827–1830,* University College London, London.

MS *Minutes* 23 March 1853, 30 April 1857, 27–29 October 1858. *Minute Book to be Kept by Dean in Conformity with the Bye Laws relating to his Jurisdiction Over Students,* Faculty of Medicine, Section 14 vols *22,* 1853–8, University College London, London.

MS *Minutes 26 October 1827. Minute Book of the Faculty of Medicine of the University of London, 1827–1830.* University College London, London.

MS *Minutes of Committee of Zoology and Animal Physiology* beginning 23 November 1838, Royal Society, London.

MS *Minutes of Council Meetings,* Royal Society, London.

MS *Minutes of Council,* University of London, London.

MS *Minutes of Senate,* University College London, London.

MS *Minutes of the Royal Medical Society,* University of Edinburgh, Edinburgh.

MS Note, Hall to Weld, 12 May 1847, *Archives 1847, AP.29.5–6,* Royal Society, London.

MS *Notebooks Containing a Student's Abstracts of Grant's Lectures, 1832–33 and 1833–34. Strong Rm. MS.ADD.39–38,* University College London, London.
MS Drawing of Hall's apparatus for his respirometer, *Archives 1831, PT.74.15,* Royal Society, London.

MS *Records of the Royal Medical Society,* University of Edinburgh, Edinburgh.

MS Referee's Report on a Paper by J. Lockhart Clarke on the Structure of the Spinal Cord. By Marshall Hall, 18 February 1851, *RR.2.44,* Royal Society, London.

MS Referee's Report on Dr M. Hall's two papers received 28 January, 15 April, and the Appendix 26 June 1847. *RR.1.86–89,* Royal Society, London.

MS *Researches into the Effects of Certain Physical and Chemical Agents on the Nervous System;* by Marshall Hall MD, F.R.S. etc., *Archives 1847, AP.29.5–6,* Royal Society, London.

MS *The Denis Notebook,* n.d., *Student Lecture Notebook,* St Thomas's Medical School, London.

MS *Will of Marshall Hall,* and Probate Certificate *Certificate 482,* Lord Chancellor's Department, Somerset House, London, 1896.

Index

A

Abernethy, John *71–4, 119*
 Abernethian Code *73–4*
 advice books *72–3*
Academy of Medicine, Paris *26, 171*
Academy of Sciences, Paris *26–7, 171, 272*
Acton, William *86*
Adams, John *310*
Aerometer *44*
Africa *310–11*
 See also: Negroes, Slavery
Alcock, Thomas *117*
Aldersgate School – *See:* Anatomy Schools
Alpine Club *30*
Alternative Medicine – *See:* Quackery
America – *See:* United States
American Civil War *305*
Anaemia *68, 89*
Anatomy *7, 26, 49, 58–61 passim, 72, 95, 110–35 passim, 145, 165–8, 174–84 passim, 190–213 passim, 248–58 passim, 271, 281*
Anatomy Act *1832 58–9*
Anatomy Schools (London, Private)
 Aldersgate *26, 114, 119, 127, 132*
 Blenheim Street *115–16*
 Great Windmill Street *115, 127*
 Sydenham College *50–6 passim, 114–17* passim, *127, 132, 282, 284*
 Webb Street *114, 116, 121, 132, 198, 200, 282–85 passim*
Anencephalous *269–70*
Anglican Church *17*
 See also: Church of England
Animal Electricity *205, 271–2*
 See also: Electrophysiology, Galvani, Volta
Animal Experimentation – *See:* Experimental Work on Animals, Vivisection
Annales de Chimie 2
Anti-vivisection *183, 238*
Apnoea *315–18*
Apoplexy *19, 88*
Apothecaries *6, 16, 19, 108–13 passim, 118, 137–46 passim*
 apprenticeships *5–6, 111–16 passim*
 as general practitioners *108–11*
 licensing of *108, 111, 144*
 right to prescribe *113, 145*
Apothecaries Act 1815 *108–09, 111, 118, 139–40, 144*
 1825 Amendment *111*
Apprenticeships *11, 25, 111, 112, 116*
 apothecaries – *See:* Apothecaries
 surgeons *11, 111–18 passim*
Archiv fur Anatomie und Physiologie 206, 241
Archives Générales de Médecine 96–7
Artificial Respiration – *See:* Resuscitation
Asphyxia *313–15, 317*
Asylums *129–30*
 Commissioners in Lunacy *130*
 mechanical constraints *130*
 Middlesex County Asylum *130*
 Moorcroft Asylum *130–2*
 Nottidge versus Ripley *130–1*
Autopsies/Post-mortems *57–61, 84–5*
 See also: Dissection

B

Babbage, Charles *158–68 passim, 179, 204, 212*
 calculating machine *160, 164*
 declinist views on state of science *160, 163*

threatened ejection from Royal Society *160*
Bacon, Francis *55, 233*
Baillie, Matthew *18, 46, 63, 115*
Baly, W. *117, 269*
Bancroft J. *321*
Banks, Sir Joseph – President Royal. Society. *53, 159, 161, 168–9, 320, 322*
on admission procedures to Royal Society *159*
Barker, Dr *121–3*
Barlow, Edward *139, 268*
Bartlett, Elisha *47, 55–6*
Beaumont, Gustave de – On slavery *306*
Bedside teaching xii, *48–9, 54, 110, 182–3*
Bell, Sir Charles *71, 111–15 passim, 125–35 passim, 158, 166, 173–7 passim, 190, 200, 213, 237–40 passim, 284–5, 289*
controversies with University College *128*
criticism of Magendie's vivisection, Bell-Magendie *239–40, 253*
experimental work *239, 248, 252*
sensory and motor nerves *190, 237, 251–3, 289–90*
vital nerves *252*
Bell, George *126*
Bell, John *252*
Bell, Thomas *173–81 passim, 208–11 passim*
Bellot, H. Hale *128*
Benet, Dr. *90*
Bentham, Jeremy *124*
Berlin *14, 206, 256, 281*
Bernard, Claude *27, 262, 279–81*
two forms of *sensibilité 280*
Berthollet, Claude *2*
Bichat, Xavier *9, 48, 58–61*
Birkbeck, George *126, 128*
Bishop, Mr – Royal Society *208, 211*
Black, Joseph *2, 8–9, 309*
Blane, Sir Gilbert – Dissociated sensation from reflex action *259*
Bleaching *x, 2–4, 28*
chlorine *2–3*
oxymuriatic acid *2–3*
scale in Britain *3*
Blenheim Street School – *See:* Anatomy Schools
Bloodletting/Venesection *12, 45, 64–6, 89–100, 184*
experiments on dogs *93–5*
Hall's paper on circulation *95–6*
inflammation v. irritation *91–3*
Jackson's work *98–100*
Louis's work *96–8*
rule for use *92–3*
See also: Diagnosis
Blumenbach, [D.]J. F. *14, 190*
Boerhaave, H. *47, 287*
Bostock, John *173, 180–1*
Boston Medical and Surgical Journal 277, 304
Bowman, W. *205, 208, 271*
Boyle, Robert *233*
Bradley, I. *15, 305*
Brewster, David *158–9, 161, 163*
Bridgewater Treatise – Roget P. M. *194, 196–7*
Bristol *14–15*
British and Foreign Bible Society *17*
British and Foreign Medical Review 14, 26, 182, 235, 239, 249, 256, 264, 267–8, 284
British Annals of Medicine, Pharmacy, Vital Statistics 57, 61
British Association for the Advancement of Science *157-8, 161–5, 167–70, 173, 178–81*
Annual Meetings *163–8 passim*
commissioning of research *168*
committees and membership *166*
cultivators of science v. gentlemen of science *163, 168*
Herschel's reservations *165, 168*
origins and aims *161–4*

place of and reports on physiology *166–8 passim*
British Medical Association *xii, 20, 25, 110, 139–43, 167–8*
British Medical Journal 327
British Museum *20, 158, 194*
Brodie, Sir Benjamin C. *115, 132, 158–9, 173–82 passim, 208–10, 269, 325*
See also: Physiological Committee – Royal Society
Brookes, Joshua *115, 116*
Broughton, S. D. *168*
Broussais, François *64*
Brown, John *45*
Brown-Sequard, C. E. *319*
Budd, W. *25, 117, 268*
Bullying of Hall *23–4*
Bulwer-Lytton, Sir Edward *25*
Burnside, William – Publisher *20–1*
Bynum, W. F. *335*

C

Cadavers *59, 85, 115–16*
See also: Corpses
Calmeil, L. F. *262*
Canguilhem, G. *245, 290*
Cannon, S .F. *163–4*
Carpenter, W. B. *14, 124, 141, 182, 239, 249, 256, 312*
Carrington, Lord (Smith, Robert) *4*
Cartwright, S. A. – *See:* Controversies, Vivisection
Case histories *18–19, 24, 62–5 passim, 76, 83–7 passim, 91–9 passim, 268–70, 327*
Catholics *7*
Cayrade, Jules *283*
Chadwick, Edwin *321–2, 324*
Charing Cross Hospital *118*
Charleston Medical Journal and Review 276, 304
Chatham Street School, Manchester *312*
Chemists and Druggists *5–6, 108–11 passim, 145–6*
Children, J.G. *184, 187, 191, 194–7*
Chirac, M. *232, 269*
Chlorosis-Anaemia *68, 77, 89*
See also: Women's Disorders
Cholera *84, 312*
Christian Medical Association *116*
Christie, Samuel H. *208, 211*
Christison, R. *59, 119*
Church of England *4, 7, 157*
See also: Anglican Church
Clark, F. le Gros *124, 258, 262, 267, 270*
Clark, William *166–7*
Clark-Kennedy, A. E. *120*
Clarke, Charles Mansfield *80, 86*
Clarke, Edwin *195*
Clarke, J. Lockhart *213–14*
Clerkships – Clinical, Medical, Physicians and Surgeons *7, 11–13, 49, 58, 66, 139*
Clift, William *173, 181, 188*
Collins, John – Abolitionist *305, 311*
Colonization Societies
American *308, 310*
British *311*
Comptes Rendus 26–7, 211, 318
Conolly, John *79, 125–7, 129–32, 134, 136, 139*
Consciousness *257–65 passim, 267–9, 280, 282*
See also: Bernard, Longet, Lewes, Pflüger, Whytt
Constipation *65, 77–8, 89, 91*
See also: Women's Disorders
Contagious Fever, Contagion *321*
Controversies *199, 239*
Cartwright and Dowler in America *276–8, 304*
Newport and Hall *197–201*
Pflüger-Lotze controversy *279–82*
Prochaska controversy *282–91*
Cook, John Parsons *318–20*
See also: Strychnine
Cooper, Sir Astley P. *9, 86, 174*
Cope, Z. *28, 82–3, 86, 308*
Copley Medal – *See:* Royal Society
Corpses *58, 60, 116*
See also: Cadavers

Corry, John *78*
Corvisart, J. N. *9, 63, 64*
Croonian Lectures and Lecturers – Royal College of Physicians *26, 213–14, 254, 266*
Crystal Palace *312*
Cullen, William *44–8, 50–7, 59, 63, 67, 75, 82, 90, 115, 136*
Cuvier, Baron George *161, 173, 175, 177*

D

Darwin, Charles *9, 177*
Davis, D. B. – *See:* Slavery
Davy, Sir Humphry – President Royal Society *28, 159, 169–70*
Descartes, R. du P. *232, 234–5*
Design in Nature *60, 267, 277*
Diagnosis, *14, 64–5, 71, 91, 110*
  criterion of usefulness *52–3, 67*
  definition *46, 48*
  nature of evidence *53–5, 63*
  on observation *55–62, 93*
  Tables of Diagnosis *62–3*
  See also: Nosology
Diastaltic Nervous System *26, 214, 254, 266*
  See also: Excito-motory System, True Spinal System
Dickens, Charles *322*
Diseases x, xi, xiii, *8, 9, 26,* Chapter *2 passim, 130, 136, 145, 183, 206,* Chapter *5 passim, 303, 317*
  classification – *See:* Diagnosis
  definition *46–8*
  effect of class, exercise, gender, lifestyle *76–81*
  sexually transmitted diseases – *See:* Venereal Diseases
  See also: Infectious Diseases, Nervous Diseases, Puerperal Fever, Women's Disorders
Dispensaries *111–14, 129, 135–6*
Dissection *7–8, 57–60, 112, 114–16, 248, 270*
  See also: Vivisection
Dissenters *4, 6–7, 17*
  See also: Methodists, Unitarians
Dowler – See: Controversies
Drug Resistance *52*
Duncan, Andrew *12*
Dutrochet, R. H. J. – First observation of nerve cells *256*
Duverny, C. J. *232, 269*
  See also: Chirac, M.
Dyspepsis *64, 68*
Dyspnoea *57*

E

Eckhard, C. *235–6*
Edinburgh xi–xii, *1–19 passim, 27–8, 57*
  See also: Cullen, Edinburgh Royal Infirmary, Royal Medical Society, University of Edinburgh
Edinburgh General Lying-in Hospital *72, 83*
Edinburgh Medical and Surgical Journal 9, 15, 45, 48, 124, 268
Edinburgh New Philosophical Journal 211
Edinburgh Royal Infirmary *48–50, 58–60, 65, 72, 82, 89–90, 110*
  clerkships *7, 11–13, 49, 58*
  managers' meetings *11–13*
  mortality rate *59*
Edwards, William F. *185–6, 189, 256, 315, 316, 318*
Electricity - Experiments On *185–6, 206–08 ,passim, 271–2*
  See also: Electrophysiology, Faraday
Electrophysiology *173, 186, 209, 271–5*
  Hall's Memoir *1847 186, 206–9, 211–12, 271*
  See also: Animal Electricity, Galvani, Matteucci, Volta
Emblen, D. L. *8*
Esquirol, J. E. D. *71*
Ethics
  Hall's proposed code of ethics *96*
  Percival's 'Medical Ethics' *87*
Eupaedia xi, *68*
Evangelicals *15, 305–6*
  See also: Dissenters, Slavery

Excito-motory System *124, 190–2, 195–6, 213, 242, 251–60 passim, 267, 270–83 passim, 318*
See also: Diastaltic Nervous System, True Spinal System
Experimental Medicine – *See:* Physiology
Experimental Work on Animals *21, 94–5, 184, 232–3, 236–9, 243–5, 253, 259–60, 262, 267–9, 276, 279*
See also: Anti-vivisection, Vivisection

F

Faculty of Medicine – London *143–46*
Faculty of Medicine – Paris *238*
Faraday, Michael *206, 271–2*
investigations on electric fish *271*
Farr, W. *57, 61, 84*
Fearing, Franklin *234, 241, 242, 269, 282*
Felkin, W. *19*
Female Venereal Hospital, Paris *86*
Flourens, M. J. P. *26, 132, 188, 190, 192, 203, 238-9, 248, 259, 262, 264, 279, 312*
cerebrum essential for sensation and volition *239, 259, 262*
experimental work on pigeons *238, 239, 262*
Flower, W. H. *157*
Forbes, J. D. *28*
Fordyce, George *46, 99*
Franklin, Benjamin *310*
Fyfe, Andrew *7–8*

G

Galen – Greek physiology *232, 278, 282*
Galvani, L. *271–2*
galvanism *108, 157, 271*
nerve function and animal electricity *271*
See also: Electrophysiology
Gassing of lace *x, 3*
Gaubius, H. D. *47*
Gay Lussac, J. L. *44*
Geikie, A. *30*
General Medical Council *146*
Medical Register *146*
General Practice *6, 14–20 passim, 30, 44, 66, 72–82 passim, 90–9 passim, 112, 117, 127–40 passim, 157, 179, 190, 282, 291, 303*
General Practitioners *5–6, 16, 19, 25, 47–64 passim, 72–80 passim, 108–19 passim, 134–46 passim*
Gentleman's Magazine 27
Geographical Society *30*
George III, King *18, 23, 170, 303*
George IV, King *177, 321*
George, John Durance *137, 141, 201, 246, 283–91*
See also: Controversies – Prochaska
Gilbert, Davies – President Royal Society *169*
Glisson, F. *232, 258*
Glover, S. *320*
Goldsmid, Isaac L. *125–6*
Goldsmith, Oliver *9*
Gooch, R. *66*
Government Grant – Royal Society *160–1, 163, 179–81*
Grainger, Edward *116*
Grainger, Richard D. *116, 118, 121, 122, 124, 159, 196, 237, 239, 249, 263–5, 267–8*
refutation of Mayo's claim *263–4*
supporter of Hall on reflex action *249*
Grant, Robert *49, 126, 127, 140, 159, 182, 193, 194, 196, 198–201, 254, 283*
Granville, A. B. *159, 161*
Great Exhibition, 1851 *162*
Great Windmill Street School – *See:* Anatomy Schools
Gregory, James *9, 90*
Grove, W. R. *176, 210–12*
Gulstonian Lectures – Royal College of Physicians *186*
Guthrie, George *142*
Guy's Hospital *116, 118, 122, 137*

H

Haemorrhage – *See:* Blood-letting
Hale-White, Sir William *318*
Hall, Alfred – Father of barrister *325*
Hall, Charlotte (née Green) – Wife Chapter *1 passim, 79, 95, 116, 131, 184, 208, 278,* Chapter *6 passim*
Hall, Sir Edward Marshall – Barrister *325*
Hall, John – Brother *1, 29*
Hall, Marie Boas *179, 181*
Hall, Marshall
  application for chair at University College London *132–5*
  British Medical Association *138–46*
  diagnosis *45–9*
  Edinburgh medical education *7–15*
  family and early years Chapter *1 passim*
  final illness and post-mortem *325–7*
  final works – *See:* Royal Humane Society, Sewage Disposal, Slavery, Strychnine
  Nottingham years *15–19,* Chapter *2 passim*
  proposed code of ethics *96*
  residences *15, 20, 303*
  Royal Society
    controversy - Newport *197–201*
    correspondence with Herschel *201–6*
    Fellowship *182–3, 188*
    papers; rejections *183–97, 206–9*
    Rosse letter *211–13*
  travels *304–5, 312*
  See also: Controversies, Experimental Work on Animals, Nervous System, Reflex Action, Sensation
Hall, Marshall – Son *20, 22, 27–30, 115, 303, 311, 317, 323*
Hall, Robert – Father *x, xi, 1–7, 9, 13, 20, 22, 44, 160, 325*
  prize for crane design *3, 160*
Hall, Samuel – Brother *3–4, 19, 22*
Haller, Albrecht von *188, 192, 232, 257, 258, 286*
Hamilton, Alexander *71, 72*
Hamilton, James *48, 58, 65, 66, 71–3, 83, 85*
Harcourt, Rev. William V. *161, 163–6*
  founding of British Association *161*
Harley, George *319, 320*
Harvey, William *29, 202, 258*
Hastings, Charles *xii, 139*
Hawkins, Caesar *115, 325*
Helmholtz, H. von *272, 275*
  See also: Electrophysiology
Henry, Thomas *2, 4, 5*
Henry, William C. – Son of Thomas *2, 5, 166*
Hermaphroditism *51*
Herschel, Sir John F. W. – President Royal Society *156–9 passim, 161–73 passim, 192–8 passim, 201–6, 211–12*
  correspondence with Hall *192, 197, 198, 201–6*
Hess, Rudolph *67*
Higginbottom, John – Brother-in-Law *4, 7, 16, 19, 208*
Higginbottom, M.H. – Nephew *325*
Hoblyn, Richard *117, 254*
Hodgkin, Thomas *22, 121–3, 185, 305, 309–11*
Hogg, John *129, 135, 136*
Holland, M. *132, 214*
Home, James *7, 139*
Horner, Leonard *127, 129, 132, 169*
Hospital Appointments *12, 115, 118–28 passim, 133–6, 141*
Hospital Medical Schools – *See* under name of Hospital
Hospitals – *See* under appropriate name
  Hôtel-Dieu, Paris *58*
Humoral Theory of Disease *64, 78, 89*
  See also: Blood-letting
Hunter, Charles *29, 314, 317, 318, 325*
Hunter, William *29, 46, 68, 72, 114, 115*
Huxley, T. H. *234*
Hypnosis *108*
Hypochondriasis *64, 68*

I

Iconography – Illustration of Symptoms *71*
Idiopathic Symptoms *67*
Indian Tribes – Chippewas, Sioux, Winnebagoes *304*
Industrialization Including Effects on Health *77–8, 82, 83, 90, 113, 320–1*
Infectious Diseases – Herpes, Rubeola, Tuberculosis, Typhoid, Typhus *57–62 passim, 90*
See also: Cholera
Inflammation *47, 52, 65, 67, 83, 89, 91–3, 96–9*
See also: Blood-letting

J

Jackson, J. *96, 98–100*
Jacyna, S. *195*
Jefferson, Thomas *309, 310*
Jenner, Sir William *138, 303*
Jews *7, 125, 310*
Journals – *See* under relevant title

K

Keith, Sir Arthur *17*
Kidd, John *142*
Kiernan, K. *173, 176, 181, 201–3*
Kölbe, A. W. H. *214*

L

La Charité, Paris *98*
Laennec, R. T. H. – Auscultation *61, 64*
See also: Stethoscope
Lallemand, Claude F. *86, 270*
Lancet *89, 91, 97, 98*
See also: Blood-letting
Lancet, The xii, *21, 29, 111–22 passim, 135–42 passim, 158, 182, 193, 198–200, 209–11, 239, 284, 285, 315–27 passim*
See also: Wakley
Lansdowne, Marquis of – President Royal Society *126*
La Pitié, Paris *282*
Lavoisier, Antoine *2, 5*
Lawrence, William *119, 139, 173–80 passim, 190–6, 208, 263, 269–70*
Laycock, William *284, 289*
Lee, Robert *49, 66, 159, 175, 187, 208, 209, 212*
Leeches *64, 89, 90, 97, 98*
See also: Blood-letting
Legallois, J. J. C. *188, 189, 237–9, 242, 248, 259*
Lewes, George H. *257, 262, 279–82*
consciousness and reflex action *262, 280*
imaginary physiology *281*
views on reflex action *257*
Lhermitte, J. *234*
Liebig, Justus von *3, 323*
Linnaeus, C. *49, 50*
Lister, Joseph *137, 184*
Lister, Robert *12*
London *4, 15–22 passim,* Chapter *2 passim,* Chapter *3 passim*
Great Exhibition, *1851 162*
Health of Towns Association *322*
Medical Officer of Health *321*
mortality rates *321*
number of physicians *6, 109*
population *113, 320*
Public Health Issues – drainage, water supply, sewage *113, 320–3*
Thames flood barrier *323*
transport system *320*
See also: Dispensaries, Hospitals, Institutions, Journals, Societies under appropriate name
London and Edinburgh Philosophical Magazine 241
London College of Medicine *109, 139, 140*
London Hospital *118, 121*
London Medical Gazette 109, 112, 283, 284
London University: *See* University College London, University of London

Longet, F. A. *279, 280*
Lotze, Rudolf H. *276, 279, 281, 282*
See also: Controversies
Loudon, I. *138*
Louis, P. C. A. *xii, 20, 26, 55–7, 64, 92, 96–100, 132, 141, 179, 203, 312, 325*
Lubbock, John *187, 188*
Luddite *4*
Lyell, Charles *158*

M

MacLeay W. S. *173, 174*
Macleod, R. – Professionalization of science *163*
Magendie, F. *237, 239–40, 248, 253, 279, 280, 290*
Manchester *157, 166, 312, 319, 325*
Manchester College of Arts and Science *2*
Manchester Infirmary *87*
Manchester Literary and Philosophical Society *2, 15, 157*
Martin, C. *113, 138*
Martineau, Harriet *311*
Massachusetts General Hospital *96, 98*
Matteucci, C. *207, 272–5*
galvanoscopic frog *273, 275*
See also: Electrophysiology
Maudsley, Henry *79*
Maulitz, R. C. *49, 183*
Mayo, Herbert *115, 126, 158–81 passim, 190–203*passim, *239, 262–4, 266–70 passim*
Mayow, J. *232, 234*
McMenemey, W. H. *119*
Meckel, John *127*
Medical Act, 1858 *6, 110, 146*
Medical Education Chapter *3 passim*
See also: Anatomy Schools, Apprenticeships, Medical Schools, University College London, University of Edinburgh
Medical Ethics – *See:* Ethics
Medical Officer of Health (City of London) *321*
See also: Simon
Medical Profession – *See:* Apothecaries, Physicians, Surgeons
Medical Reform *109, 110, 138–46, 206*
Medical Reform Bill *146*
See also: Hall, Williams
Medical Schools
Dispensaries – *See* Dispensaries
Hospital – *See* under name of Hospital
Private – See Anatomy Schools
Medical Students *8, 47, 49, 80, 84, 110, 118–20, 137–8*
Medico-Chirurgical Society *45, 66, 91, 92, 94, 158, 159, 193*
Medico-Chirurgical Transactions 270
Men-midwives *80, 81*
Mental Disorders; Mental Illnesses *47, 67, 130*
mania *48, 86*
melancholia *48*
Mesmerism *108*
Methodists *1, 4, 14, 21, 78*
Metropolitan Society of General Practitioners *139*
Miasma *84*
Microscope, Microscopic Studies *95, 137, 184, 213, 249, 255–6, 275*
Middlesex County Asylum – *See:* Asylums
Middlesex Hospital *115, 118, 128–9, 135*
Midland Medical and Surgical Reporter 139
Midwifery, Midwives *7, 65, 68, 71, 72, 80–2, 84*
Milne Edwards, H. *256, 315, 316, 318*
Mimoses *48, 61, 64, 65, 67–71, 77*
See also: Women's Disorders
Mollett, J. W. *90*
Monro Dynasty *8, 9, 10, 57*
Monstrosity *51, 268*
Montesquieu, C. L. de S. *306*
Moor, M. – Drug seller *5, 6, 111*

Moorcroft Asylum – See: Asylums
Morrell, J. B. *7, 163*
Mosaic Physiology *278*
Müller, Johannes *25, 117, 132, 187, 190, 192, 203, 206, 239–42, 262, 265, 269, 275, 283*
See also: Reflex Action
Munk, W. *214*

N

Naples and Vesuvius *312*
Napoleon *14, 64*
National Institute of France *238*
Negroes, Negro Emancipation *19, 25, 123, 125, 304-6 passim, 308–10*
Nepotism *140, 187*
Nervous Diseases *53, 77, 246,* Chapter *5 passim*
Nervous System Chapter *5 passim*
Arthropod nervous system *254*
British Association Reports *166–7*
Hall's Croonian Lectures *213*
Hall's Gulstonian Lectures *186*
Hall's Memoirs to Royal Society *189–201, 206–9*
history of work on nervous system *232–42*
issues of priority *239–42, 253*
nerve function *179, 192, 207, 271–5 passim*
specialist terms
Electrogenic *207–12 passim, 273–4*
Excitabilité, Motorische Kraft, Vis motoria 192
Vis nervosa 192, 257–8, 285–6
See also: Electrophysiology, Experimental Work on Animals, Reflex Action, Sensation
Neve, M. *14*
New York Times *277*
Newark *5, 111*
Newcastle, Duke of *16, 17*
Newport, George *177, 180, 182, 194, 197–201, 210–11, 249, 282*
Nichols, J. *319*
Nicholson's Journal 2
Nobili, L. *272, 273*
Noble, D. *312*
Northampton, Marquis of – President Royal Society *187, 192, 209*
Nosology *47, 48, 50–2, 64, 67, 136*
Cullen *47, 48, 50, 51, 67, 136*
Hall *48, 52, 64, 110, 136*
Nottingham *x, 1, 4, 5, 11, 15–20, 22, 44, 45, 68, 77, 82, 89, 119, 127, 312, 325, 327*
Basford – Birthplace, *x, 1*
Bridgewater – First Practice *15*
burial place *327*
Dukeries *17*
Libraries *16*
Societies *16, 17, 22*
See also: Hall, Nottingham Infirmary/ Nottingham General Hospital
Nottingham Infirmary/Nottingham General Hospital *16, 19, 82, 89*
Numerical Method/Méthode Numérique *26, 57, 97*

O

Observation in Diagnosis *55–62 passim*
Obstetrics *7, 67, 68, 72, 110, 121*
Oliffe, Sir Joseph and Lady *312*
Ollivier, C. P. *270*
Olmsted, J. M. D. *240*
Owen, Richard *173–80, 206*

P

Paget, James *25, 111, 112, 115, 117–19, 325*
Palmer, William – *See:* Strychnine
Paraplegia, Hemiplegia *268, 269*
Paris *xii, 14, 26, 27, 46–8 passim, 56, 58, 63, 86, 90, 96, 99, 171, 179, 238, 282, 312*
Clinical School, Medical School *47, 58*
See also: Louis, Named: Academies, Hospitals
Paris, John A. – On claims to priority *73, 252, 290*
Parkes, Samuel *2, 3, 135*

Pathology *16, 26, 46, 54, 58, 59, 61, 84, 133, 195, 255, 256*
Patients xii, *11–13, 16, 18–20, 25, 26,* Chapters *2* and *3 passim, 183, 251, 303*
Paton, George *266*
Pattison, Granville-Sharp *127, 128, 132*
Pearson, Frederick *138*
Peel, Sir Robert – Home Secretary *143, 173, 177*
Penitentiary System – United States *306*
See also: Beaumont, Tocqueville
Pepys, Samuel *90*
Percival, T. *87*
See also: Ethics
Perrault, C. *232*
Pestalozzi, J. H. *76*
Peterson, M. J. *5*
Pettigrew, T .J. *117*
Pflüger, Edward F.W. *246, 276, 279, 281, 282*
attack on Hall's view *re* brain and consciousness *281*
See also: Controversies
Pharmacopoeia *25, 146, 319*
Philanthropy *15, 16, 18, 112, 114, 123, 125, 309, 311*
Philip, A. P. W. *139, 166, 173, 174, 181, 186–8*
Philosophical Transactions 123, 157, 162, 171, 175, 177, 182, 186–8, 191, 194, 197, 199, 202, 206, 208, 209, 214, 263, 269, 272
Physicians *6, 18, 49, 60-79 passim, 108–18 passim, 173, 212, 232*
See also: Royal College of Physicians
Physicians' Clerks – *See:* Clerkships, Edinburgh Royal Infirmary
Physiological Committee – Royal Society *173–82, 190, 207–11, 213*
See also: Royal Society
Physiology – Experimental Medicine *x, xi, 9, 16, 44, 54, 58, 72, 73, 95,* Chapter *3 passim,* Chapter *4 passim,* Chapter *5 passim, 304, 312, 313, 316, 318, 319, 323*
See also: Experimental Work on Animals, Nervous System, Reflex Action, Royal Society – Physiological Committee, Sensation
Pierson, G. W. *306*
Pinel, P. *9, 61*
Plagiarism *96, 137, 141, 191, 194, 196, 198, 199, 276, 282–3, 286*
Pneumatometer *189, 198*
Pole, J. R. *310, 311*
Pollution *321, 324*
See also: Sewage Disposal
Porter, D. *108*
Porter, R. *108*
Power, Sir D'Arcy *96, 327*
Pregnancy *53, 65, 66, 82, 83*
See also: Women's Disorders
Preston, C. *232, 269*
See also: Anencephalous
Prevost, The Abbé *306*
Priestley, J. B. *71, 78*
Priestley, Joseph *2*
Pringle, Sir John *2*
Private Medical/Anatomy Schools, London - *See:* Anatomy Schools
Prochaska, George *141, 234, 235, 251, 282–91*
charge of plagiarism against Hall *282–91*
See also: Controversies
Professionalization of Science *156, 159, 160, 162, 163*
Prostitution *86*
See also: Women's Disorders
Provincial Medical Association *139, 145, 168*
Psychology, Psychical Activity *183, 242, 267, 278*
Puerperal Fever *xi, 65, 71, 82–5, 89, 91*
See also: Women's Disorders
Purposiveness *266, 281*

Q

Quackery *6, 16, 54, 78, 79, 108, 109*
Quain, Jones *117, 127, 133, 134*
Quakers – Society of Friends *22, 113, 121, 305, 325*
Quarterly Journal (Journal of the Royal Institution) 15, 44, 139

R

Redi, F. *196, 223*
Reflex Action xi, *117, 189–91, 208, 232–5, 237, 240–3, 246, 249, 251–5, 257, 264, 266, 267, 275, 276, 278, 279, 283, 285, 287, 288, 290*
  Bell's work on afferent and efferent nerves *251–2*
  Descartes's views *234*
  Hall's concept of distinct anatomical system *248–57*
  Hall's early experiments – newt tail, Zoological Society *190, 243–4*
  Hall's papers – Royal Society *172–3, 189–97*
  history of concept *233–9*
  issues of priority *239–42*
  mechanistic explanation *266–7*
  purposiveness *281–2*
  special principle *242, 245–8, 257, 258*
  spinal cord – crucial role of *236, 242, 260*
  See also: Bell C., Bernard, Flourens, Lewes, Longet, Magendie, Whytt Consciousness, Sensation
Reform *156, 306*
  medical reform *109–10, 138–46, 206*
  Reform Bill 1832 *25, 158*
  Royal Society *158–61*
Refrigeration, Artificial *44*
Register Thermometer *44*
Reiser, S. J. *63*
Respiration *26, 62, 183–98 passim, 200, 206, 235–46 passim, 259, 266, 313–18 passim*
Resurrectionists/Body Snatchers *60, 115–16*
Resuscitation *30, 62, 276, 312–18*
  See also: Royal Humane Society
Reymond, Emil du Bois- *272, 275*
  dismissed electrophysiology of predecessors *275*
Reynolds, J. Russell *303*
Reznek, L. *48, 50, 51*
Richardson, R. *59*
Richet, C. *234, 235*
Risse, G. B. *12, 59, 82*
Ritter, J. W. *207, 272*
Roberton, J. D. *197*
Roget, Peter Mark *9, 115, 126, 157–63 passim, 192–212 passim, 321–2*
  See also: Bridgewater Treatise, Royal Society
Rolleston, Sir Humphry *16, 19*
Rome *28, 278, 312*
Rosse, Earl of – President Royal Society *176, 177, 187, 192, 193, 210–13*
  letter from Marshall Hall *176, 177, 187, 192, 193, 211–13*
Royal College of Physicians *6, 18, 26, 109, 110, 113, 138, 141–3, 145, 157, 187, 214, 254*
  Fellowships *6, 18, 110, 141, 145, 157*
  Hall's criticism and reform *141, 145*
  Licentiates *6, 110, 141*
Royal College of Surgeons *6, 72, 109–21 passim, 137–45 passim, 190, 290*
Royal College of Surgeons, Edinburgh *146*
Royal Humane Society *313–17*
  Annual Reports *313, 314*
  See also: Resuscitation
Royal Institution London *15, 44, 127, 158, 159, 169*
Royal Medals – *See:* Royal Society
Royal Medical and Chirurgical Society *132*
Royal Medical Society, Edinburgh *9, 10*
Royal Society *119–39 passim,* Chapter *4 passim, 238–63 passim, 271–85 passim*
  charges against Royal Society *158–61*

donation fund *179–80*
government grant – *See:* Government Grant
Hall's Fellowship *96, 156*
Medals *159*
Copley Medal *5, 168, 175–9, 202*
Royal Medals *141, 175–8, 187–97 passim, 200–3, 210*
Rumford Medal *177*
new elections *209–11*
Presidents *168–9, 187, 209*
specialist committees *169–73*
See also: Hall – Royal Society – papers, Herschel, Physiological Committee, Reform, Roget
Royal Society, Edinburgh *15*
Rumford, Count Benjamin T. *159, 177*
Rumford Medal – *See:* Royal Society
Russell Literary and Scientific Institution, Bloomsbury *157*
Russell, Lord John *180*

S

Sadler, M. T. – Law of Population *75*
St Bartholomew's Hospital *72, 118, 190*
St George's Hospital *29, 118, 119, 314, 321*
St Mary's Hospital *112, 138, 319*
St Pancras Infirmary – Case of Hemiplegia *269*
St Thomas's Hospital *58, 116, 118, 120–4*
Sauvages, F. B. de *49, 50*
Schafer, E. A. *239*
Scheele, C. *2*
Scientific Literacy *See:* Society for the Diffusion of Useful Knowledge
Sea -sickness *304*
See also: Hall
Sedgwick, Adam *159*
Seeley, Mr – Publisher *20, 21*
Semmelweis, I. *84*
See also: Puerperal Fever
Sensation *199, 237*
early and contemporary views on *260–7*
organs associated with *238–9*
role in reflex action *237, 257–60*
role of will, volition *237*
sensorium commune - location of sentient principle *234, 241–2, 285–8*
sentient principle *285–6*
See also: Bernard, Controversies, Flourens, Legallois, Lewes, Hall, Magendie, Mayo, Paton, Whytt
Sensorium Commune – *See:* Sensation
Sentient Principle *260, 261, 285*
See also: Sensation, Whytt
Sewage Disposal *320–5*
fertilizer production *323–4*
pollution of Thames *323*
Sexual Abuse *86, 87*
Sexuality *66, 68, 71, 77, 78, 79, 80*
femininity and illness *66–8*
masturbation *86*
modesty and virtue *80*
perceptions of women *66, 79*
prostitution *86*
sexual abuse *87*
See also: Women's Disorders, Venereal Diseases
Shakespeare, W. *130*
Sharpey, William *9, 133, 134, 138, 182, 283*
Sherrington, Sir Charles – Integrated nervous system *238, 239, 249, 275*
Showalter, Elaine *71*
Simon, John – Medical Officer of Health *321–2*
See also: Medical Officer of Health
Slavery, Abolition *19, 20, 157, 304, 305–12, 325*
See also: Negroes
Smellie, William *68, 72*
Smith, Ginnie *72, 73*
Smith, Henry – Helped Hall with Experiments *206*
Smollett, T. G. *27*
Snow Beck, T. *175, 192, 212*
Society for the Diffusion of Useful Knowledge *157, 158*
Society of Apothecaries *6, 109, 137, 142, 144–6*

Society of Arts – Father's award *3*
Society of Friends – *See:* Quakers
Sontag, Susan *48*
Soul *60, 183, 234, 260–2*
See also: Controversies – Pflüger-Lotze, Sentient Principle, Whytt
Smith, Sir James – Astronomer *160–1*
Spallanzani, L. *188*
Spens, Thomas *12, 13, 48*
Spinal Soul – *See:* Controversies – Pflüger-Lotze
Sprigge, S. S. *137*
Statistical Method *57, 84*
Stethoscope *54, 62*
See also: Laennec
Stilling, B. *213*
Stilwell, A. – *See:* Asylums – Moorcroft
Strychnine *211, 273, 318–20*
detection; tests *318–20*
Palmer-Cook murder trial *318–19*
Student Medical Societies *136–7*
See also: Royal Medical Society
Surgeons *1, 6, 11, 12, 19, 49, 58, 67–73 passim, 108-18 passim, 120, 140, 142–5 passim, 173, 181, 182, 190*
apprenticeships *11, 111, 112, 116*
See also: Medical Schools, Royal College of Surgeons
Suspended Animation *62*
Sussex, Duke of – President Royal Society *1, 161–76 passim*
See also: Royal Society
Swammerdam, J. *271*
Sydenham, T. *50, 55, 56*
Sydenham Society *284*
Sydenham College – *See:* Anatomy Schools
Sympathy *236*
Symptoms – *See:* Nosology
Syncope – Fainting *91–8* passim
See also: Blood-letting
Szasz, T. S. *48*

T

Technology *xiii, 1, 3, 63*
Thackeray, W. *82*
Thames
bridges, transport *322–3*
sewage disposal *320, 321, 323–4*
Thames barrier *323*
Tillotson, John *23, 24*
Tocqueville, Alexis de *306*
See also: Penitentiary System, Slavery
Todd, R. B. *173, 205, 208, 209, 213, 249, 271*
Tractarians Tract Societies *17*
Transatlantic Steamships
'Arabia' – Cunard Royal Mail *303–4*
'Atlantic' – Collins' Steamer *305*
True Spinal System *54, 249–50, 253, 254, 270, 312*
See also: Diastaltic Nervous System, Excito-motory System
Turner, Edward – First Professor of Chemistry, University College London *126*

U

Underwood, Michael *68, 87*
Unitarians *4, 14, 311*
See also: Dissenters
United Hospitals – *See:* Guy's Hospital, St Thomas's Hospital
United States *xii, 1, 7, 19, 29, 183, 276–8, 303-12*
Universities – *See* under appropriate name
University College Hospital *114, 134, 135, 268*
University College London – Originally the University of London *22, 49, 117, 121, 141–2, 182–203 passim, 283–303 passim, 319*
foundation and backers *124–6*

medical education *135–8*
professorial appointments and disputes *126–32*
University of Cambridge, *7, 110, 111, 124, 125, 141, 158, 166*
University of Edinburgh *1, 5, 7–10, 45, 49, 50, 57, 71, 119, 139, 183*
Apparatus and Museum Committee *10*
Faculty of Medicine *9*
medical curriculum *7–10, 110*
student numbers *8*
See also: Royal Medical Society
University of Glasgow *44, 157*
University of London *20, 110, 111, 114, 115, 118, 124–38 passim*
See also: University College London
University of Oxford *7, 110, 111, 124, 125, 141, 163*
Unzer, Johann A. *251, 284*
See also: Controversies – Prochaska

V

Valli, E. *207*
Van Deen, I. – Dissociated sensation from reflex action *117, 203, 214, 259*
Venereal Diseases *85–7*
Venesection *89*
See also: Blood-letting
Vernon, Hanotte – On strychnine *319*
Victoria, Queen – Coronation *21, 303*
Vienna General Hospital *84*
Virchow, Rudolf *256–7*
Virginia Medical and Surgical Journal 304
Vivisection *174, 253, 277, 312*
Cartwright and Dowler's experiments – alligators *276–8*
Hall's early experiments on newt *190*
Magendie's work – Bell's criticism *239–40*
See also: Bell C., Experimental Work on Animals
Vogel, M. J. *50*
Volkmann, A. W. *239, 260, 267*
Volta *207, 271, 272*
artificial electric organ *271*
opposed Galvani *271*
voltaic-pile, -current, –influence *207, 271–5 passim*
See also: Galvani
Voltaire, A. F-M. *306*

W

Wakley, Thomas *xiii, 21, 25, 111, 127–43 passim, 182, 193, 284, 285*
Walker, Alexander *263*
Walshe, W. H. *135*
Webb Street School – *See:* Anatomy Schools
Webster, George *25, 28, 44, 139, 143, 307*
Weld, C. R. – Assistant Sec. Royal Society *207*
Wesley, John; Wesleyans *4, 21*
Westminster Review *156*
Whaley, F. *9, 12*
Wheatstone, C. *165*
Whewell, William *159–81 passim*
Whig *xi, xii, 21*
Whitfield, R. J. *122*
Whytt, Robert *188, 232–42 passim, 257–69 passim, 285–90*
Wilkie, Sir David *90*
Wilkins, George – Archdeacon, Nottingham *18*
William IV, King *21*
Williams, C. J. B. *134, 135, 142, 325*
Willis, T. *232, 234, 235*
Wilson, Erasmus *117, 208, 209*
Wollaston, William H. – President Royal Society *168, 169, 179*
Women's Disorders
associated with old age *65*
childbearing years *82–5*

constipation *77–8*
importance of clothing and exercise *77–9*
middle and later years *88–9*
mimoses *61, 65–71 passim*
See also: Constipation, Mimoses, Puerperal Fever
Worcester Infirmary *139*

Y

York Philosophical Society *164*

Z

Zeitschrift für klinische Medizin 183
Zoological Society *96, 167, 190, 240, 241, 243, 258, 285*